STUDY GUIDE

FOR

STEWART'S

MULTIVARIABLE CALCULUS

SEVENTH EDITION

Richard St. Andre
Central Michigan University

BROOKS/COLE
CENGAGE Learning

Australia • Brazil • Japan • Korea • Mexico • Singapore • Spain • United Kingdom • United States

BROOKS/COLE
CENGAGE Learning

Printer: Thomson/West

Cover Image: © M. Neugebauer/zefa/Corbis

For product information and technology assistance, contact us at **Cengage Learning Customer & Sales Support, 1-800-354-9706**

For permission to use material from this text or product, submit all requests online at **www.cengage.com/permissions**
Further permissions questions can be emailed to **permissionrequest@cengage.com**

ISBN-13: 978-0-8400-5410-4
ISBN-10: 0-8400-5410-6

Brooks/Cole
10 Davis Drive
Belmont, CA 94002-3098
USA

Cengage Learning is a leading provider of customized learning solutions with office locations around the globe, including Singapore, the United Kingdom, Australia, Mexico, Brazil, and Japan. Locate your local office at: **www.cengage.com/global**

Cengage Learning products are represented in Canada by Nelson Education, Ltd.

To learn more about Brooks/Cole, visit
www.cengage.com/brookscole

Purchase any of our products at your local college store or at our preferred online store
www.cengagebrain.com

Printed in the United States of America
1 2 3 4 5 6 7 15 14 13 12 11

Preface

This *Study Guide* is designed to supplement the multivariable chapters of *Calculus*, 7th edition, by James Stewart. It may also be used with the 7th editions of *Calculus: Early Transcendentals, Multivariable Calculus,* and *Multivariable Calculus: Early Transcendentals.*

Your text is well written in a very complete and patient style. This *Study Guide* is not intended to replace it. You should read the relevant sections of the text and work problems, lots of problems. Calculus is learned by doing it.

What this *Study Guide* does do is capture the main points and formulas of each section of your text and provide complete examples and short, concise questions that will help you understand the essential concepts. Every question has an explained answer. Some of our solutions begin with parenthetical comments offset by < . . . > and in italics to explain the approach to take to solve the problem. The two-column format allows you to cover the answer portion of a question while working on it and then uncover the given answer to check your solution. Working in this fashion leads to greater success than simply perusing the solutions. Students have found this *Study Guide* helpful for reviewing for examinations.

Technology, such as graphing calculators and computer algebra systems, can help the understanding of calculus concepts by drawing accurate graphs, solving or approximating solutions to equations, doing numerically intensive calculations, and performing symbolic manipulations. A number of sections have "Technology Plus" exercises at the conclusion of the section to help you master the calculus.

As a quick check of your understanding of a section, we have included a page of On Your Own questions located toward the back of the *Study Guide*. There are more than 270 multiple-choice-type questions—the kind you might see on an exam in a calculus class. You are "on your own" in the sense that an answer, but no solution, is provided for each question.

I hope that you find this *Study Guide* helpful in understanding the concepts and solving the exercises in *Calculus*, 7th edition.

Richard St. Andre

Table of Contents

13 Vector Functions

14 Partial Derivatives

15 Multiple Integrals

16 Vector Calculus

17 Second-Order Differential Equations

On Your Own

Answers to On Your Own

Chapter 10 — Parametric Equations and Polar Coordinates

Section 10.1   Curves Defined by Parametric Equations

As a point (x, y) moves along a curve from time $t = a$ to $t = b$, the coordinates x and y may each be described as a function of a time variable t. The idea of describing a curve with a pair of parametric equations (t is the parameter) is covered in this section. We will see that graphs of ordinary functions ($y = f(x)$) are one of many types of curves that may be defined parametrically.

Concepts to Master

A. Parameter; Parametric equations; Graphs of curves defined parametrically; Cycloid

B. Elimination of the parameter

C. Technology Plus

Summary and Focus Questions

A. A set of **parametric equations** has the form

$$x = f(t)$$
$$y = g(t),$$

where f and g are functions of a third variable t, called a **parameter.** Each value of t determines a point (x, y) in the plane. The collection of all such points is a **parametric curve.**

A curve may be described by several different pairs of equations. Here are two sets of parametric equations for the quarter of the unit circle in the first quadrant:

$$x = t$$
$$y = \sqrt{1 - t^2}$$

for $t \in [0, 1]$.

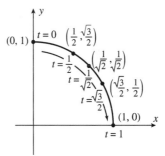

1

$$x = \cos t$$
$$y = \sin t$$
$$\text{for } t \in \left[0, \frac{\pi}{2}\right].$$

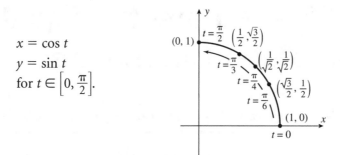

Each has the same graph (set of points). The first curve is traversed clockwise as t increases; the second graph is traversed counterclockwise.

Note that a curve defined parametrically need not be the graph of a function. The graph of

$$x = 2 \sin 3t, \ y = 2 \cos 5t, \ 0 \le t \le 2\pi,$$

given at the right, is clearly not the graph of a function $y = f(x)$.

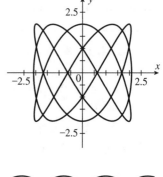

Suppose you paint a white spot on the edge of a bicycle wheel. As the bicycle rides past you the spot traces out a repeating arched curve that touches the ground once every revolution. That curve is a **cycloid**—a curve traced out by a point P on a circle as the circle rolls along a straight line. If r is the radius of the circle and the point P passes through $(0, 0)$, the parametric equations to describe the cycloid are:

$$x = r(\theta - \sin \theta)$$
$$y = r(1 - \cos \theta).$$

1) Sketch a graph of the curve given by $x = \sqrt{t}, y = t + 2$.

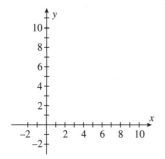

Compute some values and plot points:

t	0	1	4	9
x	0	1	2	3
y	2	3	6	11

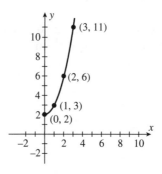

We see the curve is half a parabola since $x = \sqrt{t}$ implies $x^2 = t$.
Thus $y = x^2 + 2$, $x \geq 0$.

2) Describe the graph of $x = a + bt$, $y = c + dt$ where a, b, c, and d are constants.

If both b and d are zero, this is the single point (a, c). If $b = 0$ and $d \neq 0$ this is a vertical line through (a, c). If $b \neq 0$, then $t = \frac{x - a}{b}$ and $y = c + d\frac{(x - a)}{b}$. Thus $y = \frac{d}{b}x + c - \frac{da}{b}$, so the graph is a line through (a, c) with slope $\frac{d}{b}$.

3) Does the pair
$$x = e^t$$
$$y = e^{2t}, \ -\infty < t < \infty$$
represent the parabola $y = x^2$?

The pair represent a portion of the parabola $y = e^{2t} = (e^t)^2 = x^2$. Since $x = e^t$, $x > 0$. The pair of equations represent that portion of $y = x^2$ to the right of the y-axis.

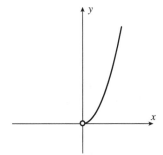

4) Use a graphing calculator to sketch
$$x = t^2 + 3t$$
$$y = 2 - t^3.$$

t	x	y
−4	4	66
−3	0	29
−2	−2	10
−1	−2	3
0	0	2
1	4	1
2	10	−6
3	18	−25
4	28	−62

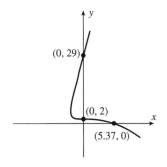

5) For a cycloid formed by a circle of radius 2:

a) Draw the cycloid for $0 \le \theta \le 4\pi$.

<Use a calculator or plot several points.>

θ	x	y
0	0	0
$\dfrac{\pi}{2}$	$\pi - 2$	2
π	2π	4
$\dfrac{3\pi}{2}$	$3\pi + 2$	2
2π	4π	0
3π	6π	4
4π	8π	0

b) Where is the slope of the tangent line to the cycloid not defined for $0 \le \theta \le 4\pi$?

The tangent line does not exist at the cusps, where $\theta = 0, 2\pi, 4\pi$.

c) For what values of θ is the curve at maximum height?

By the symmetry of the cycloid, the maximum occurs midway between the cusps. The maximum is 4 and occurs at $\theta = \pi$ and $\theta = 3\pi$.

B. Sometimes parametric equations may be combined to produce a single equation not involving the parameter t; this process is called **eliminating the parameter.** The method to do so depends greatly on the nature of the parametric equations involved. Sometimes identities need to be employed, especially if trigonometric functions are involved. Other times you can solve for t in one parametric equation and substitute the result in the other equation.

Example: Eliminate the parameter in $x = 2 + 3t, y = t^2 - 2t + 1$.

Solve $x = 2 + 3t$ for t to get $t = \frac{1}{3}(x - 2)$. Substitute this in the other equation:

$$y = \left[\tfrac{1}{3}(x - 2)\right]^2 - 2\left[\tfrac{1}{3}(x - 2)\right] + 1.$$

$$y = \tfrac{1}{9}x^2 - \tfrac{10}{9}x + \tfrac{25}{9}.$$

Example: Eliminate the parameter in $x = \sin t, y = \cot^2 t$.

$$y = \cot^2 t = \frac{\cos^2 t}{\sin^2 t} = \frac{1 - \sin^2 t}{\sin^2 t}. \text{ Thus, } y = \frac{1 - x^2}{x^2}.$$

We note from $x = \sin t, -1 \le x \le 1$ and $x \ne 0$ (since $\cot t$ is undefined when $\sin t = 0$).

6) Eliminate the parameter in each

    a) $x = e^t, y = t^2$.

Solve for t in terms of y:
$y = t^2, \sqrt{y} = t$ for $t \ge 0$.
Thus, $x = e^{\sqrt{y}}, y \ge 0$.
A solution may also be obtained by solving
for t in terms of x:
$x = e^t, t = \ln x, y = (\ln x)^2, x > 0$.

    b) $x = 1 + \cos t, y = \sin^2 t$.

$x = 1 + \cos t, x - 1 = \cos t$
$\cos^2 t = (x - 1)^2$.
Since $\sin^2 t = y$, and $\cos^2 t + \sin^2 t = 1$
we have $(x - 1)^2 + y = 1$.
Thus, $y = 1 - (x - 1)^2$.
Since $x = 1 + \cos t, 0 \le x \le 2$.

    c) $x = 2 \sec t, y = 3 \tan t$.

$\sec t = \frac{x}{2}$ and $\tan t = \frac{y}{3}$.
Hence the identity $\tan^2 t + 1 = \sec^2 t$
becomes $\left(\frac{y}{3}\right)^2 + 1 = \left(\frac{x}{2}\right)^2$ or $\frac{x^2}{4} - \frac{y^2}{9} = 1$.

7) Describe the motion of a particle moving
along a curve given by
$x = 4 \sin t, y = 3 \cos t, 0 \le t \le \frac{3\pi}{2}$.

Eliminate the parameter t:

$\frac{x}{4} = \sin t$ and $\frac{y}{3} = \cos t$. Since
$\sin^2 t + \cos^2 t = 1$

$$\left(\frac{x}{4}\right)^2 + \left(\frac{y}{3}\right)^2 = 1$$

$\frac{x^2}{16} + \frac{y^2}{9} = 1$, which is an ellipse.

Because $0 \le t \le \frac{3\pi}{2}$, the particle starts at

$(0, 3)$ and traverses $\frac{3}{4}$ of the way around the
ellipse.

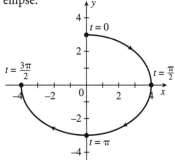

C. Technology Plus. Use a computer algebra system or a graphing calculator
to solve.

T-1) Sketch the graphs of $r = \cos(2k - 1)\theta$,
$0 \le \theta \le 2\pi$, for $k = 1, 2, 3$, and 4.
How does the graph change as k increases?

$k = 1$

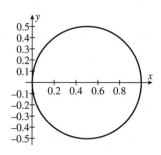

$k = 2$

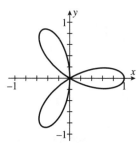

$k = 3$

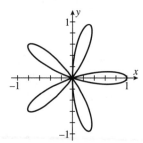

$k = 4$

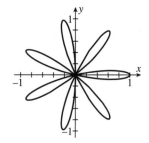

The number of leaves is $2k - 1$.

T-2) Use a graphing calculator to sketch a graph of $x = t + 2 \sin 2t$, $y = t + \sin 4t$ for $-9 \le t \le 9$.

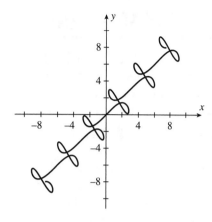

Section 10.2 Calculus with Parametric Curves

This section describes several familiar concepts for curves that are defined parametrically: slope of a tangent line to a curve, curve sketching, length of a curve, areas of regions enclosed by curves, and surfaces of solids of revolution.

Concepts to Master

A. First and second derivative of a function defined by a pair of parametric equations; applications to graphs of curves defined parametrically

B. Area of a region enclosed by curves defined parametrically

C. Length of a curve defined parametrically

D. Area of a surface of revolution for a curve defined parametrically

E. Technology Plus

Summary and Focus Questions

Page 669

A. The slope of the line tangent to a parametric curve described by $x = f(t)$, $y = g(t)$ is the first derivative of y with respect to x and is given by

$$\frac{dy}{dx} = \frac{\frac{dy}{dt}}{\frac{dx}{dt}} \quad \text{if } \frac{dx}{dt} \neq 0.$$

Likewise, the second derivative of y with respect to x is the derivative of the first derivative:

$$\frac{d^2y}{dx^2} = \frac{\frac{d}{dt}\left(\frac{dy}{dx}\right)}{\frac{dx}{dt}}.$$

The curve will have a horizontal tangent when $\frac{dy}{dt} = 0$ and $\frac{dx}{dt} \neq 0$, and will have a vertical tangent when $\frac{dx}{dt} = 0$ and $\frac{dy}{dt} \neq 0$.

1) Let $x = t^3, y = t^2 - 2t$.

a) Find $\frac{dy}{dx}$ and $\frac{d^2y}{dx^2}$.

$\frac{dx}{dt} = 3t^2$ and $\frac{dy}{dt} = 2t - 2$, so $\frac{dy}{dx} = \frac{2t - 2}{3t^2}$.

$\frac{d}{dt}\left(\frac{dy}{dx}\right) = \frac{3t^2(2) - (2t - 2)(6t)}{(3t^2)^2} = \frac{4 - 2t}{3t^3}$.

Thus $\frac{d^2y}{dx^2} = \frac{\frac{4 - 2t}{3t^3}}{3t^2} = \frac{4 - 2t}{9t^5}$.

b) Find the horizontal and vertical tangents for the curve.

$\frac{dy}{dx} = 0$ at $t = 1$ ($x = 1, y = -1$).

A horizontal tangent is at $(1, -1)$.

$\frac{dy}{dx}$ does not exist at $t = 0$ ($x = 0, y = 0$).

A vertical tangent is at $(0, 0)$.

c) Discuss the concavity of the curve.

$\dfrac{d^2y}{dx^2} > 0$ for $0 < t < 2$ $(0 < x < 8)$ and negative for $t < 0$ $(x < 0)$ and $t > 2$ $(x > 8)$. The curve is concave upward for $x \in (0, 8)$ and concave downward for $x \in (-\infty, 0)$ and $x \in (8, \infty)$.

d) Sketch the curve using the information above.

$\dfrac{d^2y}{dx^2} = \dfrac{4 - 2t}{9t^5} = 0$ at $t = 2$ $(x = 8, y = 0)$.

$(8, 0)$ and $(0, 0)$ are inflection points.

t	x	y	$\dfrac{dy}{dx}$
-3	-27	15	$-\dfrac{8}{27}$
-2	-8	8	$-\dfrac{1}{2}$
-1	-1	3	$-\dfrac{4}{3}$
0	0	0	undefined
1	1	-1	0
2	8	0	$\dfrac{1}{6}$
3	27	3	$\dfrac{4}{27}$

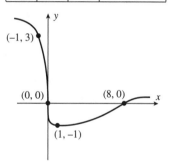

2) Find the equation of the tangent line to the curve $x = 4t^2 + 2t + 1$, $y = 7t + 2t^2$ at the point corresponding to $t = -1$.

At $t = -1$, $x = 3$ and $y = -5$.

$\dfrac{dx}{dt} = 8t + 2 = -6$ at $t = -1$.

$\dfrac{dy}{dt} = 7 + 4t = 3$ at $t = -1$.

Thus $\dfrac{dy}{dx} = \dfrac{3}{-6} = -\dfrac{1}{2}$.

The tangent line is $y + 5 = -\dfrac{1}{2}(x - 3)$.

B. Suppose the parametric equations $x = f(t)$, $y = g(t)$, $t \in [\alpha, \beta]$ define an integrable function $y = F(x) \geq 0$ over the interval $[a, b]$, where $a = f(\alpha)$, $b = f(\beta)$. The area under $y = F(x)$ is

$$\int_{\alpha}^{\beta} g(t)\, f'(t)\, dt \quad \text{if } (f(\alpha), g(\alpha)) \text{ is the left endpoint}$$

or

$$\int_{\beta}^{\alpha} g(t)\, f'(t)\, dt \quad \text{if } (f(\beta), g(\beta)) \text{ is the left endpoint.}$$

Example: Find the area bounded by the x-axis and the curve given by $x = \sqrt{t}, y = 4t - t^2, 0 \le t \le 4$.

The area is graphed at the right and is

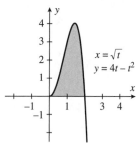

$$\int_0^4 (4t - t^2)\left(\tfrac{1}{2}t^{-1/2}\right)dt = \tfrac{1}{2}\int_0^4 \left(4t^{1/2} - t^{3/2}\right)dt$$

$$= \tfrac{1}{2}\left(4\left(\tfrac{2}{3}t^{3/2}\right) - \tfrac{2}{5}t^{5/2}\right)\Big]_0^4 = \left(\tfrac{4}{3}t^{3/2} - \tfrac{1}{5}t^{5/2}\right)\Big]_0^4$$

$$= \left(\tfrac{4}{3}(4)^{3/2} - \tfrac{1}{5}(4)^{4/2}\right) - 0 = \tfrac{32}{3} - \tfrac{32}{5} = \tfrac{64}{15}.$$

Note that we may eliminate the parameter and obtain $y = 4x^2 - x^4$,

for $0 \le x \le 2$. Thus the area is $\int_0^2 (4x^2 - x^4)\,dx = \tfrac{64}{15}$.

3) Find the area under the curve $x = t^2 + 1$, $y = e^t, 0 \le t \le 1$.

\<Set up the integral and use integration by parts to evaluate it.\>

The area is $\displaystyle\int_0^1 e^t\,(2t)dt$

$(u = 2t, dv = e^t\,dt; du = 2\,dt, v = e^t)$

$= 2te^t\Big] - \displaystyle\int_0^1 e^t 2\,dt = (2te^t - 2e^t)\Big]$

$= (2e - 2e) - (0 - 2) = 2.$

4) Find the area inside the loop of the curve given by $x = 9 - t^2$, $y = t^3 - 3t$.

\<First draw the graph to get a sense for what area is described.\>

$$\frac{dy}{dx} = \frac{\dfrac{dy}{dt}}{\dfrac{dx}{dt}} = \frac{3t^2 - 3}{-2t}.$$

$\dfrac{dy}{dx} = 0$ at $t = 1, t = -1$. $\dfrac{dy}{dx}$ is not defined at $t = 0$. There are horizontal tangents at $(8, 2)$ (where $t = -1$) and $(8, -2)$ (where $t = 1$). There is a vertical tangent at $(9, 0)$ (where $t = 0$).

t	x	y
−3	0	−18
−2	5	−2
−1	8	2
0	9	0
1	8	−2
2	5	2
3	0	18

The graph is

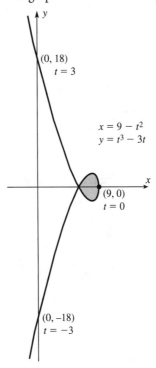

The curve crosses itself (to form a loop) at
$y = 0$.

$$t^3 - 3t = 0$$
$$t(t^2 - 3) = 0$$
$$t = 0, t = \sqrt{3}, t = -\sqrt{3}.$$

The area of the loop is twice the area under
the top portion which corresponds to
$t \in [-\sqrt{3}, 0]$.

The area is $2 \displaystyle\int_{-\sqrt{3}}^{0} (t^3 - 3t)(-2t) \, dt$

$$= 2 \int_{-\sqrt{3}}^{0} (-2t^4 + 6t^2) \, dt$$

$$= 2 \left(-\tfrac{2}{5} t^5 + 2t^3 \right) \Big]_{-\sqrt{3}}^{0}$$

$$= \frac{24\sqrt{3}}{5}.$$

**Page
672**

C. The length of a curve defined by $x = f(t), y = g(t), t \in [\alpha, \beta]$ with f', g' continuous and the curve traversed only once as t increases from α to β is

$$\int_{\alpha}^{\beta} \sqrt{\left(\frac{dx}{dt}\right)^2 + \left(\frac{dy}{dt}\right)^2}\, dt.$$

Recall that if s is the arc length function, then $(ds)^2 = (dx)^2 + (dy)^2$. Thus,

$$\int ds = \int \sqrt{\left(\frac{dx}{dt}\right)^2 + \left(\frac{dy}{dt}\right)^2}\, dt,$$ which is consistent with the previous formula for

arc length.

Example: Find the length of the curve given by $x = t^2 + 1$, $y = \frac{1}{3}t^3$

for $0 \le t \le 2$.

With $\dfrac{dx}{dt} = 2t$ and $\dfrac{dy}{dt} = t^2$, the length is

$$\int_0^2 \sqrt{(2t)^2 + (t^2)^2}\, dt = \int_0^2 \sqrt{4t^2 + t^4}\, dt$$

$$= \int_0^2 t\sqrt{4 + t^2}\, dt.$$

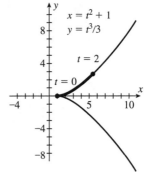

Let $u = 4 + t^2$, $du = 2t\, dt$. For $t = 0$, $u = 4$ and
for $t = 2$, $u = 8$.

The length is $\dfrac{1}{2}\displaystyle\int_4^8 u^{1/2}\, du = \dfrac{1}{2}\left(\dfrac{2}{3}u^{3/2}\right)\Big]_4^8$

$$= \frac{1}{3}\left(8^{3/2} - 4^{3/2}\right) = \frac{8\sqrt{8}}{3} - \frac{8}{3} \approx 4.88.$$

---

5) Write a definite integral for the arc length
of each:

a) the curve given by $x = \ln t$, $y = t^2$, for
$1 \le t \le 2$.

$x'(t) = \dfrac{1}{t}$ and $y'(t) = 2t$. The arc length is

$$\int_1^2 \sqrt{\left(\frac{1}{t}\right)^2 + (2t)^2}\, dt.$$

b) an ellipse given by $\dfrac{x^2}{a^2} + \dfrac{y^2}{b^2} = 1$, where
$a, b > 0$.

The ellipse may be defined parametrically
by $x = a \cos t$, $y = b \sin t$, $t \in [0, 2\pi]$. The
length of the ellipse is

$$\int_0^{2\pi} \sqrt{(-a \sin t)^2 + (b \cos t)^2}\, dt$$

$$= \int_0^{2\pi} \sqrt{a^2 \sin^2 t + b^2 \cos^2 t}\, dt.$$

c) The curve given by
$x = \cos 2t,$
$y = \sin t,$
$t \in [0, 2\pi].$

Begin with a sketch of the curve. We plot some points.

t	x	y
0	1	0
$\frac{\pi}{4}$	0	$\frac{\sqrt{2}}{2}$
$\frac{\pi}{2}$	−1	1
$\frac{3\pi}{4}$	0	$\frac{\sqrt{2}}{2}$
π	1	0
$\frac{5\pi}{4}$	0	$-\frac{\sqrt{2}}{2}$
$\frac{3\pi}{2}$	−1	−1
$\frac{7\pi}{4}$	0	$-\frac{\sqrt{2}}{2}$
2π	1	0

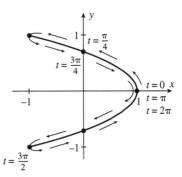

The curve is traversed once for $t \in \left[\frac{\pi}{2}, \frac{3\pi}{2}\right]$ and has length

$$\int_{\frac{\pi}{2}}^{\frac{3\pi}{2}} \sqrt{4 \sin^2 2t + \cos^2 t}\, dt.$$

Page 674

D. If a curve given by $x = f(t), y = g(t), t \in [\alpha, \beta]$, with f', g' continuous and $g(t) \geq 0$ is rotated about the x-axis, the area of the resulting surface of revolution is

$$S = \int_{\alpha}^{\beta} 2\pi y \sqrt{\left(\frac{dx}{dt}\right)^2 + \left(\frac{dy}{dt}\right)^2}\, dt.$$

Using the ds notation this is $S = \int 2\pi y\, ds$, the same formula as in the section on areas of surfaces of revolution.

6) Find the definite integral for the area of the surface of revolution about the x-axis for each curve:

a) $x = 2t + 1, y = t^3, t \in [0, 2]$

$$S = \int_0^2 2\pi(t^3)\sqrt{(2)^2 + (3t^2)^2}\ dt$$

$$= \int_0^2 2\pi t^3 \sqrt{4 + 9t^4}\ dt.$$

b) a "football" obtained from rotating the ellipse $\dfrac{x^2}{a^2} + \dfrac{y^2}{b^2} = 1, y \geq 0$ about the x-axis.

The top half of the ellipse is $x = a \cos t$,
$y = b \sin t, t \in [0, \pi]$.

$$ds = \sqrt{(dx)^2 + (dy)^2}$$
$$= \sqrt{(-a \sin t)^2 + (b \cos t)^2}$$
$$= \sqrt{a^2 \sin^2 t + b^2 \cos^2 t}.$$

The area is

$$S = \int_0^\pi 2\pi(b \sin t)\sqrt{a^2 \sin^2 t + b^2 \cos^2 t}\ dt.$$

E. Technology Plus. Use a computer algebra system or a graphing calculator to solve.

T-1) Use a CAS to evaluate the definite integral for the arc length of
$x = t + \cos t$
$y = t + \sin t$
for $0 \leq t \leq 2\pi$.

$\dfrac{dx}{dt} = 1 - \sin t, \dfrac{dy}{dt} = 1 + \cos t$

The arc length is

$$\int_0^{2\pi} \sqrt{(1 - \sin t)^2 + (1 + \cos t)^2}\ dt$$

$$\approx 10.037.$$

Section 10.3 Polar Coordinates

All our graphs have been in a rectangular coordinate system where the two coordinates are distances from perpendicular axes. In this section on polar coordinates the pair of numbers that determines a point in a plane is an angle through which to rotate from the *x*-axis together with a distance from the origin. We will see how to convert from one coordinate system to another and see that certain curves are much easier to express in polar coordinates than rectangular coordinates.

Concepts to Master

A. Points in polar coordinates; conversion to and from rectangular to polar coordinates

B. Graphs of equations in polar coordinates

C. Tangents to polar curves

Summary and Focus Questions

Page
678

A. To construct a **polar coordinate system** start with a point called the **pole** and a ray from the pole called the **polar axis.**

Pole   Polar axis

A point P has polar coordinates (r, θ) if
   $|r|$ = the distance from the pole to P, and
   θ = the measure of a directed angle with initial side the polar axis and
        terminal side the line through the pole and P.

Example: The point P with polar coordinates $(r, \theta) = (2, \frac{\pi}{4})$ is plotted by swinging an angle upward of $\frac{\pi}{4}$ radians (45°) and then moving out 2 units from the pole along the angled line.

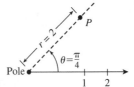

In general, plotting a point P which has polar coordinates (r, θ) depends on the signs of r and θ. Here are the possibilities:

r and *θ*	How to plot *P* (*r, θ*)	Graph
$r = 0$ θ = any value	*P* is the pole.	
$r > 0$ $\theta \geq 0$	Rotate the polar axis *counterclockwise* by the angle *θ* and locate *P* on this ray a distance *r* units from the pole.	
$r > 0$ $\theta < 0$	Rotate the polar axis *clockwise* by the angle *θ* and locate *P* on this ray a distance *r* units from the pole.	
$r < 0$ $\theta \geq 0$	Rotate the polar axis *counterclockwise* by the angle *θ* and then reflect a ray about the pole. *P* is located on this reflected ray a distance *−r* units from the pole.	
$r < 0$ $\theta < 0$	Rotate the polar axis *clockwise* by the angle *θ* and then reflect a ray about the pole. *P* is located on this reflected ray a distance *−r* units from the pole.	

Example: Plot these five points in polar coordinates:

$A(0, \pi/2)$

$B(2, \pi/3)$

$C(3, -\pi/6)$

$D(-2, \pi/6)$

$E(-1, -\pi/3)$

Note that when *θ* is negative (points C and E) we rotate clockwise.

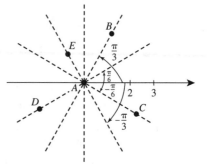

A point in a rectangular coordinate system will have a unique pair of coordinates. On the other hand, a point in a polar coordinate system will have infinitely many pairs of coordinates. In particular, the point with coordinates (r, θ) also has coordinates $(r, \theta + 2\pi)$, $(-r, \theta + \pi)$, $(-r, \theta - \pi)$, and many others.

Suppose a rectangular coordinate system is placed upon the polar coordinate system as in the figure.

To change from polar to rectangular:

$x = r \cos \theta$

$y = r \sin \theta.$

To change from rectangular to polar:

Solve $\tan \theta = \dfrac{y}{x}$ $(x \neq 0)$ for θ.

Solve $r^2 = x^2 + y^2$ for r. *Note:* $r > 0$ if the terminal side of the angle θ is in the same quadrant as P; if not, then $r \leq 0$.

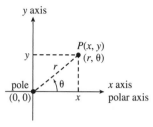

1) Plot these points in polar coordinates.

$A\left(4, \dfrac{\pi}{3}\right)$ $\qquad$ $B\left(-3, \dfrac{\pi}{4}\right)$

$C(0, 3\pi)$ $\qquad$ $D\left(2, -\dfrac{\pi}{4}\right)$

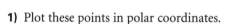

2) Find two other polar coordinates for the point with polar coordinates $\left(8, \dfrac{\pi}{3}\right)$.

There are infinitely many answers including $\left(8, \dfrac{7\pi}{3}\right), \left(-8, \dfrac{4\pi}{3}\right), \left(8, -\dfrac{5\pi}{3}\right), \ldots$

3) Sometimes, Always, or Never:

a) The polar coordinates of a point are unique.

Never.

b) $(r, \theta) = (r, \theta + 2\pi).$

Always.

c) $(r, \theta) = (-r, \theta + \pi).$

Always.

d) $(r, \theta) = (-r, \theta).$

Sometimes. (True when $r = 0$.)

4) Find polar coordinates for the point P with rectangular coordinates $P: (-3, 3\sqrt{3}).$

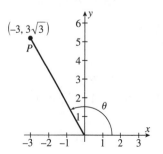

$\tan \theta = \dfrac{y}{x} = \dfrac{3\sqrt{3}}{-3} = -\sqrt{3}.$

A solution is $\theta = \dfrac{2\pi}{3}$.

$r^2 = (-3)^2 + (3\sqrt{3})^2 = 9 + 27 = 36.$

$r = \pm 6.$

Because the terminal side of $\theta = \dfrac{2\pi}{3}$ lies in the same quadrant as P, $r > 0$. Therefore, P has coordinates $\left(6, \dfrac{2\pi}{3}\right)$.

Note: $\theta = -\dfrac{\pi}{3}$ is another solution to $\tan \theta = -\sqrt{3}$, which results in $r = -6$ and coordinates $\left(-6, -\dfrac{\pi}{3}\right)$ for P.

5) Find the rectangular coordinates for the point with polar coordinates $\left(-4, \dfrac{5\pi}{6}\right)$.

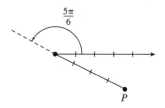

$x = -4 \cos \dfrac{5\pi}{6} = (-4)\left(\dfrac{-\sqrt{3}}{2}\right) = 2\sqrt{3}.$

$y = -4 \sin \dfrac{5\pi}{6} = (-4)\left(\dfrac{1}{2}\right) = -2.$

P has coordinates $(2\sqrt{3}, -2)$.

Page 680

B. Your first procedure for graphing a polar equation is the same as that used when first graphing functions—compute values and plot points. In some cases the graph of a polar equation is easily identified when the equation is transformed into rectangular coordinates.

Example: Graph $r = 6 \sin \theta$.

Multiply both sides of the equation by r:

$r^2 = 6r \sin \theta$
$x^2 + y^2 = 6y$
$x^2 + y^2 - 6y = 0$
$x^2 + (y - 3)^2 = 9.$

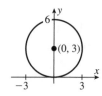

The graph is the circle of radius 3 centered at $(0, 3)$.

There are three possible symmetries that can help you draw a graph. For polar equation $r = f(\theta)$, the graph is

a) **symmetric about the polar axis** if $f(-\theta) = f(\theta)$. In other words, $r = f(\theta)$ is unchanged when θ is replaced by $-\theta$.
b) **symmetric about the pole** (unchanged when rotated 180°) if either $-f(\theta) = f(\theta)$ or $f(\theta + \pi) = f(\theta)$. In other words, $r = f(\theta)$ is unchanged when r is replaced by $-r$ or when θ is replaced by $\theta + \pi$.
c) **symmetric about the vertical line through the pole** (through the line $\theta = \frac{\pi}{2}$) if $f(\theta) = f(\pi - \theta)$. In other words, $r = f(\theta)$ is unchanged when θ is replaced by $\pi - \theta$.

Example: The graph of $r = 2 \cos 4\theta$ exhibits all three symmetries. If we use the facts that cosine has periodicity 2π, $\cos(-t) = \cos t$ for all t, and $\cos(\pi - t) = \cos t$ for all t, then

a) $2 \cos 4(-\theta) = 2 \cos(-4\theta) = 2 \cos 4\theta$.
b) $2 \cos 4(\theta + \pi) = 2 \cos(4\theta + 4\pi) = 2 \cos(4\theta + 2\pi + 2\pi) = 2 \cos 4\theta$.
c) $2 \cos 4(\pi - \theta) = 2 \cos(4\pi - 4\theta)$

$= 2 \cos(\pi - (-3\pi + 4\theta)) = 2 \cos(-3\pi + 4\theta)$
$= 2 \cos(3\pi - 4\theta) = 2 \cos(\pi - (-2\pi + 4\theta))$
$= 2 \cos(-2\pi + 4\theta) = 2 \cos(2\pi - 4\theta)$
$= 2 \cos(\pi - (-\pi + 4\theta)) = 2 \cos(-\pi + 4\theta)$
$= 2 \cos(\pi - 4\theta) = 2 \cos 4\theta$.

The graph of $r = 2 \cos 4\theta$ is an "eight-leaf rose."

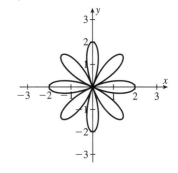

6) Sketch a graph of $r = \sin\theta - \cos\theta$.

Compute several points for values of θ.

Point	θ	r
A	0	-1
B	$\frac{\pi}{6}$	$\frac{1}{2} - \frac{\sqrt{3}}{2}$
C	$\frac{\pi}{4}$	0
D	$\frac{\pi}{2}$	1
E	$\frac{3\pi}{4}$	$\sqrt{2}$
F	π	1 (same as A)
G	$\frac{7\pi}{6}$	$\frac{-1}{2} + \frac{\sqrt{3}}{2}$ (same as B)

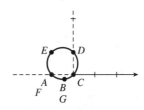

We can verify that the figure is a circle by switching to rectangular coordinates:

Multiply $r = \sin\theta - \cos\theta$ by r:

$r^2 = r\sin\theta - r\cos\theta$

$x^2 + y^2 = y - x$

$x^2 + x + y^2 - y = 0.$

Completing the square gives

$$\left(x + \tfrac{1}{2}\right)^2 + \left(y - \tfrac{1}{2}\right)^2 = \tfrac{1}{2},$$

the circle with center $\left(-\tfrac{1}{2}, \tfrac{1}{2}\right)$ and radius $\dfrac{1}{\sqrt{2}}.$

7) Sketch the graph of $r = \cos 3\theta$.

Since $\cos 3\theta = \cos(-3\theta)$, the curve is symmetric about the polar axis. Also $\cos 3\theta$ repeats every $\dfrac{2\pi}{3}$ units.

Point	θ	r	Point	θ	r
A	0	1	H	$\dfrac{7\pi}{12}$	$\dfrac{\sqrt{2}}{2}$
B	$\dfrac{\pi}{12}$	$\dfrac{\sqrt{2}}{2}$	I	$\dfrac{2\pi}{3}$	1
C	$\dfrac{\pi}{6}$	0	J	$\dfrac{3\pi}{4}$	$\dfrac{\sqrt{2}}{2}$
D	$\dfrac{\pi}{4}$	$-\dfrac{\sqrt{2}}{2}$	K	$\dfrac{5\pi}{6}$	0
E	$\dfrac{\pi}{3}$	-1	L	$\dfrac{11\pi}{12}$	$-\dfrac{\sqrt{2}}{2}$
F	$\dfrac{5\pi}{12}$	$-\dfrac{\sqrt{2}}{2}$	M	π	-1
G	$\dfrac{\pi}{2}$	0			

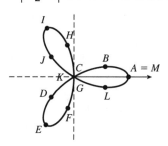

Page 683

C. For polar curve $r = f(\theta)$, we can switch to rectangular coordinates $x = f(\theta)\cos\theta,\; y = f(\theta)\sin\theta$. Treating θ as a parameter,

$$\frac{dy}{dx} = \frac{\dfrac{dy}{d\theta}}{\dfrac{dx}{d\theta}} = \frac{\dfrac{dr}{d\theta}\sin\theta + r\cos\theta}{\dfrac{dr}{d\theta}\cos\theta - r\sin\theta}.$$

This is the slope of the tangent line to $r = f(\theta)$ at (r, θ).

8) Find the slope of the tangent line to $r = e^\theta$ at $\theta = \frac{\pi}{2}$.

$x = r \cos \theta = e^\theta \cos \theta$, so

$\frac{dx}{d\theta} = e^\theta(-\sin \theta) + e^\theta(\cos \theta)$.

$y = r \sin \theta = e^\theta \sin \theta$, so

$\frac{dy}{d\theta} = e^\theta \cos \theta + e^\theta \sin \theta$.

At $\theta = \frac{\pi}{2}$,

$\frac{dx}{d\theta} = -e^{\pi/2}$ and $\frac{dy}{d\theta} = e^{\pi/2}$.

Therefore, $\frac{dy}{dx} = \dfrac{\frac{dy}{d\theta}}{\frac{dx}{d\theta}} = \dfrac{e^{\pi/2}}{-e^{\pi/2}} = -1.$

9) Find the points on $r = \sin \theta - \cos \theta$ where the tangent line is horizontal or vertical for $0 \le \theta \le \pi$.

$\frac{dy}{dx} = \dfrac{\frac{dy}{d\theta}}{\frac{dx}{d\theta}}$ is not defined when $\frac{dx}{d\theta} = 0$

and is zero when $\frac{dy}{d\theta} = 0$.

From $x = r \cos \theta$

$\qquad = (\sin \theta - \cos \theta) \cos \theta,$

$\frac{dx}{d\theta} = (\sin \theta - \cos \theta)(-\sin \theta)$

$\qquad + \cos \theta \,(\cos \theta + \sin \theta)$

$\qquad = 2 \sin \theta \cos \theta + (\cos^2 \theta - \sin^2 \theta)$

$\qquad = \sin 2\theta + \cos 2\theta.$

$\frac{dx}{d\theta} = 0$ when

$\sin 2\theta = -\cos 2\theta$

$\tan 2\theta = -1$

$2\theta = \frac{3\pi}{4}$ and $2\theta = \frac{7\pi}{4}$

$\theta = \frac{3\pi}{8}$ and $\theta = \frac{7\pi}{8}$.

At $\theta = \frac{3\pi}{8}, r = \sin \frac{3\pi}{8} - \cos \frac{3\pi}{8}$

$\qquad = \dfrac{\sqrt{\sqrt{2} + 2}}{2} - \dfrac{\sqrt{2 - \sqrt{2}}}{2} \approx .54.$

At $\theta = \frac{7\pi}{8}, r = \sin \frac{7\pi}{8} - \cos \frac{7\pi}{8}$

$\qquad = \dfrac{\sqrt{2 - \sqrt{2}}}{2} - \dfrac{-\sqrt{\sqrt{2} + 2}}{2} \approx 1.31.$

There are vertical tangents at $P\left(.54, \frac{3\pi}{8}\right)$ and $Q\left(1.31, \frac{7\pi}{8}\right)$.

From $y = r \sin \theta$

$$= (\sin \theta - \cos \theta) \sin \theta,$$

$$\frac{dy}{d\theta} = (\sin \theta - \cos \theta) \cos \theta$$
$$+ \sin \theta (\cos \theta + \sin \theta)$$
$$= 2 \sin \theta \cos \theta - (\cos^2 \theta - \sin^2 \theta)$$
$$= \sin 2\theta - \cos 2\theta = 0.$$

$$\sin 2\theta = \cos 2\theta$$

$$\tan 2\theta = 1$$

$$2\theta = \frac{\pi}{4} \text{ and } \frac{5\pi}{4}$$

$$\theta = \frac{\pi}{8} \text{ and } \frac{5\pi}{8}.$$

At $\theta = \frac{\pi}{8}, r = \sin \frac{\pi}{8} - \cos \frac{\pi}{8}$

$$= \frac{\sqrt{2 - \sqrt{2}}}{2} - \frac{\sqrt{2 + \sqrt{2}}}{2} \approx -.54.$$

At $\theta = \frac{5\pi}{8}, r = \sin \frac{5\pi}{8} - \cos \frac{5\pi}{8}$

$$= \frac{\sqrt{2 + \sqrt{2}}}{2} - \frac{-\sqrt{2 - \sqrt{2}}}{2} \approx 1.31.$$

There are horizontal tangents at $R\left(-.54, \frac{\pi}{8}\right)$ and $S\left(1.31, \frac{5\pi}{8}\right)$.

From exercise 6, the graph is a circle.

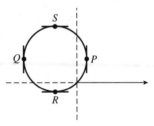

Section 10.4 Areas and Lengths in Polar Coordinates

This section develops a formula for the area of a region bounded by equations given in polar form and the arc length of a curve given by a polar equation.

Concepts to Master

A. Area of a region described by polar equations

B. Length of a curve described by a polar equation

Summary and Focus Questions

Page 689

A. Let $r = f(\theta)$ be a continuous and positive polar function for $a \le \theta \le b$, where $0 < b - a \le 2\pi$. The **area** bounded by $r = f(\theta)$, $\theta = a$ and $\theta = b$ may be approximated by partitioning the interval $[a, b]$ in n equal subintervals, each of length $\Delta\theta = \dfrac{b-a}{n}$, which correspond to n sectors. Each sector has a central angle of $\Delta\theta$.

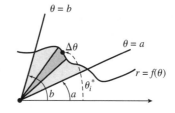

The area of a sector is approximately $\dfrac{1}{2}\left(f(\theta_i^*)\right)^2 \Delta\theta$.

The area of the region is $\displaystyle\int_a^b \dfrac{1}{2}\left[f(\theta)\right]^2 \, d\theta$.

If $0 \le f(\theta) \le g(\theta)$ for all $a \le \theta \le b$ and if f and g are continuous, then the area between f and g and between $\theta = a$ and $\theta = b$ is $\displaystyle\int_a^b \dfrac{1}{2}\left[(g(\theta))^2 - (f(\theta))^2\right] d\theta$.

Example: Find the area in the first quadrant between the polar curves $r = 1 + \cos\theta$ and $r = 2 - \dfrac{2\theta}{\pi}$.

The curves are sketched at the right. The outer curve is $r = 1 + \cos\theta$ and the inner curve is $r = 2 - \dfrac{2\theta}{\pi}$. They intersect at $\theta = 0$ and $\theta = \dfrac{\pi}{2}$.

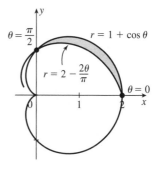

The area is $\displaystyle\int_{0}^{\pi/2} \frac{1}{2}\left[(1+\cos\theta)^2 - \left(2-\frac{2\theta}{\pi}\right)^2\right]d\theta$

$\displaystyle = \frac{1}{2}\int_{0}^{\pi/2}\left[(1+2\cos\theta+\cos^2\theta) - \left(4 - \frac{8\theta}{\pi} + \frac{4\theta^2}{\pi^2}\right)\right]d\theta$

$\displaystyle = \frac{1}{2}\int_{0}^{\pi/2}\left[\left(1+2\cos\theta+\frac{1}{2}+\frac{\cos 2\theta}{2}\right) - \left(4 - \frac{8\theta}{\pi} + \frac{4\theta^2}{\pi^2}\right)\right]d\theta$

$\displaystyle = \frac{1}{2}\int_{0}^{\pi/2}\left[-\frac{5}{2}+2\cos\theta+\frac{\cos 2\theta}{2} + \frac{8\theta}{\pi} - \frac{4\theta^2}{\pi^2}\right]d\theta = \frac{1}{2}\left(-\frac{5\theta}{2}+2\sin\theta+\frac{\sin 2\theta}{4}+\frac{4\theta^2}{\pi}-\frac{4\theta^3}{3\pi^2}\right)\Bigg]_{0}^{\pi/2}$

$\displaystyle = \frac{1}{2}\left[\left(-\frac{5\pi}{4}+2(1)+\frac{0}{4}+\pi-\frac{\pi}{6}\right)-0\right] = 1 - \frac{5\pi}{24} \approx 0.346.$

1) Find a definite integral for each region:

a) the shaded area

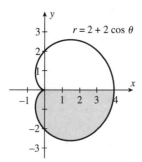

b) The region bounded by $\theta = \frac{\pi}{3}, \theta = \frac{\pi}{2},$ $r = e^{\theta}.$

c) The shaded area

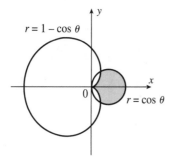

The region is bound by $\theta = \pi, \theta = 2\pi,$ $r = 2 + 2\cos\theta.$ The area is

$\displaystyle = \int_{\pi}^{2\pi}\frac{1}{2}(2+2\cos\theta)^2\,d\theta$

$\displaystyle = 2\int_{\pi}^{2\pi}(1+\cos\theta)^2\,d\theta\left(=\frac{3}{2}\pi\right).$

The area is $\displaystyle\int_{\pi/3}^{\pi/2}\frac{1}{2}e^{2\theta}\,d\theta\left(=\frac{e^{\pi}-e^{2\pi/3}}{4}\right).$

The area is between 2 curves. First determine where the curves intersect:

$1 - \cos\theta = \cos\theta,\ 1 = 2\cos\theta,\ \cos\theta = \frac{1}{2}.$

Therefore, $\theta = \frac{\pi}{3}$ and $\theta = -\frac{\pi}{3}.$

For $-\frac{\pi}{3} \leq \theta \leq \frac{\pi}{3},$

$\cos\theta \geq 1 - \cos\theta$ so the area is

$\displaystyle = \int_{-\pi/3}^{\pi/3}\frac{1}{2}[\cos^2\theta - (1-\cos\theta)^2]\,d\theta$

$\displaystyle = \int_{-\pi/3}^{\pi/3}\left(\cos\theta - \frac{1}{2}\right)d\theta\left(=\sqrt{3}-\frac{\pi}{3}\right).$

2) Find the area above the line $r = \csc\theta$ and inside the circle $r = 2$.

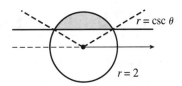

The curves intersect when

$$\csc\theta = 2$$

$$\sin\theta = \frac{1}{2}$$

$$\theta = \frac{\pi}{6} \text{ and } \frac{5\pi}{6}.$$

The area is $\displaystyle\int_{\frac{\pi}{6}}^{\frac{5\pi}{6}} \left(\frac{1}{2}(2)^2 - \frac{1}{2}\csc^2\theta\right) d\theta$

$$= \frac{1}{2}\int_{\frac{\pi}{6}}^{\frac{5\pi}{6}} (4 - \csc^2\theta)\, d\theta$$

$$= \frac{1}{2}(4\theta + \cot\theta)\Big|_{\frac{\pi}{6}}^{\frac{5\pi}{6}} = \frac{4\pi}{3} - \sqrt{3}.$$

B. A curve given in polar coordinates by $r = f(\theta)$ for $a \le \theta \le b$ has arc length

$$L = \int_a^b \sqrt{[f(\theta)]^2 + [f'(\theta)]^2}\, d\theta.$$

Page 692

3) Set up a definite integral for the length of each curve.

a) $r = e^{2\theta}$ for $0 \le \theta \le 1$.

$$\int_0^1 \sqrt{(e^{2\theta})^2 + (2e^{2\theta})^2}\, d\theta = \sqrt{5}\int_0^1 e^{2\theta}\, d\theta$$

$$\left(= \frac{\sqrt{5}}{2}(e^2 - 1)\right).$$

b) The curve sketched below.

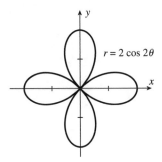

$r = 2\cos 2\theta$

Since the curve is quite symmetric, the arc length is 8 times the length of half of one leaf:

$$8\int_0^{\frac{\pi}{4}} \sqrt{[2\cos 2\theta]^2 + [-4\sin 2\theta]^2}\, d\theta$$

$$= 8\int_0^{\frac{\pi}{4}} \sqrt{4\cos^2 2\theta + 16\sin^2 2\theta}\, d\theta$$

$$= 16\int_0^{\frac{\pi}{4}} \sqrt{1 + 3\sin^2 2\theta}\, d\theta \;(\approx 19.4).$$

Section 10.5 Conic Sections

In three dimensions, a cone and a plane may intersect in a point, a line, or a curve. Conic sections are the various types of curves that result from the intersection of a plane with a cone. This section gives geometric definitions of those curves and the equations that describe them.

Concepts to Master

Focus-directrix definition of parabolas, ellipses, and hyperbolas; Vertices; Standard form of the equations of conics

Summary and Focus Questions

Page 694

Three different types of curves may result when a cone and a plane intersect. Each may be described as the set of points in a plane satisfying a certain geometric property:

Parabola: Given a line (a **directrix**) and a point (the **focus,** F), a point P is on the parabola if the distance from P to the directrix is the same as the distance from P to F.

Parabola: $d_1 = d_2$

Ellipse: Given two points (the **foci,** F_1 and F_2), a point P is on the ellipse if the *sum* of the distances from P to F_1 and from P to F_2 is a constant.

Hyperbola: Given two points (the **foci,** F_1 and F_2), a point P is on the hyperbola if the *difference* of the distances from P to F_1 and from P to F_2 is a constant.

Ellipse: $d_1 + d_2 = $ constant
Hyperbola: $d_1 - d_2 = $ constant

The equations of the conics and their graphs in rectangular coordinates are given below.

Conic Section	Equation	Properties	Graphs
Parabola	$x^2 = 4py$	$p > 0$ focus: $(0, p)$ vertex: $(0, 0)$ directrix: $y = -p$	
	$x^2 = 4py$	$p < 0$ focus: $(0, p)$ vertex: $(0, 0)$ directrix: $y = -p$	

Parabola	$y^2 = 4px$	$p > 0$ focus: $(p, 0)$ vertex: $(0, 0)$ directrix: $x = -p$	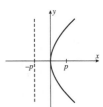
	$y^2 = 4px$	$p < 0$ focus: $(p, 0)$ vertex: $(0, 0)$ directrix: $x = -p$	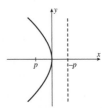

Ellipse	$\dfrac{x^2}{a^2} + \dfrac{y^2}{b^2} = 1$	$a \geq b > 0$ $c^2 = a^2 - b^2$ foci: $(c, 0), (-c, 0)$ vertices: $(a, 0), (-a, 0)$ constant sum $= 2a$ center: $(0, 0)$ major axis: line segment from $(-a, 0)$ to $(a, 0)$	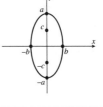
Ellipse	$\dfrac{x^2}{b^2} + \dfrac{y^2}{a^2} = 1$	$a \geq b > 0$ $c^2 = a^2 - b^2$ foci: $(0, c), (0, -c)$ vertices: $(0, a), (0, -a)$ constant sum $= 2a$ center: $(0, 0)$ major axis: line segment from $(0, -a)$ to $(0, a)$	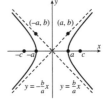

Hyperbola	$\dfrac{x^2}{a^2} - \dfrac{y^2}{b^2} = 1$	$c^2 = a^2 + b^2$ foci: $(c, 0), (-c, 0)$ vertices: $(a, 0), (-a, 0)$ constant difference $= 2a$ center $(0, 0)$ asymptotes: $y = \dfrac{b}{a}x,$ $\qquad\qquad y = -\dfrac{b}{a}x$	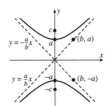
Hyperbola	$\dfrac{y^2}{a^2} - \dfrac{x^2}{b^2} = 1$	$c^2 = a^2 + b^2$ foci: $(0, c), (0, -c)$ vertices: $(0, a), (0, -a)$ constant difference $= 2a$ center: $(0, 0)$ asymptotes: $y = \dfrac{a}{b}x,$ $\qquad\qquad y = -\dfrac{a}{b}x$	

A second degree equation in x and y (with no xy term) represents a conic section whose center or vertex may be shifted from $(0, 0)$. To determine the type of conic, complete the square for x and y as in this example.

Example: What type of conic is given by the equation $x^2 + 6x + 4y^2 - 8y = 3$?
Complete the square by adding 9 and $4(1)$ to both sides:

$$x^2 + 6x \qquad + 4y^2 - 8y \qquad = 3$$
$$x^2 + 6x + 9 + 4(y^2 - 2y + 1) = 3 + 9 + 4(1).$$
$$(x + 3)^2 + 4(y - 1)^2 = 16$$
$$\frac{(x + 3)^2}{16} + \frac{(y - 1)^2}{4} = 1.$$

This is an ellipse shifted 3 units left and one unit upward.
Its center is $(-3, 1)$, $a = 4$ and $b = 2$.

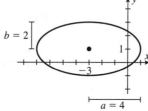

1) Find the vertices, foci, directrix (if a parabola), and asymptotes (if a hyperbola). Sketch the graph of each:

a) $\frac{x^2}{144} = 1 + \frac{y^2}{25}$.

Rewrite in standard form: $\frac{x^2}{144} - \frac{y^2}{25} = 1$.

This is a hyperbola with $a = 12, b = 5$.
$c^2 = 12^2 + 5^2 = 169, c = 13$.
Foci: $(13, 0), (-13, 0)$.
Vertices: $(12, 0), (-12, 0)$.
Asymptotes: $y = \frac{5}{12}x, y = -\frac{5}{12}x$.

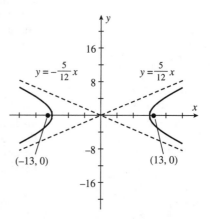

b) $x^2 = 4x + 8y - 4.$

$x^2 = 4x + 8y - 4$

$x^2 - 4x + 4 = 8y$

$(x - 2)^2 = 4(2)y$

This is a parabola, shifted two units to the right. The vertex is $(2, 0)$.

$p = 2$; the directrix is $y = -2$.

The focus is $(2, 2)$, also shifted two units.

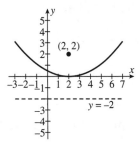

2) What is the conic given by each?

a) $2x(6 - x) = y(8 + y)$

<Put each in standard form.>

$2x(6 - x) = y(8 + y)$

$-2x^2 + 12x = y^2 + 8y$

$2x^2 - 12x + y^2 + 8y = 0$

$2(x^2 - 6x + 9) + y^2 + 8y + 16 = 18 + 16$

$2(x - 3)^2 + (y + 4)^2 = 34$

$\dfrac{(x - 3)^2}{17} + \dfrac{(y + 4)^2}{34} = 1.$

This is an ellipse, $a = \sqrt{34}, b = \sqrt{17}$.

b) $x(2 + x) - 24y = x^2 + 4y^2$

$x(2 + x) - 24y = x^2 + 4y^2$

$2x + x^2 - 24y = x^2 + 4y^2$

$2x = 4y^2 + 24y$

$x = 2y^2 + 12y = 2(y^2 + 6y)$

$x + 18 = 2(y^2 + 6y + 9) = 2(y + 3)^2$

$(y + 3)^2 = \frac{1}{2}(x + 18)$

This is a parabola opening to the right with vertex $(-18, -3)$.

Section 10.6 Conic Sections in Polar Coordinates

This section gives a polar coordinate definition for each conic section using just one focus and one directrix. This approach leads to simple forms for the equations of parabolas, ellipses, and hyperbolas in polar coordinates.

Concepts to Master

A. Eccentricity; Eccentric definition of conic sections; Equations of conics in polar form

B. Technology Plus

Summary and Focus Questions

Page 702

A. Another way to define conic sections is to specify a fixed point F (the focus), a fixed line l (the directrix), and a positive constant e called the **eccentricity***. Then the set of all points P such that

$$\frac{\text{distance from } P \text{ to } F}{\text{distance from } P \text{ to } l} = e$$

is a conic section. The value of e determines whether the conic is a parabola, ellipse, or hyperbola:

Eccentricity	Type	Graph	Ratio
$e = 1$	Parabola		$\dfrac{\|PF\|}{\|Pl\|} = 1$
$e < 1$	Ellipse		$\dfrac{\|PF\|}{\|Pl\|} = e < 1$
$e > 1$	Hyperbola		$\dfrac{\|PF\|}{\|Pl\|} = e > 1$

* This use of e for eccentricity is not to be confused with the use of e as the symbol for the base of natural logarithms.

Suppose a conic with eccentricity $e > 0$ is drawn in a rectangular coordinate system with focus F at $(0, 0)$, and one of the lines $x = d$, $x = -d$, $y = d$, or $y = -d$, where $d > 0$, is the directrix l. By superimposing a polar coordinate system the polar equation of the conic has one of these forms:

Conic	**Polar Form of Equation**			
Equation:	$r = \dfrac{ed}{1 + e\cos\theta}$	$r = \dfrac{ed}{1 - e\cos\theta}$	$r = \dfrac{ed}{1 + e\sin\theta}$	$r = \dfrac{ed}{1 - e\sin\theta}$
Directrix:	$x = d$	$x = -d$	$y = d$	$y = -d$
Parabola ($e = 1$)				
Ellipse ($0 < e < 1$)				
Hyperbola ($e > 1$)				

1) The sketch below shows one point P on a conic and its distances from the focus and directrix. What type of conic is it?

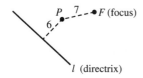

$e = \dfrac{|PF|}{|Pl|} = \dfrac{7}{6} > 1.$ The conic is a hyperbola.

2) What is the polar equation of the conic with:

a) eccentricity 2, directrix $y = 3$.

$e = 2$ and $d = 3$, so $r = \dfrac{6}{1 + 2 \sin \theta}$.

b) eccentricity $\frac{1}{2}$, directrix $x = -3$.

$e = \frac{1}{2}$ and $d = 3$, so $r = \dfrac{\frac{3}{2}}{1 - \frac{1}{2} \cos \theta}$,

$r = \dfrac{3}{2 - \cos \theta}$.

c) directrix $x = 4$ and is a parabola.

$e = 1$ and $d = 4$, so $r = \dfrac{4}{1 + \cos \theta}$.

3) What polar form does the equation of the graph below have?

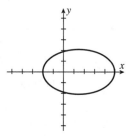

The graph is that of an ellipse. The directrix has the form $x = -d$ so the equation is

$r = \dfrac{ed}{1 - e \cos \theta}$ with $0 < e < 1$.

4) Find the eccentricity and directrix and identify the conic given by $r = \dfrac{3}{4 - 5 \cos \theta}$.

Divide numerator and denominator by 4.

$r = \dfrac{\frac{3}{4}}{1 - \frac{5}{4} \cos \theta}$. We see that $e = \dfrac{5}{4}$.

Since $ed = \frac{3}{4}$, $d = \frac{3}{4} \cdot \frac{1}{e} = \frac{3}{4} \cdot \frac{4}{5} = \frac{3}{5}$.

The trigonometric term is $-\cos \theta$ so the directrix is $x = -\frac{3}{5}$. Since $e > 1$, the conic is a hyperbola.

B.   Technology Plus. Use a computer algebra system or a graphing calculator to solve.

T-1) Sketch a graph of $r = \dfrac{1}{1 - \frac{1}{2} \cos \theta}$ on a graphing calculator. What type of conic is it?

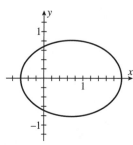

The graph is an ellipse with eccentricity $e = \frac{1}{2}$.

Chapter 11 — Infinite Sequences and Series

Section 11.1   Sequences

The main topic of this chapter is the representation of a function as a series. A series (introduced in the next section) is a sum of an infinite list of numbers— that is, the sum of a sequence of numbers. This section describes the basic concepts for sequences.

Concepts to Master

A. Sequences; Limit of a sequence; Convergence and divergence
B. Monotone sequence (increasing, decreasing); Bounded sequence; Monotonic Sequence Theorem
C. Sequences defined recursively
D. Technology Plus

Summary and Focus Questions

Page 714

A. A **sequence** is an infinite list of numbers given in a specific order:
$$\{a_n\} = \{a_1, a_2, a_3, \ldots, a_n, a_{n+1}, \ldots\}$$
a_n is called the nth **term** of the sequence.

Examples: a) The sequence $a_n = \dfrac{1}{n^2}$ has first few terms $1, \frac{1}{4}, \frac{1}{9}, \frac{1}{16}, \frac{1}{25}, \ldots$.
For $n = 10$, $a_{10} = \dfrac{1}{100}$.

b) The sequence $b_n = (-1)^n$ has alternating terms: $-1, 1, -1, 1, -1, 1, \ldots$.
Here $b_{109} = (-1)^{109} = -1$.

The sequence $\{a_n\}$ **converges** to a number L, written $\displaystyle\lim_{n\to\infty} a_n = L$, means that the values of a_n get closer and closer to L as n grows larger. If $\displaystyle\lim_{n\to\infty} a_n$ does not exist, we say $\{a_n\}$ **diverges.**

Sometimes writing out the first several terms of the sequence helps reveal whether it converges.

Examples: For the sequences above,

a) $a_n = \dfrac{1}{n^2}$ converges. The terms of a_n get smaller as n increases. Thus, $\displaystyle\lim_{n\to\infty} \frac{1}{n^2} = 0$.

33

b) $b_n = (-1)^n$ diverges because the terms of b_n do not settle in near one particular number.

The formal definition of **convergence** of $\{a_n\}$ is:

$\lim\limits_{n \to \infty} a_n = L$ means for all $\epsilon > 0$ there exists a positive integer N such that if $n > N$, then $|a_n - L| < \epsilon$.

$\lim\limits_{n \to \infty} a_n = \infty$ means the a_n terms grow without bound.

One way to evaluate $\lim\limits_{n \to \infty} a_n$ is to find a real function $f(x)$ such that $f(n) = a_n$ for all n. If $\lim\limits_{n \to \infty} f(x) = L$, then $\lim\limits_{n \to \infty} a_n = L$. The converse is false.

Example: To find $\lim\limits_{n \to \infty} \dfrac{\ln n}{n^2}$, we let $f(x) = \ln \dfrac{x}{x^2}$ and evaluate

$$\lim\limits_{x \to \infty} \frac{\ln x}{x^2} \left(\text{form } \frac{\infty}{\infty} \right) = \lim\limits_{x \to \infty} \frac{1/x}{2x} = \lim\limits_{x \to \infty} \frac{1}{2x^2} = 0. \text{ Therefore, } \frac{\ln n}{n^2} \text{ also converges to } 0.$$

Versions of the limit laws at infinity for functions are valid for convergent sequences. Thus, for example, $\lim\limits_{n \to \infty} ca_n = c \lim\limits_{n \to \infty} a_n$, where c is a constant. In addition,

if $\lim\limits_{n \to \infty} |a_n| = 0$, then $\lim\limits_{n \to \infty} a_n = 0$.

There is a version of the Squeeze Theorem for sequences:

if $a_n \le b_n \le c_n$ for all $n \ge N$ and $\lim\limits_{n \to \infty} a_n = \lim\limits_{n \to \infty} c_n = L$, then $\lim\limits_{n \to \infty} b_n = L$.

For a sequence whose nth term is a rational expression, such as $a_n = \dfrac{2n^2 - 4n + 7}{3n^2 + n + 10}$,

dividing both numerator and denominator by the highest power of n will yield an expression that may be easier to evaluate.

Example: $\lim\limits_{n \to \infty} \dfrac{2n^2 - 4n + 7}{3n^2 + n + 10} = \lim\limits_{n \to \infty} \dfrac{\dfrac{2n^2}{n^2} - \dfrac{4n}{n^2} + \dfrac{7}{n^2}}{\dfrac{3n^2}{n^2} + \dfrac{n}{n^2} + \dfrac{10}{n^2}} = \lim\limits_{n \to \infty} \dfrac{2 - \dfrac{4}{n} + \dfrac{7}{n^2}}{3 + \dfrac{1}{n} + \dfrac{10}{n^2}} = \lim\limits_{n \to \infty} \dfrac{2 - 0 + 0}{3 + 0 + 0} = \dfrac{2}{3}.$

Example: By the limit laws, $\lim\limits_{n \to \infty} \left(\dfrac{1}{n} + 3 \dfrac{n^2 - 1}{n^2 + 1} \right) = \lim\limits_{n \to \infty} \dfrac{1}{n} + 3 \lim\limits_{n \to \infty} \dfrac{n^2 - 1}{n^2 + 1} = 0 + 3(1) = 3.$

1) Find the fourth term of the sequence $a_n = \dfrac{(-1)^n}{n^2}$.

$a_4 = \dfrac{(-1)^4}{4^2} = \dfrac{1}{16}.$

2) $\lim\limits_{n \to \infty} x_n = K$ means

for all _____ there

exists _____

such that if _____ then

_____.

$\epsilon > 0$

a positive integer N

$n > N$

$|x_n - K| < \epsilon.$

3) Determine whether each converges.

    a) Let c_n be the decimal expansion of π to n places. For example, $c_5 = 3.14159$ (5 places after the decimal).

Converges. $\lim\limits_{n \to \infty} a_n = \pi$.

    b) $x_n = \dfrac{n}{n^2 + 1}$.

$\lim\limits_{n \to \infty} \dfrac{n}{n^2 + 1}$ (divide numerator and denominator by n^2)

$$= \lim_{n \to \infty} \frac{\frac{1}{n}}{1 + \frac{1}{n^2}} = \frac{0}{1 + 0} = 0.$$

Thus, $\{x_n\}$ converges to 0.

    c) $b_n = \dfrac{n^2 + 1}{2n}$.

b_n grows without bound, so $\{b_n\}$ diverges. In this case we may write

$$\lim_{n \to \infty} \frac{n^2 + 1}{2n} = \infty.$$

    d) $s_n = \dfrac{(-1)^n n}{n + 1}$.

The sequence $\dfrac{n}{n+1}$ converges to 1, so $s_n = \dfrac{(-1)^n n}{n + 1}$ alternately takes on values near 1 and -1. Thus, $\{s_n\}$ diverges.

    e) $d_n = \dfrac{n}{n + 1} + \dfrac{2}{n^2}$.

Converges.
$$\lim_{n \to \infty} d_n = \lim_{n \to \infty} \frac{n}{n + 1} + \lim_{n \to \infty} \frac{2}{n^2} = 1 + 0 = 1.$$

4) Find $\lim\limits_{n \to \infty} \dfrac{\cos n}{n}$.

Since $-1 \le \cos n \le 1$ for all n,

$-\dfrac{1}{n} \le \dfrac{\cos n}{n} \le \dfrac{1}{n}$ for all n.

Because both $\left\{-\dfrac{1}{n}\right\}$ and $\left\{\dfrac{1}{n}\right\}$ converge to 0,

$\lim\limits_{n \to \infty} \dfrac{\cos n}{n} = 0.$

5) Find $\lim\limits_{n \to \infty} \dfrac{\sin x}{n}$.

Since $\sin x$ is a constant as far as n is concerned, $\lim\limits_{n \to \infty} \dfrac{\sin x}{n} = (\sin x) \lim\limits_{n \to \infty} \dfrac{1}{n}$
$= (\sin x)\,(0) = 0.$

Silliness: Don't conclude the limit is 6 with this "computation":

$$\lim_{n \to \infty} \frac{\sin x}{n} = \lim_{n \to \infty} \frac{\sin x}{\cancel{n}} = \lim_{n \to \infty} \text{six} = 6.$$

6) Find $\lim\limits_{n\to\infty} \dfrac{1-\cos\left(\frac{1}{n}\right)}{\sin\left(\frac{1}{n}\right)}$.

Let $f(x) = \dfrac{1-\cos\frac{1}{x}}{\sin\frac{1}{x}}$. Then, by

l'Hospital's Rule,

$$\lim_{x\to\infty} \frac{1-\cos\frac{1}{x}}{\sin\frac{1}{x}}\left(\text{form }\frac{0}{0}\right) = \lim_{x\to\infty} \frac{0+\left(\sin\frac{1}{x}\right)\left(\frac{-1}{x^2}\right)}{\left(\cos\frac{1}{x}\right)\left(\frac{-1}{x^2}\right)}$$

$$= \lim_{x\to\infty} \frac{\sin\frac{1}{x}}{\cos\frac{1}{x}} = \frac{0}{1} = 0. \text{ Therefore } \lim_{n\to\infty} \frac{1-\cos\left(\frac{1}{n}\right)}{\sin\left(\frac{1}{n}\right)} = 0.$$

7) True, False:

a) If $\{a_n\}$ and $\{b_n\}$ converge, then $\{a_n + b_n\}$ converges.

True.

b) If $\{a_n\}$ and $\{b_n\}$ diverge, then $\{a_n + b_n\}$ diverges.

False. For example, $a_n = n^2$ and $b_n = -n^2$ diverge, but $\{a_n + b_n\}$ is the constant sequence of all zeros that converges to zero.

c) If $a_n \leq b_n \leq c_n$ for all n, and $\{a_n\}$ and $\{c_n\}$ converge, then $\{b_n\}$ converges.

False. Let $\{c_n\}$ be the sequence in Exercise **3a**). Let $a_n = -c_n$ and $b_n = (-1)^n$ for all n. Then $\{c_n\}$ converges to π, $\{a_n\}$ converges to $-\pi$, but we have seen that $\{b_n\}$ diverges.

8) If $\lim\limits_{n\to\infty} s_n = 4$ and $\lim\limits_{n\to\infty} t_n = 2$, then

a) $\lim\limits_{n\to\infty} (8s_n - 2t_n) = $ _____.

$8(4) - 2(2) = 28.$

b) $\lim\limits_{n\to\infty} \dfrac{3s_n}{t_n} = $ _____.

$\dfrac{3(4)}{2} = 6.$

Page 720

B. Let $\{a_n\}$ be a sequence.

$\{a_n\}$ is **increasing** means $a_{n+1} > a_n$ for all n.
$\{a_n\}$ is **decreasing** means $a_{n+1} < a_n$ for all n.
$\{a_n\}$ is **monotonic** if it is either increasing or decreasing.
$\{a_n\}$ is **bounded above** means $a_n \leq M$ for some M and all n.
$\{a_n\}$ is **bounded below** means $a_n \geq m$ for some m and all n.
$\{a_n\}$ is **bounded** if it is both bounded above and bounded below.

Examples: Let $a_n = n^2 + 1$, $b_n = \frac{1}{n}$, and $c_n = \dfrac{(-1)^n}{n^2}$. Then

$\{a_n\}$ is increasing and not bounded above; $\{a_n\}$ is bounded below by 2.
$\{b_n\}$ is decreasing, bounded above by 1 and below by 0.
$\{c_n\}$ is neither increasing nor decreasing, and is bounded above by $\frac{1}{4}$ and below by -1.

Example: $t_n = 1 + \dfrac{1}{e^n}$ is a decreasing sequence because $e > 1$ and

$$t_n = 1 + \frac{1}{e^n} > 1 + \frac{1}{e^n} \cdot \frac{1}{e} = 1 + \frac{1}{e^{n+1}} = t_{n+1}.$$

Every convergent sequence is bounded. For example, the sequence $b_n = \dfrac{3600}{n^2}$ has several large first terms ($b_1 = 3600$, $b_2 = 900$, $b_3 = 400, \ldots$), but after the 60th term, $0 < b_n < 1$. Therefore, $\{b_n\}$ is bounded below by 0 and bounded above by 3600.

Monotonic Sequence Theorem:

If $\{a_n\}$ is bounded and monotonic, then $\{a_n\}$ converges.

Example: $t_n = 1 + \dfrac{1}{e^n}$ is a monotonic (decreasing) sequence and is bounded

$(1 < t_n < 2$ for all n). Therefore, $\lim\limits_{n \to \infty} 1 + \dfrac{1}{e^n}$ exists. (The limit is 1.)

The converse of the Monotonic Sequence Theorem is false—a convergent sequence need not be monotonic. For example, $\lim\limits_{n \to \infty} \dfrac{(-1)^n}{n^2} = 0$ but $\dfrac{(-1)^n}{n^2}$ is neither increasing nor decreasing.

---

9) Is $c_n = \frac{1}{3n}$ bounded above? bounded below?

$\left\{\frac{1}{3n}\right\}$ is bounded above by $\frac{1}{3}$ and bounded below by 0. (There are many other bounds.)

10) Is $s_n = \dfrac{n}{n+1}$ an increasing sequence?

Yes, because

$$s_n = \frac{n}{n+1}$$
$$< \frac{n}{n+1} + \frac{1}{(n+1)(n+2)}$$
$$= \frac{n(n+2)+1}{(n+1)(n+2)}$$
$$= \frac{n^2 + 2n + 1}{(n+1)(n+2)}$$
$$= \frac{(n+1)^2}{(n+1)(n+2)}$$
$$= \frac{n+1}{n+2} = s_{n+1}.$$

11) True or False:

a) If $\{a_n\}$ is not bounded below then $\{a_n\}$ diverges.

True.

b) If $\{a_n\}$ is decreasing and $a_n \geq 0$ for all n then $\lim\limits_{n \to \infty} a_n$ exists.

True.

c) If $\{a_n\}$ is bounded, then $\{a_n\}$ converges.

False. For example, $a_n = (-1)^n$.

C. A sequence $\{a_n\}$ is **defined by a recurrence relation** (defined **recursively**) means:

   i) a_1 is defined.

   ii) a_{n+1}, for $n = 1, 2, 3, \ldots$, is defined in terms of previous a_i. (Often a_{n+1} is defined using only a_n.)

For example, the sequence $\{a_n\}$ given by

$$a_1 = \frac{1}{2}$$
$$a_{n+1} = \frac{a_n}{2}, n = 1, 2, 3, \ldots$$

is the sequence $\frac{1}{2}, \frac{1}{4}, \frac{1}{8}, \frac{1}{16}, \ldots$. This is a recursive definition of $a_n = \frac{1}{2^n}$.

---

12) Define the sequence $\frac{1}{2}, \frac{3}{4}, \frac{7}{8}, \frac{15}{16}, \frac{31}{32}, \ldots$ with a recurrence relation.

$a_1 = \frac{1}{2}, a_2 = \frac{3}{4} = \frac{1}{2} + \frac{1}{4} = a_1 + \frac{1}{4}$,

$a_3 = \frac{7}{8} = \frac{3}{4} + \frac{1}{8} = a_2 + \frac{1}{8}$.

The pattern shows that

$a_{n+1} = a_n + \frac{1}{2^{n+1}}$, for $n = 1, 2, 3, \ldots$

(Note: a non-recursive answer is

$a_n = \frac{2^n - 1}{2^n}$ for $n = 1, 2, 3, \ldots$)

13) Let $a_1 = 2$ and $a_{n+1} = \frac{2a_n}{3}$. Find $\lim\limits_{n \to \infty} a_n$.

$a_1 = 2, a_2 = \frac{2(2)}{3} = \frac{4}{3}$,

$a_3 = \frac{2\left(\frac{4}{3}\right)}{3} = \frac{8}{9}, a_4 = \frac{2\left(\frac{8}{9}\right)}{3} = \frac{16}{27}$,

$a_5 = \frac{2\left(\frac{16}{27}\right)}{3} = \frac{32}{81}$.

Since the denominator is growing faster than the numerator, it seems that $\lim\limits_{n \to \infty} a_n = 0$. This is so since a non-recursive formula for a_n is $\frac{2^n}{3^{n-1}} = 2\left(\frac{2}{3}\right)^{n-1}$.

14) Suppose $\{a_n\}$ is defined as $a_1 = 2$, $a_2 = 1$ and $a_{n+2} = a_{n+1} - a_n$ for $n = 1, 2, 3, \ldots$. Find $\lim\limits_{n \to \infty} a_n$.

Calculate a few terms to understand the pattern.

$a_1 = 2, a_2 = 1$

$a_3 = 1 - 2 = -1$

$a_4 = -1 - 1 = -2$

$a_5 = -2 - (-1) = -1$

$a_6 = -1 - (-2) = 1$

$a_7 = 1 - (-1) = 2$

$a_8 = 2 - 1 = 1$

$a_9 = 1 - 2 = -1$

$\vdots$

The terms repeat the pattern $1, -1, 2, -1$, $1, 2$ every six terms. $\lim\limits_{n \to \infty} a_n$ does not exist.

D. Technology Plus. Use a computer algebra system or a graphing calculator to solve.

T-1) Using a spreadsheet or calculator, find the first 20 partial sums of $\sum_{n=1}^{\infty} \dfrac{1}{n^4 + 1}$. Estimate the sum of the series.

n	$\dfrac{1}{n^4 + 1}$	$\sum_{k=1}^{n} \dfrac{1}{k^4 + 1}$
1	0.50000	0.50000
2	0.05882	0.55882
3	0.01220	0.57102
4	0.00389	0.57491
5	0.00160	0.57651
6	0.00077	0.57728
7	0.00042	0.57770
8	0.00024	0.57794
9	0.00015	0.57809
10	0.00010	0.57819
11	0.00007	0.57826
12	0.00005	0.57831
13	0.00004	0.57835
14	0.00003	0.57838
15	0.00002	0.57840
16	0.00002	0.57842
17	0.00001	0.57843
18	0.00001	0.57844
19	0.00001	0.57845
20	0.00001	0.57846

$$\sum_{n=1}^{\infty} \dfrac{1}{n^4 + 1} \approx 0.57846.$$

Section 11.2   Series

A series is the sum of all the terms of an infinite sequence. To determine whether (and if so, to what) a series sums, we build a second sequence consisting of the first term, the sum of the first two terms, the sum of the first three terms, the sum of the first four terms, etc. The sum of the series exists (that is, the series converges) if the limit of this sequence of partial sums exists. In this section we define these concepts precisely and look at some specific series which converge and others which do not.

Concepts to Master

A. Infinite series, Partial sums; Convergent and divergent series; Convergent series laws

B. Geometric series; Value of a converging geometric series; Harmonic series

Summary and Focus Questions

Page 727

A. Adding up all the terms of a sequence $\{a_n\}$ is an (**infinite**) **series**:

$$\sum_{k=1}^{\infty} a_k = a_1 + a_2 + a_3 + \ldots + a_n + \ldots$$

If we add up just the first n terms, we have the **nth partial sum** of the series:

$$s_n = \sum_{k=1}^{n} a_k = a_1 + a_2 + a_3 + \ldots + a_n.$$

A series **converges** if the limit of its sequence of partial sums exists. In other words $\sum_{n=1}^{\infty} a_n$ converges (to a number s) means $\lim_{n \to \infty} s_n$ exists (and is s).

If $\lim_{n \to \infty} s_n$ does not exist, then $\sum_{n=1}^{\infty} a_n$ **diverges.**

Example: Let $a_n = \frac{1}{2^n}$. Then $s_1 = a_1 = \frac{1}{2}$, $s_2 = a_1 + a_2 = \frac{1}{2} + \frac{1}{4} = \frac{3}{4}$, $s_3 = a_1 + a_2 + a_3 = \frac{1}{2} + \frac{1}{4} + \frac{1}{8} = \frac{7}{8}$. In general, s_n is $\frac{2^n - 1}{2^n}$, which is a sequence that converges to 1. Thus, $\sum_{n=1}^{\infty} \frac{1}{2^n} = 1$.

Test for Divergence:

If $\lim_{n \to \infty} a_n$ does not exist or $\lim_{n \to \infty} a_n \neq 0$, then $\sum_{n=1}^{\infty} a_n$ **diverges.**

Example: The series $\sum_{n=1}^{\infty} 2^n$ diverges since $\lim_{n \to \infty} 2^n \neq 0$.

Important: Just because the terms a_n approach zero is not enough to conclude that $\sum_{n=1}^{\infty} a_n$ converges. We will see an example in part B.

If $\sum\limits_{n=1}^{\infty} a_n$ converges to L and $\sum\limits_{n=1}^{\infty} b_n$ converges to M then:

$\sum\limits_{n=1}^{\infty} (a_n + b_n)$ converges to $L + M$.

$\sum\limits_{n=1}^{\infty} (a_n - b_n)$ converges to $L - M$.

For any constant c, $\sum\limits_{n=1}^{\infty} ca_n = c \sum\limits_{n=1}^{\infty} a_n$ converges to cL.

1) A series converges if the _____

of the sequence of _____ exists.

limit

partial sums

2) Find the first six partial sums of $\sum\limits_{n=1}^{\infty} \frac{1}{n^2}$.

$s_1 = \frac{1}{1^2} = 1.$

$s_2 = \frac{1}{1^2} + \frac{1}{2^2} = 1 + \frac{1}{4} = \frac{5}{4} = 1.25$

$s_3 = \frac{1}{1^2} + \frac{1}{2^2} + \frac{1}{3^2} = \frac{5}{4} + \frac{1}{9} = \frac{49}{36} \approx 1.3611$

$s_4 = \frac{49}{36} + \frac{1}{16} = \frac{205}{144} \approx 1.4236$

$s_5 = \frac{205}{144} + \frac{1}{5^2} = \frac{5269}{3600} \approx 1.4636.$

$s_6 = \frac{5269}{3600} + \frac{1}{6^2} = \frac{5369}{3600} \approx 1.4914.$

3) From the sequence of partial sums in

question 2), estimate $\sum\limits_{n=1}^{\infty} \frac{1}{n^2}$.

This is very difficult to guess. Perhaps

$\sum\limits_{n=1}^{\infty} \frac{1}{n^2}$ is near 1.5. The famous mathematician

Euler showed that $\sum\limits_{n=1}^{\infty} \frac{1}{n^2} = \frac{\pi^2}{6} \approx 1.6449$.

(The series converges very slowly.) Later we
will see methods to help with estimations.

4) Sometimes, Always, or Never:

a) If $\lim\limits_{n \to \infty} a_n = 0$, $\sum\limits_{n=1}^{\infty} a_n$ converges.

Sometimes.

b) If $\lim\limits_{n \to \infty} a_n \neq 0$, $\sum\limits_{n=1}^{\infty} a_n$ diverges.

Always.

5) Suppose $\sum\limits_{n=1}^{\infty} a_n = 3$ and $\sum\limits_{n=1}^{\infty} b_n = 4$.

Evaluate each of the following:

a) $\sum\limits_{n=1}^{\infty} (a_n + 2b_n)$.

$3 + 2(4) = 11.$

b) $\sum\limits_{n=1}^{\infty} \frac{a_n}{5}$.

$\frac{3}{5}.$

c) $\sum_{n=1}^{\infty} \dfrac{1}{a_n}.$

This series diverges. Because $\sum_{n=1}^{\infty} a_n = 3$, $\lim_{n\to\infty} a_n = 0$. Thus, $\lim_{n\to\infty} \dfrac{1}{a_n} \neq 0$, so $\sum_{n=1}^{\infty} \dfrac{1}{a_n}$ diverges.

6) Sometimes, Always, or Never:

a) $\sum_{n=1}^{\infty} a_n$ converges, but $\lim_{n\to\infty} a_n$ does not exist.

Never. If $\sum_{n=1}^{\infty} a_n$ converges, then $\lim_{n\to\infty} a_n$ exists and is 0.

b) $\sum_{n=1}^{\infty} a_n$ converges, $\sum_{n=1}^{\infty} (a_n + b_n)$ converges, but $\sum_{n=1}^{\infty} b_n$ diverges.

Never. $\sum_{n=1}^{\infty} b_n = \sum_{n=1}^{\infty} (a_n + b_n) - \sum_{n=1}^{\infty} a_n$ is the difference of two convergent series and must converge.

7) Does $\sum_{n=1}^{\infty} \cos\left(\dfrac{1}{n}\right)$ converge?

No, the nth term does not approach 0.

Page 730

B. We need to build up a library of series that are known to converge or diverge. Here are the first two.

A **geometric series** has the form

$$\sum_{n=1}^{\infty} ar^{n-1} = a + ar + ar^2 + ar^3 + \dots$$

and converges (for $a \neq 0$) to $\dfrac{a}{1-r}$ if and only if $-1 < r < 1$.

Geometric series are important because not only do we always know whether the series converges, when it does converge we know exactly what it converges to.

Example: The series $\sum_{n=1}^{\infty} 4\left(\dfrac{1}{3}\right)^{n-1}$ is geometric with $a = 4$ and $r = \dfrac{1}{3}$. Since $\dfrac{1}{3} < 1$,

$$\sum_{n=1}^{\infty} 4\left(\dfrac{1}{3}\right)^{n-1} \text{ converges and } \sum_{n=1}^{\infty} 4\left(\dfrac{1}{3}\right)^{n-1} = \dfrac{4}{1-\frac{1}{3}} = \dfrac{4}{\frac{2}{3}} = 6.$$

The **harmonic series** has the form

$$\sum_{n=1}^{\infty} \dfrac{1}{n} = 1 + \dfrac{1}{2} + \dfrac{1}{3} + \dots + \dfrac{1}{n} + \dots$$

This series diverges and is an example where $\lim_{n\to\infty} a_n = 0$ but $\sum_{n=1}^{\infty} a_n$ diverges.

8) a) $\sum_{n=1}^{\infty} \dfrac{2^n}{100}$ is a geometric series in which

$a = \underline{\quad}$ and $r = \underline{\quad}$.

b) Does the series converge?

$\sum_{n=1}^{\infty} \dfrac{2^n}{100} = \dfrac{2}{100} + \dfrac{4}{100} + \dfrac{8}{100} + \dots$

$a = \dfrac{2}{100}$ and $r = 2$.

No, since $r \geq 1$.

9) $\displaystyle\sum_{n=1}^{\infty} \left(\frac{-2}{9}\right)^n =$ _____.

This is a geometric series with $a = r = -\frac{2}{9}$

which converges to $\dfrac{-\frac{2}{9}}{1 - \left(-\frac{2}{9}\right)} = -\frac{2}{11}$.

10) Does $\displaystyle\sum_{n=1}^{\infty} \frac{6}{n}$ converge?

The series diverges, since it is a multiple of the harmonic series:

$$\sum_{n=1}^{\infty} \frac{6}{n} = 6 \sum_{n=1}^{\infty} \frac{1}{n}.$$

11) $\displaystyle\sum_{n=1}^{\infty} \frac{2^n}{3^{n+1}} =$ _____.

$\displaystyle\sum_{n=1}^{\infty} \frac{2^n}{3^{n+1}} = \sum_{n=1}^{\infty} \frac{2}{9}\left(\frac{2}{3}\right)^{n-1}$, which is a

geometric series with $a = \frac{2}{9}, r = \frac{2}{3}$ that

converges to $\dfrac{\frac{2}{9}}{1 - \frac{2}{3}} = \frac{2}{3}$.

12) In a race between Achilles and a tortoise Achilles gives the tortoise a 100 m head start. If Achilles runs at 5 m/s and the tortoise moves at $\frac{1}{2}$ m/s how far has the tortoise traveled by the time Achilles catches him?

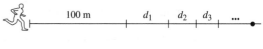

Let d_1 = distance tortoise traveled while Achilles was running the 100 m to the tortoise's starting point.

Since Achilles' velocity is 5 m/s,

$$d_1 = \left(\frac{100 \text{ m}}{5 \text{ m/s}}\right)\left(\frac{1}{2} \text{ m/s}\right) = 10 \text{ m}.$$

Let d_2 = distance tortoise traveled while Achilles was running the distance d_1.

Since $d_1 = 10, d_2 = \left(\frac{10 \text{ m}}{5 \text{ m/s}}\right)\left(\frac{1}{2} \text{ m/s}\right) = 1 \text{ m}.$

In general, for each n,

$$d_n = \left(\frac{d_{n-1}}{5}\right)\left(\frac{1}{2}\right) = \frac{d_{n-1}}{10}.$$

The total distance traveled by the tortoise is

$\displaystyle\sum_{n=1}^{\infty} d_n = 10 + 1 + \frac{1}{10}.$ This is a geometric

series with $a = 10, r = \frac{1}{10}.$

This series converges to $\dfrac{10}{1 - \frac{1}{10}} = \frac{100}{9}$ m.

Section 11.3   The Integral Test and Estimates of Sums

This section and the next three sections provide tests to determine whether a series converges without explicitly finding the sum. (It is a good first step to determine whether something exists before trying to calculate or estimate it.) The Integral Test in this section is a natural first choice, for it relates infinite series $\left(\sum\limits_{n=1}^{\infty} \ldots \right)$ to improper integrals $\left(\int_{1}^{\infty} \ldots \, dx \right)$. However, the test only applies to certain series of positive terms. The test, when applicable, also permits us to estimate the sum of the convergent series.

Concepts to Master

A. Integral Test for convergence; *p*-series

B. Estimate of the sum of a convergent series using the Integral Test

Summary and Focus Questions

Page 738

A. Integral Test: Let $\{a_n\}$ be a sequence of positive, decreasing terms ($0 < a_{n+1} < a_n$ for all n). Suppose $f(x)$ is a positive, continuous, and decreasing function on $[1, \infty)$ such that $a_n = f(n)$ for all n. Then

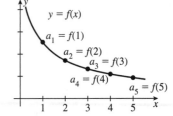

$$\sum_{n=1}^{\infty} a_n \text{ converges if and only if } \int_{1}^{\infty} f(x)dx \text{ converges.}$$

The Integral Test works well when a_n has the form of a function whose antiderivative is easily found. It is one of the few general tests that gives necessary and sufficient conditions for convergence.

Note that the Integral Test does not say $\sum\limits_{n=1}^{\infty} a_n = \int_{1}^{\infty} f(x)dx$. When these converge, they may be different numbers.

Example: To determine whether $\sum\limits_{n=1}^{\infty} \dfrac{2n}{n^2+1}$ converges, let $f(x) = \dfrac{2x}{x^2+1}$. Then

$$\int_{1}^{\infty} \frac{2x}{x^2+1}dx = \lim_{t\to\infty} \int_{1}^{t} \frac{2x}{x^2+1}dx = \lim_{t\to\infty} \ln(x^2+1)\Big]_{1}^{t} = \lim_{t\to\infty}\left(\ln(t^2+1) - \ln 2 \right) = \infty.$$

Since $\int_{1}^{\infty} \dfrac{2x}{x^2+1}dx$ diverges, $\sum\limits_{n=1}^{\infty} \dfrac{2n}{n^2+1}$ diverges by the Integral Test.

A *p*-series is a series of the form $\sum_{n=1}^{\infty} \frac{1}{n^p}$, where p is a constant.

If $p > 1$, then $\sum_{n=1}^{\infty} \frac{1}{n^p}$ converges.

If $p \leq 1$, then $\sum_{n=1}^{\infty} \frac{1}{n^p}$ diverges.

Examples:

The harmonic series $\sum_{n=1}^{\infty} \frac{1}{n}$ diverges because it is a *p*-series ($p = 1$).

The series $\sum_{n=1}^{\infty} \frac{1}{n^4}$ converges because it is a *p*-series ($p = 4$).

1) True or False:
The Integral Test will determine to what value a series converges.

False. The test only indicates whether a series converges.

2) Test $\sum_{n=1}^{\infty} \frac{n}{e^n}$ for convergence.

$\frac{n}{e^n}$ suggests the function $f(x) = xe^{-x}$.
On the interval $[1, \infty]$, f is continuous and decreasing, and $f(n) = \frac{n}{e^n}$ for $n = 1, 2, 3, \ldots$

$$\int_1^{\infty} xe^{-x} = \lim_{t \to \infty} \int_1^t xe^{-x} \, dx.$$

Using integration by parts,
($u = x$, $dv = e^{-x} \, dx$)

$$\int_1^t xe^{-x} \, dx = \left(-xe^{-x} - e^{-x}\right)\Big|_1^t = \frac{2}{e} - \frac{t+1}{e^t}.$$

$$\lim_{t \to \infty} \left(\frac{2}{e} - \frac{t+1}{e^t}\right) = \frac{2}{e} - 0 = \frac{2}{e}.$$

(By l'Hospital's Rule, $\frac{t+1}{e^t} \to 0$.)

Thus $\int_1^{\infty} xe^{-x} \, dx = \frac{2}{e}$, so $\sum_{n=1}^{\infty} \frac{n}{e^n}$ converges.

Note: We can *not* conclude that $\sum_{n=1}^{\infty} \frac{n}{e^n} = \frac{2}{e}$.

3) Which of these converge?

a) $\sum_{n=1}^{\infty} \frac{1}{n^2}$.

Converges (*p*-series with $p = 2$).

b) $\sum_{n=1}^{\infty} \frac{1}{\sqrt{n}}$.

Diverges $\left(\text{*p*-series with } p = \frac{1}{2}\right)$.

c) $\displaystyle\sum_{n=1}^{\infty} \frac{3}{2n^3}.$

Converges. $\displaystyle\sum_{n=1}^{\infty} \frac{3}{2n^3} = \frac{3}{2}\sum_{n=1}^{\infty} \frac{1}{n^3}$ is a multiple of a p-series with $p = 3$.

d) $\displaystyle\sum_{n=1}^{\infty} \left(\frac{1}{n^3} + \frac{1}{8^n}\right).$

Converges.

$$\sum_{n=1}^{\infty} \left(\frac{1}{n^3} + \frac{1}{8^n}\right) = \sum_{n=1}^{\infty} \frac{1}{n^3} + \sum_{n=1}^{\infty} \frac{1}{8^n},$$

the sum of a convergent p-series ($p = 3$) and a convergent geometric series $\left(r = \frac{1}{8}\right)$.

e) $\displaystyle\sum_{n=1}^{\infty} n^{-e}.$

Converges (p-series with $p = e$).

f) $\displaystyle\sum_{n=1}^{\infty} e^{-n}.$

Converges (This is not a p-series. It is a geometric series with $a = r = \frac{1}{e}$.)

Page 742

B. If $\displaystyle\sum_{n=1}^{\infty} a_n = s$ has been found to converge by the Integral Test using $f(x)$, then the **error** $R_n = s - s_n$ between the series value and the nth partial sum satisfies

$$\int_{n+1}^{\infty} f(x)\, dx \le R_n \le \int_{n}^{\infty} f(x)\,dx.$$

Example: The p-series $\displaystyle\sum_{n=1}^{\infty} \frac{1}{n^4}$ converges. (Remarkably, Euler showed that this series

converges to $\dfrac{\pi^4}{90} \approx 1.08232$.) Let $s = \displaystyle\sum_{n=1}^{\infty} \frac{1}{n^4}$ and $f(x) = \dfrac{1}{x^4}$. For $n = 5$,

$$\int_{6}^{\infty} \frac{1}{x^4}\, dx = \lim_{t\to\infty} \int_{6}^{t} \frac{1}{x^4}\,dx = \lim_{t\to\infty} \frac{-1}{3x^3}\Big]_{6}^{t} = \lim_{t\to\infty}\left(\frac{-1}{3t^3} - \frac{-1}{3(6)^3}\right) = \lim_{t\to\infty}\left(\frac{1}{648} - \frac{1}{3t^3}\right) = \frac{1}{648} \approx 0.00154.$$

Likewise, $\displaystyle\int_{5}^{\infty} \frac{1}{x^4}\,dx = \frac{1}{375} \approx 0.00267.$ Therefore, R_5, the difference between the infinite series sum s and the fifth partial sum s_5, satisfies:

$$0.00154 \le R_5 \le 0.00267.$$

To 5 decimals, the fifth partial sum is

$$s_5 = \sum_{n=1}^{5} \frac{1}{n^4} = \frac{1}{1^4} + \frac{1}{2^4} + \frac{1}{3^4} + \frac{1}{4^4} + \frac{1}{5^4} = \frac{1}{1} + \frac{1}{16} + \frac{1}{81} + \frac{1}{256} + \frac{1}{625} = \frac{14001361}{12960000} \approx 1.08035.$$

The actual value of $R_5 = s - s_5$ is $1.08232 - 1.08035 = 0.00197.$

4) In exercise **2** of Section 11.2, we found that the sixth partial sum of $\sum_{n=1}^{\infty} \frac{1}{n^2}$ is

$$s_6 = \frac{1}{1} + \frac{1}{4} + \frac{1}{9} + \frac{1}{16} + \frac{1}{25} + \frac{1}{36}$$

$$= \frac{5369}{3600} \approx 1.4914.$$

a) Estimate the difference R_6 between s_6 and the exact value of

$$s = \sum_{n=1}^{\infty} \frac{1}{n^2}.$$

We know that $\sum_{n=1}^{\infty} \frac{1}{n^2}$ converges (p-series, $p = 2$). Let $f(x) = \frac{1}{x^2}$.

$$\int_{7}^{\infty} \frac{1}{x^2}\, dx \le R_6 \le \int_{6}^{\infty} \frac{1}{x^2}\, dx.$$

$$\int_{7}^{\infty} \frac{1}{x^2}\, dx = \lim_{t \to \infty} \int_{7}^{t} x^{-2}\, dx$$

$$= \lim_{t \to \infty} \left(\frac{1}{7} - \frac{1}{t} \right) = \frac{1}{7}.$$

Likewise $\int_{6}^{\infty} \frac{1}{x^2}\, dx = \frac{1}{6}$.

Therefore, $\frac{1}{7} \le R_6 \le \frac{1}{6}$.

b) Estimate $s = \sum_{n=1}^{\infty} \frac{1}{n^2}$ using the results above.

From part a) $\frac{1}{7} \le s - s_6 \le \frac{1}{6}$

$$\frac{1}{7} + s_6 \le s \le \frac{1}{6} + s_6$$

$$\frac{1}{7} + \frac{5369}{3600} \le s \le \frac{1}{6} + \frac{5369}{3600}$$

$$\frac{41183}{25200} \le s \le \frac{5969}{3600}$$

$$1.6342 \le s \le 1.6581.$$

It turns out that $s = \frac{\pi^2}{6} \approx 1.6449$.

c) Calculate R_6 and compare it to the error bounds $\frac{1}{7}$ and $\frac{1}{6}$.

$R_6 = s - s_6 = 1.6449 - 1.4914 = 0.1535.$

R_6 is between $\frac{1}{7} \approx 0.1429$ and $\frac{1}{6} \approx 0.1667$.

5) How many terms of $\sum_{n=1}^{\infty} \frac{1}{n^2+1}$ are necessary to find the sum to within 0.01?

Let $f(x) = \frac{1}{x^2 + 1}$.

$\sum_{n=1}^{\infty} \frac{1}{n^2 + 1}$ converges because $\int_{1}^{\infty} \frac{1}{x^2 + 1}\, dx$

$$= \lim_{t \to \infty} \int_{1}^{t} \frac{1}{x^2 + 1}\, dx$$

$$= \lim_{t \to \infty} \left(\tan^{-1}(t) - \tan^{-1}(1) \right) = \frac{\pi}{2} - \frac{\pi}{4} = \frac{\pi}{4}.$$

Since $s - s_n \le \int_{n}^{\infty} f(x)\, dx$, it is sufficient to find n such that $\int_{n}^{\infty} f(x)\, dx \le 0.01$.

From above, we see that

$$\int_{n}^{\infty} \frac{1}{x^2 + 1}\, dx = \frac{\pi}{2} - \tan^{-1}(n).$$

$\frac{\pi}{2} - \tan^{-1}(n) < 0.01$

$\tan^{-1}(n) > \frac{\pi}{2} - 0.01$

$n > \tan\left(\frac{\pi}{2} - 0.01 \right) \approx 99.997.$

The first 100 terms are sufficient.

Section 11.4　The Comparison Tests

This section deals only with series of positive terms. If a series with positive terms converges, then any other positive series that is term for term smaller than the convergent series will also converge. This section makes precise this notion of comparing series of positive terms and discusses estimating limits.

Concepts to Master

A. Comparison Test; Limit Comparison Test

B. Estimate the sum of a convergent series

Summary and Focus Questions

Page 746

A. Let $\sum\limits_{n=1}^{\infty} a_n$ and $\sum\limits_{n=1}^{\infty} b_n$ be series whose terms are all positive.

Comparison Test:

　i)　If $a_n \leq b_n$ for all n and $\sum\limits_{n=1}^{\infty} b_n$ converges, then $\sum\limits_{n=1}^{\infty} a_n$ converges.

　ii)　If $a_n \geq b_n$ for all n and $\sum\limits_{n=1}^{\infty} b_n$ diverges, then $\sum\limits_{n=1}^{\infty} a_n$ diverges.

Limit Comparison Test:

Let c be a real number. If $\lim\limits_{n\to\infty} \dfrac{a_n}{b_n} = c > 0$, then

$\sum\limits_{n=1}^{\infty} a_n$ and $\sum\limits_{n=1}^{\infty} b_n$ either both converge or both diverge.

When $\lim\limits_{n\to\infty} \dfrac{a_n}{b_n} = 0$, if $\sum\limits_{n=1}^{\infty} b_n$ converges, then $\sum\limits_{n=1}^{\infty} a_n$ converges.

When $\lim\limits_{n\to\infty} \dfrac{a_n}{b_n} = \infty$, if $\sum\limits_{n=1}^{\infty} b_n$ diverges, then $\sum\limits_{n=1}^{\infty} a_n$ diverges.

For a given series $\sum\limits_{n=1}^{\infty} a_n$, success using either comparison test to determine whether $\sum\limits_{n=1}^{\infty} a_n$ converges depends on coming up with another series $\sum\limits_{n=1}^{\infty} b_n$ (geometric, p-series, …) whose convergence is known and for which a_n and b_n may be compared.

Example: Determine whether $\sum\limits_{n=1}^{\infty} \dfrac{1}{\sqrt{n}+4}$ converges by the Comparison Test.

To start, you need a hunch whether $\sum\limits_{n=1}^{\infty} \dfrac{1}{\sqrt{n}+4}$ converges. Because

$\dfrac{1}{\sqrt{n}+4}$ is "like" $\dfrac{1}{\sqrt{n}}$ and $\sum\limits_{n=1}^{\infty} \dfrac{1}{\sqrt{n}}$ is a divergent p-series $\left(p = \tfrac{1}{2}\right)$, we have

reason to believe that $\sum\limits_{n=1}^{\infty} \dfrac{1}{\sqrt{n}+4}$ diverges. We also have an idea of the type

of b_n to look for—one for which $a_n \geq b_n$:

Let $b_n = \dfrac{1}{3\sqrt{n}}$. Then for all n, $\dfrac{1}{\sqrt{n}+4} \geq \dfrac{1}{3\sqrt{n}}$. We see that $\sum\limits_{n=1}^{\infty} \dfrac{1}{3\sqrt{n}} = \dfrac{1}{3}\sum\limits_{n=1}^{\infty} n^{-\frac{1}{2}}$

is a divergent p-series. Therefore, $\sum\limits_{n=1}^{\infty} \dfrac{1}{\sqrt{n}+4}$ diverges.

Example: Determine whether $\sum\limits_{n=1}^{\infty} \dfrac{1}{\sqrt{n}+4}$ converges by the Limit Comparison Test.

Let $a_n = \dfrac{1}{\sqrt{n}+4}$. We choose $b_n = \dfrac{1}{\sqrt{n}}$ for all n because it is similar to a_n. We know that

$\sum\limits_{n=1}^{\infty} \dfrac{1}{\sqrt{n}}$ diverges (a p-series with $p = \tfrac{1}{2}$). Then

$$\lim_{n\to\infty} \dfrac{a_n}{b_n} = \lim_{n\to\infty} \dfrac{\frac{1}{\sqrt{n}+4}}{\frac{1}{\sqrt{n}}} = \lim_{n\to\infty} \dfrac{\sqrt{n}}{\sqrt{n}+4} = \lim_{n\to\infty} \sqrt{\dfrac{n}{n+4}} = 1.$$

Since $1 > 0$ and $\sum\limits_{n=1}^{\infty} \dfrac{1}{\sqrt{n}}$ diverges, $\sum\limits_{n=1}^{\infty} \dfrac{1}{\sqrt{n}+4}$ also diverges.

1) Determine whether each of the following converge. Find a series $\sum\limits_{n=1}^{\infty} b_n$ to use with the Comparison Test or Limit Comparison Test.

a) $\sum\limits_{n=1}^{\infty} \dfrac{1}{n^2+2n}$, $b_n = $ _____.

Because $\dfrac{1}{n^2+2n}$ is "like" $\dfrac{1}{n^2}$ for large n, and $\sum\limits_{n=1}^{\infty} \dfrac{1}{n^2}$ is a convergent p-series, we suspect that the given series converges. Use $b_n = \dfrac{1}{n^2}$. Since $\dfrac{1}{n^2+2n} < \dfrac{1}{n^2}$ and $\sum\limits_{n=1}^{\infty} \dfrac{1}{n^2}$ converges, $\sum\limits_{n=1}^{\infty} \dfrac{1}{n^2+2n}$ converges by the Comparison Test.

b) $\displaystyle\sum_{n=1}^{\infty} \frac{\sqrt[3]{n}}{n+4}$, $b_n = $ _____.

Since $\dfrac{\sqrt[3]{n}}{n+4}$ is "like" $\dfrac{\sqrt[3]{n}}{n} = \dfrac{1}{\sqrt[3]{n^2}}$,

use $b_n = \dfrac{1}{\sqrt[3]{n^2}}$.

Then $\displaystyle\lim_{n\to\infty} \frac{a_n}{b_n} = \lim_{n\to\infty} \frac{\dfrac{\sqrt[3]{n}}{n+4}}{\dfrac{1}{\sqrt[3]{n^2}}} = \lim_{n\to\infty} \frac{n}{n+4} = 1.$

Since $\displaystyle\sum_{n=1}^{\infty} \frac{1}{\sqrt[3]{n^2}}$ diverges $\left(\text{a } p\text{-series with}\right.$

$\left. p = \frac{2}{3}\right)$, $\displaystyle\sum_{n=1}^{\infty} \frac{\sqrt[3]{n}}{n+4}$ diverges by the Limit

Comparison Test.

c) $\displaystyle\sum_{n=1}^{\infty} \frac{1}{n+2^n}$, $b_n = $ _____.

Use $b_n = \dfrac{1}{2^n}$. Then for $n \ge 1$, $\dfrac{1}{n+2^n} \le \dfrac{1}{2^n}$.

Since $\displaystyle\sum_{n=1}^{\infty} \frac{1}{2^n}$ converges $\left(\text{a geometric series}\right.$

with $\left. r = \frac{1}{2}\right)$, $\displaystyle\sum_{n=1}^{\infty} \frac{1}{n+2^n}$ converges by the

Comparison Test.

2) Suppose $\{a_n\}$ and $\{b_n\}$ are positive sequences. Sometimes, Always, Never:

a) If $\displaystyle\lim_{n\to\infty} \frac{a_n}{b_n} = 0$ and $\displaystyle\sum_{n=1}^{\infty} a_n$ converges,

then $\displaystyle\sum_{n=1}^{\infty} b_n$ converges.

Sometimes. True for $a_n = \dfrac{1}{n^3}$ and $b_n = \dfrac{1}{n^2}$.

False for $a_n = \dfrac{1}{n^2}$ and $b_n = \dfrac{1}{n}$.

b) If $\displaystyle\lim_{n\to\infty} \frac{a_n}{b_n} = \infty$ and $\displaystyle\sum_{n=1}^{\infty} b_n$ diverges,

then $\displaystyle\sum_{n=1}^{\infty} a_n$ diverges.

Always.

Page 749

B. If $\displaystyle\sum_{n=1}^{\infty} a_n = s$ converges by the Comparison Test using $\displaystyle\sum_{n=1}^{\infty} b_n = t$, then

$$s - s_n \le t - t_n, \text{ for } n = 1, 2, 3, \dots.$$

This means that $T_n = t - t_n$ (which may be easier to calculate) is an upper bound for an estimate of $R_n = s - s_n$.

Example: Show that $s = \displaystyle\sum_{n=1}^{\infty} \frac{5+n}{n \cdot 4^n}$ converges by the Comparison Test and find an

upper bound for the difference between $\displaystyle\sum_{n=1}^{\infty} \frac{5+n}{n \cdot 4^n}$ and its fifth partial sum s_5.

We see that $0 \le \dfrac{5+n}{n \cdot 4^n} \le \dfrac{6n}{n \cdot 4^n} = \dfrac{6}{4^n}$ and $\displaystyle\sum_{n=1}^{\infty} \frac{6}{4^n}$ converges to $t = \dfrac{\frac{6}{4}}{1 - \frac{1}{4}} = \dfrac{\frac{6}{4}}{\frac{3}{4}} = 2$

because it is a geometric series (with $a = \frac{6}{4}$ and $r = \frac{1}{4}$). Therefore $\sum_{n=1}^{\infty} \frac{5+n}{n \cdot 4^n}$ converges.

The fifth partial sum is $t_5 = \frac{6}{4} + \frac{6}{4^2} + \frac{6}{4^3} + \frac{6}{4^4} + \frac{6}{4^5} \approx 1.99805$. Therefore

$$T_5 = 2.00000 - 1.99805 = 0.00195$$

is an upper bound for $R_5 = s - s_5$.

3) In part **c)** of question **1)** we saw that

$\sum_{n=1}^{\infty} \frac{1}{n + 2^n}$ converges.

a) Find s_4, the fourth partial sum.

$$s_4 = \frac{1}{1+2} + \frac{1}{2+4} + \frac{1}{3+8} + \frac{1}{4+16}$$
$$= \frac{1}{3} + \frac{1}{6} + \frac{1}{11} + \frac{1}{20} = \frac{141}{220}.$$

b) Estimate the difference between the sum of the series and its fourth partial sum.

From **1c)**, $b_n = \frac{1}{2^n}$ may be used to show

$\sum \frac{1}{n + 2^n}$ converges. Let $s = \sum_{n=1}^{\infty} \frac{1}{n + 2^n}$ and

$t = \sum_{n=1}^{\infty} \frac{1}{2^n}$. Then $s - s_4 \le t - t_4$.

Since t is the sum of a geometric series we can calculate it:

$$t = \frac{\frac{1}{2}}{1 - \frac{1}{2}} = 1.$$

$t_4 = \frac{1}{2} + \frac{1}{4} + \frac{1}{8} + \frac{1}{16} = \frac{15}{16}.$

Thus, $t - t_4 = 1 - \frac{15}{16} = \frac{1}{16}$ and

$s - s_4 \le \frac{1}{16}$. Therefore, we know $s_4 = \frac{141}{220}$

is within $\frac{1}{16}$ of the value of s.

Section 11.5   Alternating Series

An alternating series is one for which consecutive terms have opposite signs. This section describes a test for convergence of these kinds of series and a simple method to estimate the sum of a convergent alternating series.

Concepts to Master

A. Alternating Series Test

B. Estimating the sum of a convergent alternating series

C. Technology Plus

Summary and Focus Questions

A. An **alternating series** has successive terms of opposite sign—that is, it has one of these forms:

$$\sum_{n=1}^{\infty} (-1)^{n-1} a_n \ \text{ or } \ \sum_{n=1}^{\infty} (-1)^n a_n, \text{ where } a_n > 0.$$

Here are two examples of alternating series:

$$\sum_{n=1}^{\infty} \frac{(-1)^n}{n} = -\frac{1}{1} + \frac{1}{2} - \frac{1}{3} + \frac{1}{4} - \frac{1}{5} + \cdots$$

and $\displaystyle\sum_{n=1}^{\infty} (-1)^{n-1} n! = 1 - 2! + 3! - 4! + 5! - \ldots$.

The Alternating Series Test:

If $\{a_n\}$ is a decreasing sequence with $a_n > 0$ for all n and $\lim\limits_{n \to \infty} a_n = 0$,

then $\displaystyle\sum_{n=1}^{\infty} (-1)^{n-1} a_n$ converges.

Example: The series $\displaystyle\sum_{n=1}^{\infty} \frac{(-1)^n}{2^n}$ is an alternating series whose nth term is

$(-1)^n \frac{1}{2^n}$. Since $\lim\limits_{n \to \infty} \frac{1}{2^n} = 0$, the series converges.

Note: The Alternating Series Test is another "existence test"—it confirms convergence but does not identify the actual sum of the series.

1) Is $\displaystyle\sum_{n=1}^{\infty} \frac{\sin n}{n}$ an alternating series?

No, although some terms are positive and others negative.

2) Use the Alternating Series Test to determine whether each converge:

a) $\displaystyle\sum_{n=1}^{\infty} \frac{(-1)^{n-1}}{e^n}$.

This is an alternating series with $a_n = \dfrac{1}{e^n}$.

$\dfrac{1}{e^{n+1}} \le \dfrac{1}{e^n}$, so the terms decrease.

Since $\lim\limits_{n\to\infty} \dfrac{1}{e^n} = 0$, $\displaystyle\sum_{n=1}^{\infty} \frac{(-1)^{n-1}}{e^n}$ converges by the Alternating Series Test.

b) $\displaystyle\sum_{n=1}^{\infty} \frac{(-1)^n n}{n+1}$.

This is an alternating series but the other conditions for the test do not hold.

Since $\lim\limits_{n\to\infty} \dfrac{n}{n+1} \ne 0$, $\displaystyle\sum_{n=1}^{\infty} \frac{(-1)^n n}{n+1}$ diverges by the Divergence Test.

c) $\displaystyle\sum_{n=3}^{\infty} \frac{(-1)^{n-1} \ln n}{n}$. (Note that the series starts at $n = 3$.)

It may not be obvious that $a_n = \dfrac{\ln n}{n}$ decreases. Let $f(x) = \dfrac{\ln x}{x}$.

Then $f'(x) = \dfrac{x\left(\frac{1}{x}\right) - \ln x\,(1)}{x^2} = \dfrac{1 - \ln x}{x^2}$.

Since $n \ge 3$, $x \ge 3 > e$. Thus, $f'(x) < 0$ and f is decreasing. Therefore, $\{a_n\}$, $n \ge 3$, is decreasing. Finally,

$$\lim_{n\to\infty} \frac{\ln n}{n} = \lim_{n\to\infty} \frac{\frac{1}{n}}{1} = 0 \text{ by l'Hospital's Rule.}$$

Therefore, $\displaystyle\sum_{n=3}^{\infty} \frac{(-1)^{n-1} \ln n}{n}$ converges.

d) $\displaystyle\sum_{n=1}^{\infty} (-1)^{n-1} n!$

This series diverges. It is an alternating series but $\lim\limits_{n\to\infty} (-1)^{n-1} n!$ does not exist.

(Remember that for a series $\displaystyle\sum_{n=1}^{\infty} a_n$ to converge we must have $\lim\limits_{n\to\infty} a_n = 0$.)

B. Alternation Series Estimation Theorem:

If $\displaystyle\sum_{n=1}^{\infty} (-1)^{n-1} a_n$ is an alternating series with $a_{n+1} \le a_n$ and $\lim\limits_{n\to\infty} a_n = 0$ then $|s - s_n| \le a_{n+1}$.

If $R_n = s - s_n$, the theorem says s_n may be used to approximate s to within a_{n+1}. In other words, $|R_n| \leq a_{n+1}$ for all n.

3) a) Show that the series $\displaystyle\sum_{n=1}^{\infty} \frac{(-1)^n}{\sqrt{n} + 1}$ converges.

Since the series alternates and $\dfrac{1}{\sqrt{n} + 1}$ decreases to 0, the series converges.

b) Estimate the error between the sum of the series and its fifteenth partial sum.

The error $|s - s_{15}|$ does not exceed

$$a_{16} = \frac{1}{\sqrt{16} + 1} = 0.2.$$

4) For what value of n is s_n, the nth partial sum, within 0.001 of $s = \displaystyle\sum_{n=1}^{\infty} \frac{(-1)^{n+1}}{2n + 5}$?

Let $a_n = \dfrac{1}{2n + 5}$.

Since $|s - s_n| \leq a_{n+1}$ we need to find n such that $a_{n+1} < 0.001$.

$$a_{n+1} = \frac{1}{2(n + 1) + 5} = \frac{1}{2n + 7}.$$

$$\frac{1}{2n + 7} \leq 0.001$$

$$2n + 7 \geq 1000$$

$$2n \geq 993$$

$$n \geq 496.5.$$

Let $n = 497$. Then s_{497} is within 0.001 of s.

5) Approximate $\displaystyle\sum_{n=1}^{\infty} \frac{(-1)^{n+1}}{n^3}$ to within 0.005.

Let $a_n = \dfrac{1}{n^3}$. Then

$$|s - s_n| \leq a_{n+1} = \frac{1}{(n + 1)^3} \leq 0.005 = \frac{1}{200}.$$

$$(n + 1)^3 \geq 200$$

$$n + 1 \geq \sqrt[3]{200} \approx 5.84$$

$$n \geq 4.84. \text{ Let } n = 5.$$

Therefore, s_5 is within 0.005 of s.

$$s_5 = \frac{1}{1^3} - \frac{1}{2^3} + \frac{1}{3^3} - \frac{1}{4^3} + \frac{1}{5^3}$$

$$= 1 - \frac{1}{8} + \frac{1}{27} - \frac{1}{64} + \frac{1}{125} \approx 0.9044.$$

We do not know to what number the series converges, but we do know that 0.9044 is within 0.005 of the series sum.

C. Technology Plus. Use a computer algebra system or a graphing calculator to solve.

T-1) Use the 15th partial sum to estimate $\displaystyle\sum_{n=1}^{\infty} \frac{(-1)^{n-1}}{n^3}$. How accurate is the estimate?

n	$\dfrac{(-1)^{n-1}}{n^3}$	$\displaystyle\sum_{k=1}^{n} \frac{(-1)^{k-1}}{k^3}$
1	1.00000	1.00000
2	-0.12500	0.87500
3	0.03704	0.91204
4	-0.01563	0.89641
5	0.00800	0.90441
6	-0.00463	0.89978
7	0.00292	0.90270
8	-0.00195	0.90075
9	0.00137	0.90212
10	-0.00100	0.90112
11	0.00075	0.90187
12	-0.00058	0.90129
13	0.00046	0.90175
14	-0.00036	0.90139
15	0.00030	0.90169

The 15th partial sum is

$$s_{15} = \sum_{n=1}^{15} \frac{(-1)^{n-1}}{n^3} \approx 0.90169.$$

The absolute value of the 16th term is

$$\left| \frac{(-1)^{16-1}}{16^3} \right| \approx 0.00024.$$

The 15th partial sum is within approximately 0.00024 of the sum of the series.

Section 11.6 Absolute Convergence and the Ratio and Root Tests

The concept of convergence of an infinite series may be split into two separate subconcepts: absolute convergence and conditional convergence. This section gives a precise definition of each. It also describes the Root and Ratio Tests, which may be used to check for absolute convergence.

Concepts to Master

A. Absolute convergence; Conditional convergence

B. Ratio Test; Root Test

C. Rearrangement of terms of a series

Summary and Focus Questions

Page 756

A. For any series $\sum_{n=1}^{\infty} a_n$ (with a_n not necessarily positive or alternating) two types of convergence may be defined:

$\sum_{n=1}^{\infty} a_n$ **converges absolutely** means that $\sum_{n=1}^{\infty} |a_n|$ converges.

$\sum_{n=1}^{\infty} a_n$ **converges conditionally** means that $\sum_{n=1}^{\infty} a_n$ converges but $\sum_{n=1}^{\infty} |a_n|$ diverges.

Examples: a) $\sum_{n=1}^{\infty} \frac{(-1)^{n-1}}{n^3}$ is absolutely convergent because the series

$\sum_{n=1}^{\infty} \left| \frac{(-1)^{n-1}}{n^3} \right| = \sum_{n=1}^{\infty} \frac{1}{n^3}$ converges (p-series, $p = 3$).

b) $\sum_{n=1}^{\infty} \frac{(-1)^{n-1}}{\sqrt[3]{n}}$ is conditionally convergent because it converges (by the

Alternating Series Test) but $\sum_{n=1}^{\infty} \left| \frac{(-1)^{n-1}}{\sqrt[3]{n}} \right| = \sum_{n=1}^{\infty} \frac{1}{\sqrt[3]{n}}$ diverges

$\left(p\text{-series}, p = \frac{1}{3} \right)$.

Either one of absolute convergence or conditional convergence implies (ordinary) convergence. Conversely, convergence implies either absolute or conditional convergence (but not both!).

Thus *every* series must behave in exactly one of these three ways: diverge, converge absolutely, or converge conditionally.

1) Sometimes, Always, or Never:

a) If $\sum\limits_{n=1}^{\infty} |a_n|$ diverges, then $\sum\limits_{n=1}^{\infty} a_n$ diverges.

Sometimes. True for $a_n = n$ but false for $a_n = \dfrac{(-1)^n}{n}$.

b) If $\sum\limits_{n=1}^{\infty} a_n$ diverges, then $\sum\limits_{n=1}^{\infty} |a_n|$ diverges.

Always.

c) If $a_n \geq 0$ for all n, then $\sum\limits_{n=1}^{\infty} a_n$ is not conditionally convergent.

Always. Since $|a_n| = a_n$ for all n, we cannot have $\sum\limits_{n=1}^{\infty} a_n$ converges and $\sum\limits_{n=1}^{\infty} |a_n|$ diverges.

2) Determine whether each converges conditionally, converges absolutely, or diverges.

a) $\sum\limits_{n=1}^{\infty} \dfrac{(-1)^n}{\sqrt{n}}$.

The series is alternating with $\dfrac{1}{\sqrt{n}}$ decreasing to 0 so it converges. It remains to check for absolute convergence:

$$\sum_{n=1}^{\infty} \left| \frac{(-1)^n}{\sqrt{n}} \right| = \sum_{n=1}^{\infty} \frac{1}{\sqrt{n}} \text{ diverges } \left(p\text{-series} \right.$$

with $p = \frac{1}{2}$). Thus, $\sum\limits_{n=1}^{\infty} \dfrac{(-1)^n}{\sqrt{n}}$ converges conditionally.

b) $\sum\limits_{n=1}^{\infty} \dfrac{\sin n + \cos n}{n^3}$.

Check for absolute convergence first:

$$\sum_{n=1}^{\infty} \left| \frac{\sin n + \cos n}{n^3} \right| = \sum_{n=1}^{\infty} \frac{|\sin n + \cos n|}{n^3}.$$

Since $|\sin n + \cos n| \leq 2$ and $\sum\limits_{n=1}^{\infty} \dfrac{2}{n^3}$ converges (it is a p-series with $p = 3$), $\sum\limits_{n=1}^{\infty} \dfrac{|\sin n + \cos n|}{n^3}$ converges by the Comparison Test. Thus, $\sum\limits_{n=1}^{\infty} \dfrac{\sin n + \cos n}{n^3}$ converges absolutely.

3) Does $\displaystyle\sum_{n=1}^{\infty} \frac{\sin e^n}{e^n}$ converge?

Yes. The series is not alternating but does contain both positive and negative terms. Check for absolute convergence:

$$\sum_{n=1}^{\infty} \left| \frac{\sin e^n}{e^n} \right| = \sum_{n=1}^{\infty} \frac{|\sin e^n|}{e^n}.$$

Since $|\sin e^n| \leq 1$, $\dfrac{|\sin e^n|}{e^n} \leq \dfrac{1}{e^n}$.

$\displaystyle\sum_{n=1}^{\infty} \frac{1}{e^n}$ converges $\Big($geometric series with

$r = \dfrac{1}{e}\Big)$ so by the Comparison Test

$\displaystyle\sum_{n=1}^{\infty} \left| \frac{\sin e^n}{e^n} \right|$ converges. Therefore, $\displaystyle\sum_{n=1}^{\infty} \frac{\sin e^n}{e^n}$

converges absolutely and hence converges.

Page 758

B. The following tests are very useful for determining whether a series converges absolutely. Both tests decide whether $|a_n|$ approaches 0 fast enough so that $\displaystyle\sum_{n=1}^{\infty} a_n$ converges. Neither test involves another series or function, which makes them relatively easy to use. However, each has cases where it fails to provide any information.

The Ratio Test works well when the nth term, a_n, contains exponentials or factorials or when a_n is defined recursively. It will fail when a_n is a rational function of n.

Ratio Test: Let a_n be a sequence of nonzero terms.

i) If $\displaystyle\lim_{n\to\infty} \left| \frac{a_{n+1}}{a_n} \right| = L < 1$, then $\displaystyle\sum_{n=1}^{\infty} a_n$ converges absolutely.

ii) If $\displaystyle\lim_{n\to\infty} \left| \frac{a_{n+1}}{a_n} \right| = L > 1$ or $\displaystyle\lim_{n\to\infty} \left| \frac{a_{n+1}}{a_n} \right| = \infty$, then $\displaystyle\sum_{n=1}^{\infty} a_n$ diverges.

iii) If $\displaystyle\lim_{n\to\infty} \left| \frac{a_{n+1}}{a_n} \right| = 1$, the Ratio Test fails: $\displaystyle\sum_{n=1}^{\infty} a_n$ may converge or diverge.

The Root Test works well when a_n contains an expression to the nth power.

Root Test:

i) If $\displaystyle\lim_{n\to\infty} \sqrt[n]{|a_n|} = L < 1$, then $\displaystyle\sum_{n=1}^{\infty} a_n$ converges absolutely.

ii) If $\displaystyle\lim_{n\to\infty} \sqrt[n]{|a_n|} = L > 1$ or $\displaystyle\lim_{n\to\infty} \sqrt[n]{|a_n|} = \infty$, $\displaystyle\sum_{n=1}^{\infty} a_n$ diverges.

iii) If $\displaystyle\lim_{n\to\infty} \sqrt[n]{|a_n|} = 1$, the Root Test fails: $\displaystyle\sum_{n=1}^{\infty} a_n$ may converge or diverge.

The limit $\displaystyle\lim_{n\to\infty} \sqrt[n]{n} = 1$ is sometimes useful in applying the Root Test.

Examples: a) $\displaystyle\sum_{n=1}^{\infty} \frac{n^2}{n!}$ converges absolutely by the Ratio Test:

$$\lim_{n\to\infty} \left|\frac{a_{n+1}}{a_n}\right| = \lim_{n\to\infty} \left|\frac{\frac{(n+1)^2}{(n+1)!}}{\frac{n^2}{n!}}\right| = \lim_{n\to\infty} \frac{\frac{(n+1)^2}{(n+1)}}{n^2} = \lim_{n\to\infty} \frac{n+1}{n^2} = 0. \text{ Since this limit is less}$$

than 1 the series converges absolutely.

b) $\displaystyle\sum_{n=1}^{\infty} \frac{n}{e^{2n}}$ converges absolutely by the Root Test:

$$\lim_{n\to\infty} \sqrt[n]{|a_n|} = \lim_{n\to\infty} \sqrt[n]{\left|\frac{n}{e^{2n}}\right|} = \lim_{n\to\infty} \frac{\sqrt[n]{n}}{\sqrt[n]{e^{2n}}} = \lim_{n\to\infty} \frac{\sqrt[n]{n}}{e^2} = \frac{1}{e^2}. \text{ Since this limit is less}$$

than 1 the series converges absolutely.

4) Determine whether each converges by the Ratio Test.

a) $\displaystyle\sum_{n=1}^{\infty} \frac{n}{4^n}$

$$\lim_{n\to\infty} \left|\frac{a_{n+1}}{a_n}\right| = \lim_{n\to\infty} \frac{\frac{n+1}{4^{n+1}}}{\frac{n}{4^n}} = \lim_{n\to\infty} \frac{n+1}{4n} = \frac{1}{4}.$$

Thus, $\displaystyle\sum_{n=1}^{\infty} \frac{n}{4^n}$ converges absolutely and therefore converges.

b) $\displaystyle\sum_{n=1}^{\infty} \frac{(-4)^n}{n!}$

$$\lim_{n\to\infty} \left|\frac{a_{n+1}}{a_n}\right| = \lim_{n\to\infty} \left|\frac{\frac{(-4)^{n+1}}{(n+1)!}}{\frac{(-4)^n}{n!}}\right| = \lim_{n\to\infty} \frac{4}{n+1} = 0.$$

Thus, $\displaystyle\sum_{n=1}^{\infty} \frac{(-4)^n}{n!}$ converges absolutely and therefore converges.

c) $\displaystyle\sum_{n=1}^{\infty} \frac{(-1)^n}{\sqrt[3]{n}}$

$$\lim_{n\to\infty} \left|\frac{a_{n+1}}{a_n}\right| = \lim_{n\to\infty} \left|\frac{\frac{(-1)^{n+1}}{\sqrt[3]{n+1}}}{\frac{(-1)^n}{\sqrt[3]{n}}}\right|$$

$$= \lim_{n\to\infty} \sqrt[3]{\frac{n}{n+1}} = 1. \text{ The Ratio Test fails.}$$

The Alternating Series Test shows that this series converges. It converges conditionally because $\displaystyle\sum_{n=1}^{\infty} \frac{1}{\sqrt[3]{n}}$ diverges (*p*-series, $p = \frac{1}{3}$.)

d) $\sum\limits_{n=1}^{\infty} a_n$, where $a_1 = 4$ and

$$a_{n+1} = \frac{3a_n}{2n + 1}.$$

$$\lim_{n\to\infty} \left| \frac{a_{n+1}}{a_n} \right| = \lim_{n\to\infty} \frac{\dfrac{3a_n}{2n+1}}{a_n} = \lim_{n\to\infty} \frac{3}{2n+1} = 0.$$

Thus, $\sum\limits_{n=1}^{\infty} a_n$ converges absolutely and therefore converges.

5) Determine whether each converges by the Root Test:

a) $\sum\limits_{n=1}^{\infty} \frac{1}{(n+1)^n}$

$$\lim_{n\to\infty} \sqrt[n]{|a_n|} = \lim_{n\to\infty} \sqrt[n]{\frac{1}{(n+1)^n}} = \lim_{n\to\infty} \frac{1}{n+1} = 0.$$

Thus, $\sum\limits_{n=1}^{\infty} \frac{1}{(n+1)^n}$ converges absolutely and therefore converges.

b) $\sum\limits_{n=1}^{\infty} \frac{3^n}{n^3}$

$$\lim_{n\to\infty} \sqrt[n]{|a_n|} = \lim_{n\to\infty} \sqrt[n]{\frac{3^n}{n^3}} = \lim_{n\to\infty} \frac{3}{n^{3/n}} = \frac{3}{1} = 3.$$

Therefore by the Root Test $\sum\limits_{n=1}^{\infty} \frac{3^n}{n^3}$ diverges.

To show $\lim\limits_{n\to\infty} n^{3/n} = 1$ let $y = \lim\limits_{n\to\infty} n^{3/n}$.

Then $\ln y = \lim\limits_{n\to\infty} \ln n^{3/n} = \lim\limits_{n\to\infty} \frac{3 \ln n}{n}$

$$\left(\text{form } \frac{\infty}{\infty} \right) = \lim_{n\to\infty} \frac{\dfrac{3}{n}}{1} = 0.$$

Thus, $y = e^0 = 1$.

c) $\sum\limits_{n=1}^{\infty} \frac{(-1)^n}{n}$

$$\lim_{n\to\infty} \sqrt[n]{\left| \frac{(-1)^n}{n} \right|} = \lim_{n\to\infty} \frac{1}{\sqrt[n]{n}} = 1.$$

The Root Test fails, but we know this series converges because it is the alternating harmonic series.

C. A **rearrangement** of an infinite series is another series obtained from the series by changing the order of the terms. For example, $\frac{1}{2} + 1 + \frac{1}{8} + \frac{1}{4} + \frac{1}{32} + \frac{1}{16} + \cdots$ is a rearrangement of the geometric series $1 + \frac{1}{2} + \frac{1}{4} + \frac{1}{8} + \frac{1}{16} + \frac{1}{32} + \cdots$.

If $\sum\limits_{n=1}^{\infty} a_n$ converges absolutely to s, then all rearrangements converge to the same sum s.

If $\sum\limits_{n=1}^{\infty} a_n$ converges conditionally, its terms can be rearranged so that their sum is a different number.

6) Do all rearrangements of $\sum\limits_{n=1}^{\infty} \frac{(-1)^{n-1}}{n^4}$ converge to the same value?

Yes. The series is absolutely convergent since $\sum\limits_{n=1}^{\infty} \frac{1}{n^4}$ is a p-series ($p = 4$).

Section 11.7 **Strategy for Testing Series**

There are no new techniques in this section—just a summary of the tests to use for determining whether a given series converges.

Concepts to Master

Apply the various tests of convergence of a series.

Summary and Focus Questions

Page 763

To decide whether a given series $\sum\limits_{n=1}^{\infty} a_n$ converges you should determine the form of the nth term—whether it matches a certain type (p-series, geometric or alternating), resembles a known form (comparison tests or Integral Test), or is amenable using the Ratio or Root Tests. Here is a brief summary of each test together with a representative example of its use:

Test	Form/Conditions	Conclusion(s)	Example
Divergence	$\lim\limits_{n\to\infty} a_n \neq 0$	$\sum a_n$ diverges	$\sum\limits_{n=1}^{\infty} \dfrac{n}{n+1}$ diverges. $\lim\limits_{n\to\infty} \dfrac{n}{n+1} = 1 \neq 0$.
Integral	Find $f(x)$, continuous, decreasing with $f(n) = a_n \geq 0$.	$\sum a_n$ converges if and only if $\int_1^{\infty} f(x)\,dx$ converges.	$\sum\limits_{n=1}^{\infty} \dfrac{\ln n}{n}$ diverges. $\int_1^{\infty} \dfrac{\ln x}{x}\,dx$ does not exist.
p-series	$a_n = \dfrac{1}{n^p}$	$\sum \dfrac{1}{n^p}$ converges if and only if $p > 1$.	$\sum\limits_{n=1}^{\infty} \dfrac{1}{n^3}$ converges. $p = 3$.
Geometric Series	$a_n = ar^{n-1}$	$\sum ar^{n-1}$ converges if and only if $|r| < 1$.	$\sum\limits_{n=1}^{\infty} \dfrac{2}{3^n}$ converges. $r = \dfrac{1}{3}$.
Alternating Series	$b_n = (-1)^n a_n$ or $b_n = (-1)^{n-1} a_n$, where $a_n > 0$.	If $\lim\limits_{n\to\infty} a_n = 0$ and a_n is decreasing, $\sum b_n$ converges.	$\sum\limits_{n=1}^{\infty} \dfrac{(-1)^n}{n}$ converges. $a_n = \dfrac{1}{n}$.

Test	Form/Conditions	Conclusion(s)	Example
Comparison	$0 \le a_n \le b_n$	If $\sum b_n$ converges then $\sum a_n$ converges. If $\sum a_n$ diverges then $\sum b_n$ diverges.	$\sum\limits_{n=1}^{\infty} \dfrac{1}{n^3 + 1}$ converges. $b_n = \dfrac{1}{n^3}.$
Limit Comparison	$\lim\limits_{n\to\infty} \dfrac{a_n}{b_n} = c$	If $0 < c < \infty$, $\sum a_n$ converges if and only if $\sum b_n$ converges. If $c = 0$, then $\sum b_n$ converges implies $\sum a_n$ converges. If $c = \infty$, then $\sum a_n$ diverges implies $\sum b_n$ diverges.	$\sum\limits_{n=1}^{\infty} \dfrac{1}{2n^3 + 1}$ converges. $b_n = \dfrac{1}{n^3}.$ $\lim\limits_{n\to\infty} \dfrac{a_n}{b_n} = \dfrac{1}{2}.$
Ratio	$\lim\limits_{n\to\infty} \left| \dfrac{a_{n+1}}{a_n} \right| = L$	If $L > 1$, $\sum a_n$ diverges. If $L < 1$, $\sum a_n$ converges absolutely.	$\sum\limits_{n=1}^{\infty} \dfrac{n^2}{3^n}$ converges absolutely. $\lim\limits_{n\to\infty} \left| \dfrac{a_{n-1}}{a_n} \right| = \dfrac{1}{3}.$
Root	$\lim\limits_{n\to\infty} \sqrt[n]{|a_n|} = L$	If $L > 1$, $\sum a_n$ diverges. If $L < 1$, $\sum a_n$ converges absolutely.	$\sum\limits_{n=1}^{\infty} \left(\dfrac{2n+1}{3n+4}\right)^n$ converges absolutely. $\lim\limits_{n\to\infty} \sqrt[n]{|a_n|} = \dfrac{2}{3}.$

For each series, what is an appropriate test to apply for each?

1) $\sum\limits_{n=1}^{\infty} \dfrac{(n+1)^n}{4^n}.$

Root Test, because the $(n+1)^n$ and 4^n terms have n in the exponent.
$$\lim_{n\to\infty} \sqrt[n]{|a_n|} = \lim_{n\to\infty} \frac{n+1}{4} = \infty.$$
This series diverges.

2) $\sum\limits_{n=1}^{\infty} \dfrac{5}{n^2 + 6n + 3}.$

Comparison Test, compare to $\sum \dfrac{1}{n^2}$.
$$\frac{5}{n^2 + 6n + 3} \le 5 \cdot \frac{1}{n^2} \quad (\text{p-series, } p = 2).$$
This series converges.

3) $\displaystyle\sum_{n=1}^{\infty} \frac{(-1)^n n}{n^6 + 1}$.

Alternating Series Test for convergence.

$\displaystyle\lim_{n \to \infty} \frac{n}{n^6 + 1} = 0$ so the series converges.

Comparison Test (compare to $\sum \frac{1}{n^5}$) for absolute convergence.

$$\left| \frac{(-1)^n n}{n^6 + 1} \right| = \frac{n}{n^6 + 1} < \frac{1}{n^5}.$$

$\displaystyle\sum_{n=1}^{\infty} \frac{1}{n^5}$ converges (p-series, $p = 5$).

Therefore $\displaystyle\sum_{n=1}^{\infty} \frac{n}{n^6 + 1}$ converges.

This series converges absolutely.

4) $\displaystyle\sum_{n=1}^{\infty} \frac{2^n n^3}{n!}$.

Ratio Test, because of the $n!$ term.

$$\lim_{n \to \infty} \left| \frac{a_{n+1}}{a_n} \right| = \lim_{n \to \infty} \frac{2(n+1)^2}{n^3} = 0.$$

This series converges absolutely.

5) $\displaystyle\sum_{n=1}^{\infty} \frac{n}{e^{n^2}}$.

The exponential e^{n^2} suggests the Ratio Test (which will work) but the function $f(x) = \frac{x}{e^{x^2}} = xe^{-x^2}$ may be used in the Integral Test.

$$\int_1^{\infty} xe^{-x^2}\, dx = \frac{1}{2e}.$$

This series converges.

6) $\displaystyle\sum_{n=1}^{\infty} \frac{n!}{2^n}$.

Divergence Test.

$\displaystyle\lim_{n \to \infty} \frac{n!}{2^n} = \infty$. Since the nth term does not go to zero, this series diverges.

Section 11.8 Power Series

This section introduces a special type of function called a power series—a function f whose functional values are infinite series. As such, there may be values for its variable x for which the power series $f(x)$ is defined (for that x, the resulting series converges) and others for which $f(x)$ does not exist (resulting series diverges). The tests of convergence will help determine the domain of a power series.

Concepts to Master

Power series; Interval of Convergence; Radius of convergence

Summary and Focus Questions

Page 765

A **power series in** $(x-a)$ is an expression of the form

$$\sum_{n=0}^{\infty} c_n (x-a)^n, \text{ where } c_n \text{ and } a \text{ are constants.}$$

When a particular value of x is specified, the power series becomes an infinite series. Thus, for some values of x the expression converges and for other values of x it may diverge.

The domain of a power series in $(x-a)$ is all values of x for which $\sum_{n=0}^{\infty} c_n (x-a)^n$

converges and is called the **interval of convergence.** It consists of all real numbers or all real numbers from $a-R$ to $a+R$ for some number R, where $R \geq 0$. The power series diverges for all x outside the interval of convergence. The number a is called the **center** and R is the **radius** of convergence. When $R = 0$, the interval is the single point $\{a\}$. We use the convention $R = \infty$ to mean the interval of convergence is all real numbers.

The value of R is found by applying the Ratio Test to $\sum_{n=0}^{\infty} c_n (x-a)^n$ and solving for $|x-a|$ in the resulting inequality:

$$\lim_{n\to\infty} \left| \frac{c_{n+1}(x-a)^{n+1}}{c_n(x-a)^n} \right| = \lim_{n\to\infty} \left| \frac{c_{n+1}}{c_n} \right| |x-a| < 1.$$

(Sometimes the Root Test may be used.)

When R is a positive number, the two endpoints $a-R$ and $a+R$ pose special problems because the Ratio Test fails. Each endpoint may or may not be in the interval of convergence. You must use tests other than the Ratio Test to individually check the power series for convergence when $x = a-R$ and when $x = a+R$.

Example: The power series $\sum\limits_{n=0}^{\infty} \dfrac{(x-4)^n}{n+3} = \sum\limits_{n=0}^{\infty} \dfrac{1}{n+3}(x-4)^n$ has nth coefficient $\dfrac{1}{n+3}$.

The center of the interval of convergence is $a = 4$. To find the radius (by the Ratio Test) we solve $\lim\limits_{n\to\infty} \left| \dfrac{c_{n+1}}{c_n} \right| |x-4| < 1$.

$$\lim_{n\to\infty} \left| \frac{c_{n+1}}{c_n} \right| |x-4| = \lim_{n\to\infty} \left| \frac{\frac{1}{n+4}}{\frac{1}{n+3}} \right| |x-4| = \lim_{n\to\infty} \frac{n+3}{n+4}|x-4| = |x-4| < 1.$$

Thus, $-1 < x - 4 < 1$, or $3 < x < 5$. The radius of convergence is 1.

For $x = 3$, $\sum\limits_{n=0}^{\infty} \dfrac{(3-4)^n}{n+3} = \sum\limits_{n=0}^{\infty} \dfrac{(-1)^n}{n+3}$, which converges by the Alternating Series Test.

For $x = 5$, $\sum\limits_{n=0}^{\infty} \dfrac{(5-4)^n}{n+3} = \sum\limits_{n=0}^{\infty} \dfrac{1}{n+3}$, which diverges (it is all but the first three terms of the harmonic series).

The interval of convergence is $[3, 5)$.

Think of a power series as an infinitely long polynomial. (In fact every polynomial is a power series in which all coefficients except the first few are zero!) Just as polynomial function values are easy to calculate, we have seen that power series values are sometimes easy to approximate within a given degree of accuracy.

1) True or False:
  A power series is an infinite series.

False. A power series is an expression for a function $f(x)$. For any x in the domain of f, $f(x)$ is the sum of a convergent series.

2) For what value of x does $\sum\limits_{n=0}^{\infty} c_n(x-a)^n$ always converge?

At $x = a$, $\sum\limits_{n=0}^{\infty} c_n(x-a)^n = c_0$.
The power series always converges when x is the center a.

3) Compute the value of

  a) $\sum\limits_{n=0}^{\infty} n^2(x-3)^n$ at $x = 4$.

When $x = 4$ the power series is
$\sum\limits_{n=0}^{\infty} n^2 1^n = \sum\limits_{n=0}^{\infty} n^2$. This infinite series
diverges so there is no value for $x = 4$.

b) $\displaystyle\sum_{n=0}^{\infty} 2^n(x-1)^n$ at $x = \frac{5}{6}$.

When $x = \frac{5}{6}$ the power series is

$$\sum_{n=0}^{\infty} 2^n\left(-\frac{1}{6}\right)^n = \sum_{n=0}^{\infty} \left(-\frac{1}{3}\right)^n$$

$$= 1 - \frac{1}{3} + \frac{1}{9} - \frac{1}{27} + \cdots$$

This is a geometric series with $a = 1$,

$r = -\frac{1}{3}$. It converges to $\dfrac{1}{1 - \left(-\frac{1}{3}\right)} = \dfrac{3}{4}$.

4) Find the center, radius, and interval of convergence for:

a) $\displaystyle\sum_{n=0}^{\infty} (2n)!(x-1)^n$.

Use the Ratio Test:

$$\lim_{n\to\infty} \left| \frac{(2(n+1))!(x-1)^{n+1}}{(2n)!(x-1)^n} \right|$$

$$= \left(\lim_{n\to\infty} (2n+1)(2n+2) \right)|x-1| = \infty,$$

for all x except when $x = 1$. Thus, the center is 1, the radius is 0, and the interval of convergence is $\{1\}$.

b) $\displaystyle\sum_{n=1}^{\infty} \frac{(x-4)^n}{2^n n}$. (Although this series

starts at $n = 1$, we proceed in the same way.)

Use the Ratio Test:

$$\lim_{n\to\infty} \left| \frac{(x-4)^{n+1}}{2^{n+1}(n+1)} \cdot \frac{2^n n}{(x-4)^n} \right|$$

$$= \left(\lim_{n\to\infty} \frac{n}{2(n+1)} \right) |x-4| = \frac{1}{2}|x-4|.$$

From $\frac{1}{2}|x-4| < 1$ we conclude

$|x-4| < 2$. The center is 4 and radius is 2.

When $|x-4| = 2$, $x = 2$ or $x = 6$.

At $x = 2$, the power series becomes

$\displaystyle\sum_{n=1}^{\infty} \frac{(-1)^n}{n}$ which converges by the

Alternating Series Test.

At $x = 6$, the power series is the divergent

harmonic series $\displaystyle\sum_{n=1}^{\infty} \frac{1}{n}$.

The interval of convergence is $[2, 6)$.

Note that the Root Test could have been used to start the solution:

$$\lim_{n\to\infty} \sqrt[n]{\left| \frac{(x-4)^n}{2^n n} \right|} = \left(\lim_{n\to\infty} \sqrt[n]{\frac{1}{n}} \right) \frac{1}{2}|x-4|$$

$$= 1 \cdot \frac{1}{2}|x-4|$$

$$= \frac{1}{2}|x-4| < 1.$$

c) $\displaystyle\sum_{n=1}^{\infty} \frac{x^n}{n!}$.

Use the Ratio Test:

$$\lim_{n\to\infty} \left| \frac{\dfrac{x^{n+1}}{(n+1)!}}{\dfrac{x^n}{n!}} \right| = \lim_{n\to\infty} \frac{|x|}{n+1} = 0$$

for all values of x. Thus, the interval of convergence is all real numbers, $(-\infty, \infty)$.

5) Suppose $\displaystyle\sum_{n=0}^{\infty} c_n (x - 5)^n$ converges for $x = 8$. For what other values must it converge?

The interval of convergence is $5 - R$ to $5 + R$.

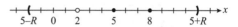

Because 8 is in this interval $8 \le 5 + R$. Thus, $3 \le R$ which means the interval at least contains all numbers between $2(= 5 - 3)$ and $8(= 5 + 3)$. We do not have enough information about the endpoint 2. The power series converges for at least all x such that $2 < x \le 8$.

6) True, False: The two power series

$$\sum_{n=0}^{\infty} c_n(x - a)^n \text{ and } \sum_{n=0}^{\infty} 2c_n (x - a)^n \text{ have}$$

exactly the same interval of convergence.

True.

Section 11.9   Representations of Functions as Power Series

In the last section we saw that a power series is an expression that may be used to define a function whose domain is either all real numbers or includes the interval $(a - R, a + R)$ and possibly either or both endpoints (where $R \geq 0$). In this section we turn things around and begin with a function f and show that it is possible to rewrite $f(x)$ as a power series for at least some of the elements in the domain of f.

Concepts to Master

A. Obtaining a power series expression for a function by manipulating known series

B. Differentiation and integration of power series

C. Power series expansions by differentiation and integration

Summary and Focus Questions

Page 770

A. The geometric series $\sum\limits_{n=1}^{\infty} ar^{n-1}$ converges to $\dfrac{a}{1-r}$ for $a \neq 0$ and $-1 < r < 1$.

If we use $a = 1$, $r = x$ and simplify $\sum\limits_{n=1}^{\infty} 1(x^{n-1})$ by rewriting it as $\sum\limits_{n=0}^{\infty} x^{n}$ (starting at $n = 0$), then

$$\frac{1}{1-x} = \sum_{n=0}^{\infty} x^{n}.$$

Even though the domain of $f(x) = \dfrac{1}{1-x}$ is all real numbers except $x = 1$, this expression is valid only for $-1 < x < 1$. Nevertheless, we have written the function f as a power series for some of the values in its domain.

In this section, to write a given function as a power series, we start with a known power series expression for a similar function (such as $f(x) = \dfrac{1}{1-x}$) and modify it by substitutions (such as x^2 for x) and multiplication by constants or powers of x. Here are two examples:

Example: Find a power series for $f(x) = \dfrac{3}{1 - 8x^3}$.

We start with the similar function $\dfrac{1}{1-x} = \sum\limits_{n=0}^{\infty} x^{n}$. Substitute $8x^3$ for x:

$$\frac{1}{1 - 8x^3} = \frac{1}{1 - (8x^3)} = \sum_{n=0}^{\infty} (8x^3)^{n} = \sum_{n=0}^{\infty} 8^{n} x^{3n}.$$

Then multiply by 3:

$$\frac{3}{1 - 8x^3} = 3\sum_{n=0}^{\infty} 8^n x^{3n} = \sum_{n=0}^{\infty} (3)8^n x^{3n}.$$

Since the interval of convergence for the original power series is $|x| < 1$, $|8x^3| < 1$. Thus, $|x^3| < \frac{1}{8}$, so $|x| < \frac{1}{2}$. The interval of convergence for f is $\left(-\frac{1}{2}, \frac{1}{2}\right)$.

Example: Find a power series for $f(x) = \dfrac{2x}{1 + x^2}$.

We start with $\dfrac{1}{1 - x} = \sum_{n=0}^{\infty} x^n$ with interval of convergence $(-1, 1)$.

Then substitute $-x^2$ for x:

$$\frac{1}{1 + x^2} = \sum_{n=0}^{\infty} (-x^2)^n = \sum_{n=0}^{\infty} (-1)^n x^{2n}.$$

Then multiply by $2x$:

$$\frac{2x}{1 + x^2} = \sum_{n=0}^{\infty} (-1)^n (2x) x^{2n} = \sum_{n=0}^{\infty} (-1)^n 2x^{2n+1}.$$

Since the interval of convergence for $\dfrac{1}{1 - x} = \sum_{n=0}^{\infty} x^n$ is $(-1, 1)$, $|x| < 1$. The substitution $-x^2$ for x yields $\left|-x^2\right| < 1$, which is still $|x| < 1$. Therefore, the interval of convergence for $\dfrac{2x}{1 + x^2} = \sum_{n=0}^{\infty} (-1)^n 2x^{2n+1}$ is $(-1, 1)$.

1) Given $\dfrac{1}{1 - x} = \sum_{n=0}^{\infty} x^n, |x| < 1$, find a power series expression for:

a) $\dfrac{1}{2 - x}$.

$$\frac{1}{2 - x} = \frac{1}{2\left(1 - \frac{x}{2}\right)} = \frac{1}{2}\left(\frac{1}{1 - \frac{x}{2}}\right).$$

Thus, $\dfrac{1}{2 - x} = \dfrac{1}{2}\sum_{n=0}^{\infty} \left(\dfrac{x}{2}\right)^n = \sum_{n=0}^{\infty} \dfrac{1}{2} \cdot \dfrac{x^n}{2^n}$

$$= \sum_{n=0}^{\infty} \frac{x^n}{2^{n+1}}.$$

From $\left|\dfrac{x}{2}\right| < 1$, $|x| < 2$, so $(-2, 2)$ is the interval of convergence.

b) $\dfrac{x}{x^4 - 1}$.

We note that $\dfrac{x}{x^4 - 1} = -\dfrac{x}{1 - x^4}$.

First substitute x^4 for x:

$$\frac{1}{1 - x^4} = \sum_{n=0}^{\infty} (x^4)^n = \sum_{n=0}^{\infty} x^{4n}.$$

Now multiply by $-x$:

$$\frac{-x}{1 - x^4} = -x \sum_{n=0}^{\infty} x^{4n} = -\sum_{n=0}^{\infty} x^{4n+1}.$$

Since $|x| < 1$, $|x^4| < 1$. Thus, the interval of convergence is still $(-1, 1)$.

B. If $f(x) = \displaystyle\sum_{n=0}^{\infty} c_n (x - a)^n$ has interval of convergence with radius $R \geq 0$ then:

Page 772

i) f is continuous on $(a - R, a + R)$.

ii) For all $x \in (a - R, a + R)$, f may be differentiated term by term:

$$f'(x) = \sum_{n=1}^{\infty} nc_n (x - a)^{n-1}.$$

Note that when the series for f starts at $n = 0$, we may write the series for f' starting at $n = 1$ since the derivative of c_0 is zero.

iii) For all $x \in (a - R, a + R)$, f may be integrated term by term:

$$\int f(x)dx = C + \sum_{n=0}^{\infty} \frac{c_n(x - a)^{n+1}}{n + 1} \text{ where } C \text{ is the constant of integration.}$$

Both $f'(x)$ and $\displaystyle\int f(x)dx$ have radius of convergence R but the endpoints $a - R$ and $a + R$ must be checked individually.

2) Find $f'(x)$ for

$$f(x) = \sum_{n=0}^{\infty} 2^n(x - 1)^n.$$

$$f'(x) = \sum_{n=0}^{\infty} 2^n n(x - 1)^{n-1}$$
$$= \sum_{n=1}^{\infty} 2^n n(x - 1)^{n-1}.$$

3) Find $\int f(x)\,dx$ where

$$f(x) = \sum_{n=1}^{\infty} \frac{(x-3)^n}{n!}.$$

$$\int f(x)\,dx = C + \sum_{n=0}^{\infty} \frac{1}{n!}\frac{(x-3)^{n+1}}{n+1}$$
$$= C + \sum_{n=0}^{\infty} \frac{(x-3)^{n+1}}{(n+1)!}.$$

4) Is $f(x) = \sum_{n=0}^{\infty} \frac{x^n}{4^n}$ continuous at $x = 2$?

Yes. 2 is in $(-4, 4)$, the interval of convergence for $f(x)$.

C. Term by term integration and differentiation is another technique to help find power series for functions.

Page 772

Example: To find a power series for $\ln(1 + x^2)$ we first note

that $\ln(1 + x^2) = \int \frac{2x}{1+x^2}\,dx$. Recall from part **A** that

$$\frac{2x}{1+x^2} = \sum_{n=0}^{\infty} (-1)^n 2x^{2n+1}.$$

Thus $\ln(1+x^2) = \int \frac{2x}{1+x^2}\,dx = \int \sum_{n=0}^{\infty} (-1)^n 2x^{2n+1}\,dx$

$$= \sum_{n=0}^{\infty} (-1)^n 2 \int x^{2n+1}\,dx = C + \sum_{n=0}^{\infty} (-1)^n 2 \frac{x^{2n+2}}{2n+2}$$

$$= C + \sum_{n=0}^{\infty} (-1)^n \frac{x^{2n+2}}{n+1}.$$

To determine C, let $x = 0$. Then $0 = \ln(1 + 0) = C + \sum_{n=0}^{\infty} (-1)^n \frac{0^{2n+2}}{n+1} =$

$C + 0$. Thus $C = 0$ and $\ln(1 + x^2) = \sum_{n=0}^{\infty} (-1)^n \frac{x^{2n+2}}{n+1}$.

This power series has the same interval of convergence $(-1, 1)$ as the power

series for $\frac{2x}{1+x^2}$ but we need to check the endpoints. At both $x = 1$ and

$x = -1$, $x^{2n+2} = 1$ and the resulting series $\sum_{n=0}^{\infty} \frac{(-1)^n}{n+1}$ converges. Therefore,

the interval of convergence for $\ln(1 + x^2) = \sum_{n=0}^{\infty} (-1)^n \frac{x^{2n+2}}{n+1}$ is $[-1, 1]$.

5) Given $\frac{1}{1-x} = \sum_{n=0}^{\infty} x^n$, $|x| < 1$, find a power series for $\frac{1}{(1+x)^2}$.

Substitute $-x$ for x:

$$\frac{1}{1+x} = \sum_{n=0}^{\infty} (-x)^n = \sum_{n=0}^{\infty} (-1)^n x^n.$$

Differentiate term by term:

$$\frac{-1}{(1+x)^2} = \sum_{n=1}^{\infty} (-1)^n n x^{n-1}.$$

$$\frac{1}{(1+x)^2} = (-1) \sum_{n=1}^{\infty} (-1)^n n x^{n-1}$$

$$= \sum_{n=1}^{\infty} (-1)^{n+1} n x^{n-1}.$$

Since $|-x| < 1$ is the same as $|x| < 1$, the radius of convergence is $R = 1$.

At $x = 1$ we have $\sum_{n=1}^{\infty} (-1)^{n+1} n$.

At $x = -1$ we have

$$\sum_{n=1}^{\infty} (-1)^{n+1} n (-1)^{n-1} = \sum_{n=1}^{\infty} n.$$

Both series diverge so the interval of convergence is $(-1, 1)$.

6) Find $\int \frac{1}{1+x^5} \, dx$ as a power series.

From $\frac{1}{1-x} = \sum_{n=0}^{\infty} x^n$, substitute $-x^5$ for x:

$$\frac{1}{1+x} = \sum_{n=0}^{\infty} (-x^5)^n$$

$$= \sum_{n=0}^{\infty} (-1)^n x^{5n}.$$

Integrate term by term:

$$\int \frac{1}{1+x^5} \, dx = \sum_{n=0}^{\infty} \int (-1)^n x^{5n} \, dx$$

$$= \sum_{n=0}^{\infty} \frac{(-1)^n}{5n+1} x^{5n+1} + C.$$

For the power series for $\frac{1}{1-x}$ the interval of convergence is $|x| < 1$. Substituting $-x^5$ for x, $|-x^5| < 1$ is still $|x| < 1$. We individually check the endpoints: At $x = 1$ the power series becomes

$$\sum_{n=0}^{\infty} \frac{(-1)^n}{5n+1} 1^{5n+1} + C = \sum_{n=0}^{\infty} \frac{(-1)^n}{5n+1} + C,$$

which is a converging alternating series.

At $x = -1$ the power series becomes

$$\sum_{n=0}^{\infty} \frac{(-1)^n}{5n+1}(-1)^{5n+1} + C$$

$$= \sum_{n=0}^{\infty} \frac{(-1)^{6n+1}}{5n+1} + C,$$

which is also a converging alternating series. The interval of convergence is $[-1, 1]$.

7) From $\dfrac{1}{1-x} = \displaystyle\sum_{n=0}^{\infty} x^n$, find the sum of the

series $\displaystyle\sum_{n=1}^{\infty} n\left(\frac{1}{3}\right)^{n-1}$.

By differentiation, $\left(\dfrac{1}{1-x}\right)' = \displaystyle\sum_{n=1}^{\infty} nx^{n-1}$.

Since $\left(\dfrac{1}{1-x}\right)' = \dfrac{1}{(1-x)^2}$,

$$\sum_{n=1}^{\infty} nx^{n-1} = \frac{1}{(1-x)^2}.$$

At $x = \dfrac{1}{3}$, $\displaystyle\sum_{n=1}^{\infty} n\left(\frac{1}{3}\right)^{n-1} = \dfrac{1}{\left(1 - \frac{1}{3}\right)^2} = \dfrac{9}{4}$.

Section 11.10   Taylor and Maclaurin Series

This section describes which functions may be written as power series and how to find such representations. The second partial sum of this representation will turn out to be the equation of the tangent line. Partial sums of the series for a function will be useful for approximating values of the function.

Concepts to Master

A. Taylor series; Maclaurin series

B. Taylor polynomials of degree n

C. Remainder of a Taylor series; Taylor's Inequality

D. Binomial coefficient; Binomial series for $(1 + x)^k$

E. Multiplication and division of power series

F. Technology Plus

Summary and Focus Questions

**Page
777**

A. Not every function can be written as a power series. However, if a function $y = f(x)$ can be expressed as a power series in $(x - a)$, then that series must have the form:

$$f(x) = \sum_{n=0}^{\infty} \frac{f^{(n)}(a)}{n!}(x - a)^n$$

This is called the **Taylor series for f at a.** In the special case of $a = 0$, it is called the **Maclaurin series for f.**

To find a Taylor series for a given function $y = f(x)$, you first find a formula for $f^{(n)}(a)$, usually in terms of n. Computing $f(a)$ and the first few derivatives, $f'(a), f''(a), f'''(a), f^{(4)}(a), \ldots$ often helps to see what the general term $f^{(n)}(a)$ looks like.

Example: Find the Maclaurin series for $f(x) = 3^x$.

First find the expression for the nth derivative at $x = 0$.

$$\begin{aligned}
f(x) &= 3^x & f(0) &= 1 \\
f'(x) &= 3^x (\ln 3) & f'(0) &= \ln 3 \\
f''(x) &= 3^x (\ln 3)^2 & f''(0) &= (\ln 3)^2 \\
f'''(x) &= 3^x (\ln 3)^3 & f'''(0) &= (\ln 3)^3
\end{aligned}$$

In general, $f^{(n)}(0) = (\ln 3)^n$. Therefore, the Maclaurin series is

$$3^x = \sum_{n=0}^{\infty} \frac{(\ln 3)^n}{n!} x^n.$$

By the Ratio Test, $\displaystyle\lim_{n\to\infty} \left| \frac{\frac{(\ln 3)^{n+1}}{(n+1)!}}{\frac{(\ln 3)^n}{n!}} \right| = \lim_{n\to\infty} \left| \frac{\ln 3}{n+1} \right| = 0$. Thus, the radius of

convergence is $R = \infty$, which means $3^x = \displaystyle\sum_{n=0}^{\infty} \frac{(\ln 3)^n}{n!} x^n$ for all x.

Here are some basic Maclaurin series:

Maclaurin Series	Domain
$\dfrac{1}{1-x} = \displaystyle\sum_{n=0}^{\infty} x^n = 1 + x + x^2 + x^3 + \dots$	$(-1, 1)$
$e^x = \displaystyle\sum_{n=0}^{\infty} \frac{x^n}{n!} = 1 + \frac{x}{1!} + \frac{x^2}{2!} + \frac{x^3}{3!} + \dots$	all real numbers
$\ln(1+x) = \displaystyle\sum_{n=1}^{\infty} \frac{(-1)^{n-1}}{n} x^n = x - \frac{x^2}{2} + \frac{x^3}{3} - \frac{x^4}{4} + \dots$	$(-1, 1]$
$\sin x = \displaystyle\sum_{n=0}^{\infty} \frac{(-1)^n}{(2n+1)!} x^{2n+1} = x - \frac{x^3}{3!} + \frac{x^5}{5!} - \frac{x^7}{7!} + \dots$	all real numbers
$\cos x = \displaystyle\sum_{n=0}^{\infty} \frac{(-1)^n}{(2n)!} x^{2n} = 1 - \frac{x^2}{2!} + \frac{x^4}{4!} - \frac{x^6}{6!} + \dots$	all real numbers
$\tan^{-1} x = \displaystyle\sum_{n=0}^{\infty} (-1)^n \frac{x^{2n+1}}{2n+1} = x - \frac{x^3}{3} + \frac{x^5}{5} - \frac{x^7}{7} + \dots$	$[-1, 1]$
$(1+x)^k = \displaystyle\sum_{n=0}^{\infty} \binom{k}{n} x^n = 1 + kx + \frac{k(k-1)}{2!} x^2 + \dots + \binom{k}{n} x^n + \dots$	Depends on k. See part D.

For some functions f it is easier to find the Taylor series for $f'(x)$ or $\int f(x)dx$ then integrate or differentiate term by term to obtain the series for f. In some other cases, substitutions in the basic Taylor series and algebra can be used to find the Taylor series for f.

1) Find the Taylor series for $f(x) = \frac{1}{x}$ at $a = 1$.

a) Directly from the definition.

First find the general form of $f^{(n)}(1)$:

$$f(x) = x^{-1} \qquad\qquad f(1) = 1$$
$$f'(x) = -x^{-2} \qquad\quad f'(1) = -1$$
$$f''(x) = 2x^{-3} \qquad\quad f''(1) = 2$$
$$f'''(x) = -6x^{-4} \qquad f'''(1) = -6$$
$$f^{(4)}(x) = 24x^{-5} \qquad f^{(4)}(1) = 24.$$

In general, $f^{(n)}(1) = (-1)^n n!$.

Thus, the Taylor series is

$$\frac{1}{x} = \sum_{n=0}^{\infty} \frac{(-1)^n n!}{n!} (x - 1)^n$$

$$= \sum_{n=0}^{\infty} (-1)^n (x - 1)^n = \sum_{n=0}^{\infty} (1 - x)^n.$$

b) Using substitution in a geometric series $\left(\text{Hint: } \frac{1}{x} = \frac{1}{1 - (1 - x)}\right)$.

In the form $\frac{1}{1 - (1 - x)}$, this is the value of a geometric series with $a = 1, r = 1 - x$.

$$\frac{1}{x} = \sum_{n=1}^{\infty} 1(1 - x)^{n-1} = \sum_{n=0}^{\infty} (1 - x)^n.$$

c) Using the Maclaurin series for $\ln(1 + x)$ and then substituting $x - 1$ for x.

Differentiate $\ln(1 + x) = \sum_{n=1}^{\infty} \frac{(-1)^{n-1}}{n} x^n$:

$$\frac{1}{1 + x} = \sum_{n=1}^{\infty} \frac{(-1)^{n-1} n}{n} x^{n-1} = \sum_{n=1}^{\infty} (-1)^{n-1} x^{n-1}$$

(Note the derivative also starts at $n = 1$.)

$$= \sum_{n=1}^{\infty} (-x)^{n-1} = \sum_{n=0}^{\infty} (-x)^n.$$

(Change the sum to start at $n = 0$.)

Now substitute $x - 1$ for x:

$$\frac{1}{1 + (x - 1)} = \sum_{n=0}^{\infty} (-(x - 1))^n = \sum_{n=0}^{\infty} (1 - x)^n.$$

Therefore, $\frac{1}{x} = \sum_{n=0}^{\infty} (1 - x)^n$.

d) What is the interval of convergence for your answer to part **a)**?

Apply the Ratio Test:

$$\lim_{n \to \infty} \left| \frac{(1 - x)^{n+1}}{(1 - x)^n} \right| = \lim_{n \to \infty} |1 - x| = |1 - x|.$$

$|1 - x| < 1$ is equivalent to $0 < x < 2$.

At $x = 0$ and $x = 2$ the terms of

$$\sum_{n=0}^{\infty} (1 - x)^n$$ do no approach zero so both

those series diverge. The interval of convergence is $(0, 2)$.

2) Find directly the Maclaurin series for $f(x) = e^{4x}$.

$$f(x) = e^{4x} \qquad\qquad f(0) = 1$$
$$f'(x) = 4e^{4x} \qquad\quad f'(0) = 4$$
$$f''(x) = 16e^{4x} \qquad f''(0) = 16.$$

In general $f^{(n)}(0) = 4^n$.

$$e^{4x} = \sum_{n=0}^{\infty} \frac{4^n}{n!} x^n.$$

(Using the basic series we can write

$$e^{4x} = \sum_{n=0}^{\infty} \frac{(4x)^n}{n!} = \sum_{n=0}^{\infty} \frac{4^n}{n!} x^n.)$$

By the Ratio Test,

$$\lim_{n\to\infty} \left| \frac{\frac{4^{n+1}}{(n+1)!} x^{n+1}}{\frac{4^n}{n!} x^n} \right| = \lim_{n\to\infty} \frac{4}{n+1} |x| = 0.$$

The interval of convergence is $(-\infty, \infty)$.

3) Obtain the Maclaurin series for $\dfrac{x}{1 + x^2}$ from the series for $\dfrac{1}{1-x} = \displaystyle\sum_{n=0}^{\infty} x^n$.

$$\frac{1}{1-x} = 1 + x + x^2 + \ldots x^n + \ldots$$

Substitute $-x^2$ for x:

$$\frac{1}{1+x^2} = 1 - x^2 + x^4 - x^6 + \ldots$$
$$+ (-1)^n x^{2n} + \ldots$$

Now multiply by x:

$$\frac{x}{1+x^2} = x - x^3 + x^5 - x^7 + \ldots$$
$$+ (-1)^n x^{2n+1} + \ldots$$
$$= \sum_{n=0}^{\infty} (-1)^n x^{2n+1}.$$

4) Using $e^x = \displaystyle\sum_{n=0}^{\infty} \frac{x^n}{n!}$ find the Maclaurin series for $\sinh x$.

$\left(\text{Remember, } \sinh x = \dfrac{e^x - e^{-x}}{2}.\right)$

$$e^x = 1 + x + \frac{x^2}{2!} + \ldots \frac{x^n}{n!} + \ldots$$
$$e^{-x} = 1 - x + \frac{x^2}{2!} + \ldots \frac{(-1)^n x^n}{n!} + \ldots$$

Subtracting term by term (the even terms cancel): $e^x - e^{-x}$

$$= 2x + \frac{2x^3}{3!} + \ldots + \frac{2x^{2n+1}}{(2n+1)!} + \ldots$$

Now divide by 2:

$$\sinh x = \frac{e^x - e^{-x}}{2}$$
$$= x + \frac{x^3}{3!} + \ldots + \frac{x^{2n+1}}{(2n+1)!} + \ldots$$
$$= \sum_{n=0}^{\infty} \frac{x^{2n+1}}{(2n+1)!}.$$

5) Evaluate $\int \dfrac{1+e^x}{x}\, dx$ using a series for the integrand.

Since $e^x = \displaystyle\sum_{n=0}^{\infty} \dfrac{x^n}{n!} = 1 + \dfrac{x}{1!} + \dfrac{x^2}{2!} + \dfrac{x^3}{3!} + \dots,$

$1 + e^x = 2 + \dfrac{x}{1!} + \dfrac{x^2}{2!} + \dfrac{x^3}{3!} + \dots$ and

$\dfrac{1+e^x}{x} = \dfrac{2}{x} + \dfrac{1}{1!} + \dfrac{x}{2!} + \dfrac{x^2}{3!} + \dots.$

Therefore,

$$\int \dfrac{1+e^x}{x}\, dx = \int \left(\dfrac{2}{x} + \dfrac{1}{1!} + \dfrac{x}{2!} + \dfrac{x}{3!} + \dots \right) dx$$

$$= 2\ln x + \dfrac{x}{1!} + \dfrac{x^2}{2\cdot 2!} + \dfrac{x^3}{3\cdot 3!} + \dots + C$$

$$= \ln x^2 + \sum_{n=1}^{\infty} \dfrac{x^n}{n\cdot n!} + C.$$

6) Find the sum of the series

$$\sum_{n=1}^{\infty} \dfrac{(-0.5)^n}{n}.$$

$$\sum_{n=1}^{\infty} \dfrac{(-0.5)^n}{n} = \sum_{n=1}^{\infty} \dfrac{(-1)(-1)^{n-1}(0.5)^n}{n}$$

$$= -\sum_{n=1}^{\infty} \dfrac{(-1)^{n-1}(0.5)^n}{n}.$$

The power series for $\ln(1+x)$

$$= \sum_{n=1}^{\infty} \dfrac{(-1)^{n-1}x^n}{n}.$$

Therefore,

$$\sum_{n=1}^{\infty} \dfrac{(-0.5)^n}{n} = -\ln(1+0.5) = -\ln 1.5$$

$$\approx -0.4054.$$

7) Find the infinite series expression for

$$\int_0^{1/2} \dfrac{1}{1-x^3}\, dx.$$

$\dfrac{1}{1-x} = \displaystyle\sum_{n=0}^{\infty} x^n.$ Thus $\dfrac{1}{1-x^3} = \displaystyle\sum_{n=0}^{\infty} x^{3n}.$

$$\int \dfrac{1}{1-x^3}\, dx = \int \sum_{n=0}^{\infty} x^{3n}\, dx$$

$$= \sum_{n=0}^{\infty} \dfrac{x^{3n+1}}{3n+1} + C.$$

$$\int_0^{1/2} \dfrac{1}{1-x^3}\, dx = \sum_{n=0}^{\infty} \dfrac{x^{3n+1}}{3n+1} \Bigg]_0^{1/2}$$

$$= \sum_{n=0}^{\infty} \dfrac{\left(\frac{1}{2}\right)^{3n+1}}{3n+1} - 0$$

$$= \sum_{n=0}^{\infty} \dfrac{1}{(6n+2)8^n}.$$

8) Find $\displaystyle\int_0^1 xe^{-x}\,dx$ using a series to within 0.01.

$$e^x = \sum_{n=0}^{\infty} \frac{x^n}{n!},\quad e^{-x} = \sum_{n=0}^{\infty} \frac{(-1)^n x^n}{n!},$$

so $\displaystyle xe^{-x} = \sum_{n=0}^{\infty} \frac{(-1)^n x^{n+1}}{n!}$. Thus,

$$\int_0^1 xe^{-x}\,dx = \int_0^1 \sum_{n=0}^{\infty} \frac{(-1)^n x^{n+1}}{n!}\,dx$$

$$= \sum_{n=0}^{\infty} \frac{(-1)^n x^{n+2}}{n!(n+2)}\Big]_0^1 = \sum_{n=0}^{\infty} \frac{(-1)^n}{n!(n+2)} - 0$$

$$= \sum_{n=0}^{\infty} \frac{(-1)^n}{n!(n+2)}$$

$$= \frac{1}{0!2} - \frac{1}{1!3} + \frac{1}{2!4} - \frac{1}{3!5} + \frac{1}{4!6} - \frac{1}{5!7} + \cdots$$

$$= \frac{1}{2} - \frac{1}{3} + \frac{1}{8} - \frac{1}{30} + \frac{1}{144} + \cdots.$$

This is an alternating series and the fifth term $\dfrac{1}{144}$ is less than 0.01. Thus, we may use

$$\int_0^1 xe^{-x}\,dx \approx \frac{1}{2} - \frac{1}{3} + \frac{1}{8} - \frac{1}{30} \approx 0.2583.$$

Note: We could use integration by parts:

$$\int_0^1 xe^{-x}\,dx = (-xe^{-x} - e^{-x})\Big] = 1 - \frac{2}{e}$$

$$\approx 0.2642$$

The approximation 0.2583 is within 0.01 of the actual value.

B. If $f^{(n)}(a)$ exists, the **Taylor polynomial of degree *n* for *f* about *a*** is the *n*th partial sum of the Taylor series:

Page
778

$$T_n(x) = f(a) + \frac{f'(a)}{1!}(x-a) + \frac{f''(a)}{2!}(x-a)^2 + \frac{f^{(3)}(a)}{3!}(x-a)^3 + \cdots \frac{f^{(n)}(a)}{n!}(x-a)^n.$$

The Taylor polynomial of degree one, $T_1(x) = f(a) + f'(a)(x-a)$, is the familiar equation for the tangent line to $y = f(x)$ at a point a. This means f and T_1 have the same functional value and same first derivative value at $x = a$: $f(a) = T_1(a)$ and $f'(a) = T_1'(a)$.

$T_2(x) = f(a) + f'(a)(x-a) + \frac{1}{2}f''(a)(x-a)^2$ is the "tangent parabola" to $y = f(x)$ at $x = a$. $T_2(x)$ has the same value, the same first derivative value, and the same second derivative value as does f at $x = a$.

$T_3(x)$ would be the "tangent cubic" and so on.

Since the first n derivatives of $T_n(x)$ at a are equal to the corresponding first n derivatives of $f(x)$ at a, $T_n(x)$ is a very good approximation for $f(x)$ if x is near a.

Example: Let $f(x) = -2 + 2x + \dfrac{1}{x}$. Then $f'(x) = 2 - \dfrac{1}{x^2}, f''(x) = \dfrac{2}{x^3}$, and

$f'''(x) = \dfrac{-6}{x^4}$. At $a = 1$, $f(1) = -2 + 2(1) + \dfrac{1}{1} = 1$, $f'(1) = 2 - \dfrac{1}{1^2} = 1$, $f''(1) = \dfrac{2}{1^3} = 2$,

and $f'''(1) = \dfrac{-6}{1^4} = -6$. Therefore, the first three Taylor polynomials for f are:

$T_1(x) = 1 + \dfrac{1}{1!}(x - 1) = 1 + (x - 1)$

$T_2(x) = 1 + \dfrac{1}{1!}(x - 1) + \dfrac{2}{2!}(x - 1)^2 = 1 + (x - 1) + (x - 1)^2$

$T_3(x) = 1 + \dfrac{1}{1!}(x - 1) + \dfrac{2}{2!}(x - 1)^2 + \dfrac{-6}{3!}(x - 1)^3 = 1 + (x - 1) + (x - 1)^2 - (x - 1)^3.$

The three figures below contain the graphs of f and T_1, of f and T_2, and f and T_3. Notice that near $a = 1$, the graph of T_n becomes an increasingly better approximation to the graph of $y = f(x)$ as n increases.

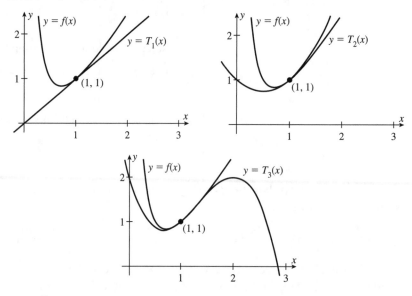

9) Let $f(x) = \sqrt{x}$. Construct the Taylor polynomial of degree 3 for $f(x)$ about $x = 1$.

$f(x) = x^{1/2}$ $f(1) = 1$

$f'(x) = \frac{1}{2}x^{-1/2}$ $f'(1) = \frac{1}{2}$

$f''(x) = -\frac{1}{4}x^{-3/2}$ $f''(1) = -\frac{1}{4}$

$f^{(3)}(x) = \frac{3}{8}x^{-5/2}$ $f^{(3)}(1) = \frac{3}{8}.$

Therefore, $T_3(x)$

$= 1 + \frac{1}{2}(x - 1) - \frac{1/4}{2!}(x - 1)^2 + \frac{3/8}{3!}(x - 1)^3$

$= 1 + \frac{1}{2}(x - 1) - \frac{1}{8}(x - 1)^2 + \frac{1}{16}(x - 1)^3.$

10) Find the Taylor polynomial of degree 6 for $f(x) = \cos x$ about $x = 0$.

$$f(x) = \cos x \qquad\qquad f(0) = 1$$
$$f'(x) = -\sin x \qquad\quad f'(0) = 0$$
$$f''(x) = -\cos x \qquad\quad f''(0) = -1$$
$$f^{(3)}(x) = \sin x \qquad\quad f^{(3)}(0) = 0$$
$$f^{(4)}(x) = \cos x \qquad\quad f^{(4)}(0) = 1$$
$$f^{(5)}(x) = -\sin x \qquad f^{(5)}(0) = 0$$
$$f^{(6)}(x) = -\cos x \qquad f^{(6)}(0) = -1.$$

$$T_6(x) = 1 + \frac{-1}{2!}x^2 + \frac{1}{4!}x^4 - \frac{1}{6!}x^6$$
$$= 1 - \frac{x^2}{2} + \frac{x^4}{24} - \frac{x^6}{720}.$$

11) a) Use your answer to question 9) to approximate $\sqrt{\frac{3}{2}}$ using the Taylor polynomial of degree 3.

$$\sqrt{\frac{3}{2}} \approx T_3\left(\frac{3}{2}\right) = 1 + \frac{1}{2}\left(\frac{1}{2}\right) - \frac{1}{8}\left(\frac{1}{2}\right)^2 + \frac{1}{16}\left(\frac{1}{2}\right)^3$$
$$= \frac{157}{128} \approx 1.2266.$$

(To 4 decimals, $\sqrt{\frac{3}{2}} = 1.2247$.)

b) Use your answer to question 10) to approximate $\cos 1$.

$$\cos 1 \approx T_6(1) = 1 - \frac{(1)^2}{2} + \frac{(1)^4}{24} - \frac{(1)^6}{720}$$
$$\approx 1 - 0.5 + 0.041667 - 0.001389$$
$$\approx 0.540278.$$
($\cos 1 = 0.540302$ to 6 decimal places.)

C. If $f^{(n)}(x)$ exists for $|x - a| < d$, the **nth remainder of the Taylor series of degree n for f** is:

Page 778

$$R_n(x) = f(x) - T_n(x).$$

(The R used in $R_n(x)$ should not be confused with the R used earlier in the chapter for the radius of convergence of an infinite series.)

$|R_n(x)|$ is the error resulting from using $T_n(x)$ to estimate $f(x)$.

Example: Let $f(x) = \cos 2x$. Then $f'(x) = -2 \sin 2x$, $f''(x) = -4 \cos 2x$, $f'''(x) = 8 \sin 2x$, and $f^{(4)}(x) = 16 \cos 2x$. At $a = 0, f(0) = 1$, $f'(0) = -2 \cdot 0 = 0$, $f''(0) = -4 \cdot 1 = -4, f'''(0) = 8 \cdot 0 = 0$, and $f^{(4)}(0) = 16 \cdot 1 = 16$.

The Taylor polynomial of degree 4 for f is

$$T_4(x) = 1 + \frac{0}{1!}(x - 0) + \frac{-4}{2!}(x - 0)^2 + \frac{0}{3!}(x - 0)^3 + \frac{16}{4!}(x - 0)^4$$

$$= 1 - 2x^2 + \frac{2}{3}x^4.$$

Choosing any x, say $x = 0.4$, $T_4(0.4) = 1 - 2 \cdot 0.4^2 + \frac{2}{3} \cdot 0.4^4 \approx 0.69707$ is an approximation of $f(0.4)$. To five decimals, $f(0.4) = \cos(0.8) = 0.69671$. Therefore, the error in this approximation (the remainder) is $|R_3(0.4)| = |0.69671 - 0.69707| = 0.00036$.

In general, $R_n(x)$ will approach 0 as n gets larger. Taylor's Inequality, stated next, is a way to establish an upper bound for $|R_n(x)|$.

Taylor's Inequality: If $|f^{(n+1)}(x)| \le M$ for $|x - a| \le d$, then the nth remainder $R_n(x)$ of the Taylor series satisfies the inequality

$$|R_n(x)| \le \frac{M}{(n+1)!}|x - a|^{n+1} \text{ for } |x - a| \le d.$$

Example: From the pattern in the example on the previous page, the form of the nth derivative of $f(x) = \cos 2x$ is

$$f^{(n)}(x) = \pm 2^n \sin 2x \text{ or } \pm 2^n \cos 2x.$$

Since sine and cosine have ranges from -1 to 1, in each case we have $|f^{(n+1)}(x)| \le 2^{n+1}$ for all x. Therefore, using $M = 2^{n+1}$ in Taylor's Inequality, an upper bound for $|R_n(x)|$ for all x is

$$|R_n(x)| \le \frac{2^{n+1}}{(n+1)!}|x|^{n+1}.$$

In the specific case where $n = 4$ and $x = 0.4$,

$$|R_4(0.4)| \le \frac{2^5}{(4+1)!}|0.4|^5 \approx 0.00273.$$

This is certainly true since we know from the example that, to five decimals, $|R_4(0.4)|$ is 0.00036.

If f has derivatives of all orders and $\lim_{n \to \infty} R_n(x) = 0$ for $|x - a| \le d$, then $f(x)$ is equal to its Taylor series for $|x - a| \le d$. This is one way to show that a function is equal to its Taylor series—show $\lim_{n \to \infty} R_n(x) = 0$. You will often need to use this limit:

$$\lim_{n \to \infty} \frac{t^n}{n!} = 0 \text{ for all real numbers } t.$$

Example: In the example above, where $f(x) = \cos 2x$, $|R_n(x)| \le \frac{2^{n+1}}{(n+1)!}|x|^{n+1}$. We then have

$$0 \le \lim_{n \to \infty} |R_n(x)| \le \lim_{n \to \infty} \frac{2^{n+1}}{(n+1)!}|x|^{n+1} = |x|^{n+1}\left(\lim_{n \to \infty} \frac{2^{n+1}}{(n+1)!} \right) = |x|^{n+1}(0) = 0 \text{ for all } x.$$

By the Squeeze Theorem, $\lim_{n \to \infty} |R_n(x)| = 0$, which proves for all x that $f(x) = \cos 2x$ is equal to its Maclaurin series; that is,

$$\cos 2x = \sum_{n=0}^{\infty} \frac{(-1)^n 2^{2n}}{(2n)!} x^{2n} \text{ for all } x.$$

12) True or False:

In general, the larger the number n, the closer the Taylor polynomial approximation is to the actual functional value.

True.

13) Let $f(x) = 2^x$.

a) Find an expression for the nth Taylor polynomial $T_n(x)$ for $f(x)$ at $x = 0$.

$$f(x) = 2^x \qquad\qquad f(0) = 1$$
$$f'(x) = (\ln 2)\, 2^x \qquad f'(0) = \ln 2$$
$$f''(x) = (\ln 2)^2\, 2^x \qquad f''(0) = (\ln 2)^2$$
and, in general,
$$f^n(x) = (\ln 2)^n\, 2^x \qquad f^n(0) = (\ln 2)^n.$$
$$T_n(x) = 1 + (\ln 2)x + \frac{(\ln 2)^2}{2!}x^2 + \dots$$
$$+ \frac{(\ln 2)^n}{n!}x^n.$$

b) Find an upper bound for $\left|f^{(n+1)}(x)\right|$ for $|x| \le 1$.

For $|x| \le 1$,
$$\left|f^{(n+1)}(x)\right| = \left|(\ln 2)^{n+1} \cdot 2^x\right|$$
$$= 2^x(\ln 2)^{n+1} \le 2(\ln 2)^{n+1}$$

c) Show that $f(x) = 2^x$ is equal to its Taylor series for $|x| < 1$.

We need to show that $\lim\limits_{n\to\infty} R_n(x) = 0$ for $|x| \le 1$. From part **b)**
$$\left|f^{(n+1)}(x)\right| \le 2\,(\ln 2)^{n+1}.$$
By Taylor's Inequality
$$0 \le |R_n(x)| \le \frac{2(\ln 2)^{n+1}}{(n+1)!}|x|^{n+1} \le \frac{2(\ln 2)^{n+1}}{(n+1)!}$$
since $|x| \le 1$.
$$\lim_{n\to\infty} 2\frac{(\ln 2)^{n+1}}{(n+1)!} = 2\lim_{n\to\infty}\frac{(\ln 2)^{n+1}}{(n+1)!} = 2(0) = 0.$$
Therefore, by the Squeeze Theorem for limits at infinity, $\lim\limits_{n\to\infty} R_n(x) = 0$.

D. For any real number k and a non-negative integer n, the **binomial coefficient** is

$$\binom{k}{n} = \begin{cases} 1 & \text{if } n = 0 \\ \dfrac{k(k-1)(k-2)\ldots(k-n+1)}{n!} & \text{if } n \geq 0 \end{cases}.$$

Example: $\quad\binom{\frac{3}{2}}{4} = \dfrac{\frac{3}{2}\cdot\frac{1}{2}\cdot\frac{-1}{2}\cdot\frac{-3}{2}}{4!} = \dfrac{\frac{9}{16}}{24} = \dfrac{3}{128}.$

Example: Evaluate $\binom{5}{n}$ for all non-negative integers n.

By definition, for $n = 0$, $\binom{5}{0} = 1$. For $n = 1, 2,$ and 3, $\binom{5}{1} = \dfrac{5}{1!} = 5$,

$\binom{5}{2} = \dfrac{5\cdot4}{2!} = \dfrac{5\cdot4}{2\cdot1} = 10$, and $\binom{5}{3} = \dfrac{5\cdot4\cdot3}{3!} = \dfrac{5\cdot4\cdot3}{3\cdot2\cdot1} = 10$. With similar

calculations, we see $\binom{5}{4} = 5$ and $\binom{5}{5} = 1$. For $n = 6$, $\binom{5}{6} = \dfrac{5\cdot4\cdot3\cdot2\cdot1\cdot0}{6!} = 0$.

In fact, $\binom{5}{n} = 0$ for all integers $n > 5$.

For any real number k the **binomial series for** $(1 + x)^k$ is the Maclaurin series for $(1 + x)^k$:

$$(1 + x)^k = \sum_{n=0}^{\infty}\binom{k}{n}x^n = 1 + kx + \frac{k(k-1)}{2!}x^2 + \ldots + \binom{k}{n}x^n + \ldots\,.$$

Examples: a) We found above that $\binom{\frac{3}{2}}{4} = \dfrac{3}{128}$ is the coefficient for x^4 in the

binomial series for $(1 + x)^{\frac{3}{2}}$. The binomial series is

$$(1 + x)^{\frac{3}{2}} = \sum_{n=0}^{\infty}\binom{\frac{3}{2}}{n}x^n = 1 + \tfrac{3}{2}x + \tfrac{3}{16}x^2 - \tfrac{1}{16}x^3 + \tfrac{3}{128}x^4 + \ldots\,.$$

b) The binomial series for $(1 + x)^5$ is finite because $\binom{5}{n} = 0$ for all integers $n > 5$.

In fact, the series $(1 + x)^5 = \sum_{n=0}^{\infty}\binom{5}{n}x^n = 1 + 5x + 10x^2 + 10x^3 + 5x^4 + x^5$

is the binomial expansion of $(1 + x)^5$ you have seen in algebra.

For $k = 0, 1, 2, 3, \ldots$, the binomial series for $(1 + x)^k$ is a polynomial of degree k. When k is any number other than a non-negative integer, the binomial series for $(1 + x)^k$ is an infinite series.

The intervals of convergence of the binomial series for various values of k are given in the table at the right.

Condition on k	Interval of Convergence
$k \leq -1$	$(-1, 1)$
$-1 < k < 0$	$(-1, 1]$
$k = 0, 1, 2, 3, \ldots$	$(-\infty, \infty)$
$k > 0$, but not an integer	$[-1, 1]$

14) a) Find $\binom{6}{4}$.

$$\binom{6}{4} = \frac{6(5)(4)(3)}{4!} = 15.$$

b) Find $\begin{pmatrix} \frac{4}{3} \\ 5 \end{pmatrix}$.

$$\begin{pmatrix} \frac{4}{3} \\ 5 \end{pmatrix} = \frac{\left(\frac{4}{3}\right)\left(\frac{1}{3}\right)\left(-\frac{2}{3}\right)\left(-\frac{5}{3}\right)\left(-\frac{8}{3}\right)}{5 \cdot 4 \cdot 3 \cdot 2 \cdot 1} = \frac{-8}{3^6} = -\frac{8}{729}.$$

c) Expand $(1 + x)^6$.

$$\binom{6}{0} = 1, \binom{6}{1} = 6, \binom{6}{2} = 15, \binom{6}{3} = 20,$$
$$\binom{6}{4} = 15, \binom{6}{5} = 6, \binom{6}{6} = 1.$$

Thus $(1 + x)^6$
$$= 1 + 6x + 15x^2 + 20x^3 + 15x^4 + 6x^5 + x^6.$$

d) Find the coefficient of x^5 in the binomial expansion of $\sqrt{1 + x}$.

$\sqrt{1 + x} = (1 + x)^{1/2}.$
The coefficient of x^5 is

$$\begin{pmatrix} \frac{1}{2} \\ 5 \end{pmatrix} = \frac{\left(\frac{1}{2}\right)\left(-\frac{1}{2}\right)\left(-\frac{3}{2}\right)\left(-\frac{5}{2}\right)\left(-\frac{7}{2}\right)}{5 \cdot 4 \cdot 3 \cdot 2 \cdot 1}$$

$$= \frac{1(-1)(-3)(-5)(-7)}{2^5 \cdot 5 \cdot 4 \cdot 3 \cdot 2 \cdot 1}$$

$$= \frac{7}{2^6} = \frac{7}{64}.$$

Therefore,
$$\sqrt{1 + x} = 1 + \cdots + \frac{7}{64}x^5 + \cdots.$$

15) True or False:

If k is a non-negative integer, then the binomial series for $(1 + x)^k$ is a finite sum.

True, only the first $k + 1$ terms of the infinite series may have non-zero coefficients.

16) Find the binomial series for $\sqrt[3]{1 + x}$.

Since $\sqrt[3]{1 + x} = (1 + x)^{\frac{1}{3}}$, let $k = \frac{1}{3}$. Then

$$\sqrt[3]{1 + x} = \sum_{n=0}^{\infty} \binom{\frac{1}{3}}{n} x^n.$$

17) Find a power series for $f(x) = \sqrt{4 + x}$.

Let $k = \frac{1}{2}$. Then $\sqrt{4 + x}$

$$= 2\left(1 + \frac{x}{4}\right)^{1/2} = 2\sum_{n=0}^{\infty} \binom{\frac{1}{2}}{n} \left(\frac{x}{4}\right)^n$$

$$= 2\sum_{n=0}^{\infty} \binom{\frac{1}{2}}{n} \frac{x^n}{4^n}.$$

18) Find an infinite series for $\sin^{-1} x$.
(Hint: Start with $(1 + x)^{-1/2}$.)

$$\frac{1}{\sqrt{1 + x}} = (1 + x)^{-1/2} = \sum_{n=0}^{\infty} \binom{-\frac{1}{2}}{n} x^n.$$

Substitute $-x$ for x:

$$\frac{1}{\sqrt{1 - x}} = \sum_{n=0}^{\infty} \binom{-\frac{1}{2}}{n} (-x)^n$$

$$= \sum_{n=0}^{\infty} (-1)^n \binom{-\frac{1}{2}}{n} x^n.$$

Substitute x^2 for x:

$$\frac{1}{\sqrt{1 - x^2}} = \sum_{n=0}^{\infty} (-1)^n \binom{-\frac{1}{2}}{n} x^{2n}.$$

Since the derivative of $\sin^{-1} x$ is $\dfrac{1}{\sqrt{1 - x^2}}$,

integrate term by term:

$$\sin^{-1} x = \sum_{n=0}^{\infty} \frac{(-1)^n \binom{-\frac{1}{2}}{n} x^{2n+1}}{2n + 1} + C.$$

Page
787

E. Convergent series may be multiplied and divided just like polynomials. Often finding the first few terms of the result is sufficient.

Example: Find the first three terms of the Maclaurin series for $e^x \sin x$.

$$e^x = 1 + x + \frac{x^2}{2} + \frac{x^3}{6} + \ldots$$

$$\sin x = x - \frac{x^3}{3!} + \frac{x^5}{5!} - \ldots$$

$$e^x \sin x = x\left(1 + x + \frac{x^2}{2} + \frac{x^3}{6} + \ldots\right) - \frac{x^3}{6}\left(1 + x + \frac{x^2}{2} + \frac{x^3}{6} + \ldots\right) + \frac{x^5}{120}\left(1 + x + \frac{x^2}{2} + \frac{x^3}{6} + \ldots\right) + \ldots$$

$$= \left(x + x^2 + \frac{x^3}{2} + \frac{x^4}{6} + \ldots\right) - \left(\frac{x^3}{6} + \frac{x^4}{6} + \frac{x^5}{12} + \frac{x^6}{36} + \ldots\right) + \left(\frac{x^5}{120} + \frac{x^6}{120} + \frac{x^7}{240} + \ldots\right)$$

$$= x + x^2 + \frac{x^3}{3} + \ldots$$

19) Use the power series

$$\cos x = 1 - \frac{x^2}{2} + \frac{x^4}{24} - \cdots \text{ and}$$

$$e^x = 1 + x + \frac{x^2}{2} + \frac{x^3}{6} + \frac{x^4}{24} + \cdots$$

to find the first four terms of the power series for $e^{-x} \cos x$.

First, substitute $-x$ for x:

$$e^{-x} = 1 - x + \frac{x^2}{2} - \frac{x^3}{6} + \frac{x^4}{24} + \ldots$$

$$e^{-x} \cos x$$

$$= 1\left(1 - \frac{x^2}{2} + \frac{x^4}{24} - \ldots\right) - x\left(1 - \frac{x^2}{2} + \frac{x^4}{24} - \ldots\right)$$

$$+ \frac{x^2}{2}\left(1 - \frac{x^2}{2} + \frac{x^4}{24} - \ldots\right) - \frac{x^3}{6}\left(1 - \frac{x^2}{2} + \frac{x^4}{24} - \ldots\right) + \ldots$$

$$= \left(1 - \frac{x^2}{2} + \frac{x^4}{24} - \ldots\right) - \left(x - \frac{x^3}{2} + \frac{x^5}{24} - \ldots\right)$$

$$+ \left(\frac{x^2}{2} - \frac{x^4}{4} + \frac{x^6}{48} - \ldots\right) - \left(\frac{x^3}{6} - \frac{x^5}{12} + \frac{x^7}{144} - \ldots\right) + \ldots$$

Therefore, $e^{-x} \cos x = 1 - x + \frac{1}{3}x^3 - \frac{5}{24}x^4 + \ldots$

20) Use the series in question 19) to find the first three terms of the power series for $\dfrac{e^{-x}}{\cos x}$.

$$
\require{enclose}
\begin{array}{r}
1 - x \\
1 - \frac{x^2}{2} + \frac{x^4}{24} - \ldots \enclose{longdiv}{1 - x + \frac{x^2}{2} - \frac{x^3}{6} + \frac{x^4}{24} - \frac{x^5}{120} + \ldots}
\end{array}
$$

$$1 \qquad - \frac{x^2}{2} \qquad + \frac{x^4}{24}$$

$$- x + x^2 \quad - \frac{x^3}{6} + 0 \quad - \frac{x^5}{120} + \ldots$$

$$- x \qquad + \frac{x^3}{2} \qquad - \frac{x^5}{24} + \ldots$$

$$x^2 - \frac{2x^3}{3} \qquad + \frac{5x^5}{30} + \ldots$$

Therefore, $\dfrac{e^{-x}}{\cos x} = 1 - x + x^2 + \ldots$

F. Technology Plus. Use a computer algebra system or a graphing calculator to solve.

T-1) On the same screen sketch a graph of $f(x) \frac{1}{\sqrt{x}}$ and its first 3 Taylor polynomials about $x = 1$.

$f(x) = x^{-1/2}, \ f(1) = 1$

$f'(x) = -\frac{1}{2}x^{-3/2}, \ f'(1) = -\frac{1}{2}$

$f''(x) = \frac{3}{4}x^{-5/2}, \ f''(1) = \frac{3}{4}$

$f'''(x) = -\frac{15}{8}x^{-7/2}, \ f'''(1) = -\frac{15}{8}.$

$T_1(x) = 1 + (-\frac{1}{2})(x - 1) = 1 - \frac{1}{2}(x - 1).$

$T_2(x) = 1 - \frac{1}{2}(x - 1) + \frac{\frac{3}{4}}{2!}(x - 1)^2$

$\qquad = 1 - \frac{1}{2}(x - 1) + \frac{3}{8}(x - 1)^2.$

$T_3(x) = 1 - \frac{1}{2}(x - 1) + \frac{3}{8}(x - 1)^2$

$\qquad + \frac{-\frac{15}{8}}{3!}(x - 1)^3$

$\qquad = 1 - \frac{1}{2}(x - 1) + \frac{3}{8}(x - 1)^2 - \frac{5}{16}(x - 1)^3.$

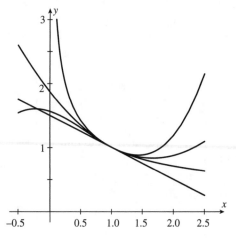

Section 11.11   Applications of Taylor Polynomials

This last section uses Taylor polynomials to estimate values of functions and uses Taylor's Inequality to measure the accuracy of such approximations.

Concepts to Master

A. Approximate functional values using a Taylor polynomial
B. Estimate the error of a Taylor polynomial approximation

Summary and Focus Questions

Page 792

A. The Taylor polynomial of degree n for f about a is

$$T_n(x) = f(a) + \frac{f'(a)}{1!}(x - a) + \frac{f''(a)}{2!}(x - a)^2 + \frac{f^{(3)}(a)}{3!}(x - a)^3 + \ldots \frac{f^{(n)}(a)}{n!}(x - a)^n.$$

$T_n(x)$ may be used to approximate $f(x)$ for any given x in the interval of convergence for the Taylor series for f about a.

1) a) For $f(x) = \frac{1}{3 + x}$, find $T_3(x)$ where $a = 2$.

$$f(x) = (3 + x)^{-1} \qquad f(2) = \frac{1}{5}$$
$$f'(x) = -(3 + x)^{-2} \qquad f'(2) = \frac{-1}{25}$$
$$f''(x) = 2(3 + x)^{-3} \qquad f''(2) = \frac{2}{125}$$
$$f'''(x) = -6(3 + x)^{-4} \qquad f'''(2) = \frac{-6}{625}$$

$$T_3(x) = \frac{1}{5} - \frac{1}{25}(x - 2) + \frac{2/125}{2!}(x - 2)^2$$
$$- \frac{6/625}{3!}(x - 2)^3$$
$$= \frac{1}{5} - \frac{1}{25}(x - 2) + \frac{1}{125}(x - 2)^2$$
$$- \frac{1}{625}(x - 2)^3.$$

b) Estimate $f(3)$ with $T_3(3)$.

$$T_3(3) = \frac{1}{5} - \frac{(3 - 2)}{25} + \frac{(3 - 2)^2}{125} - \frac{(3 - 2)^3}{625}$$
$$= \frac{1}{5} - \frac{1}{25} + \frac{1}{125} - \frac{1}{625}$$
$$= \frac{104}{625} = 0.1664.$$

We note that this is a lot of work to approximate $f(3) = \frac{1}{6}$ but understanding the method is important.

2) a) For $f(x) = x^{5/2}$, find $T_3(x)$ where $a = 4$.

$f(x) = x^{5/2}$ $\qquad$ $f(4) = 32$

$f'(x) = \frac{5}{2}x^{3/2}$ $\qquad$ $f'(4) = 20$

$f''(x) = \frac{15}{4}x^{1/2}$ $\qquad$ $f''(4) = \frac{15}{2}$

$f'''(x) = \frac{15}{8}x^{-1/2}$ $\qquad$ $f'''(4) = \frac{15}{16}$.

$T_3(x) = 32 + 20(x - 4) + \frac{15/2}{2!}(x - 4)^2$
$\qquad + \frac{15/16}{3!}(x - 4)^3$
$\qquad = 32 + 20(x - 4) + \frac{15}{4}(x - 4)^2$
$\qquad + \frac{5}{32}(x - 4)^3$.

b) Estimate $(5)^{5/2}$ with $T_3(x)$.

$T_3(5) = 32 + 20(1)^1 + \frac{15}{4}(1)^2 + \frac{5}{32}(1)^3$

$\qquad = \frac{1789}{32} \approx 55.9063$.

To four decimals $(5)^{5/2}$ is approximately 55.9017.

B. The accuracy of the approximation $f(x) \approx T_n(x)$ will depend on the values of n, x, and a. In general:

i) The larger the value of n, the better the estimate.

ii) The closer that x is to a, the better the estimate.

From the previous section, we may use Taylor's Inequality to estimate the error:

If $\left|f^{(n+1)}(x)\right| \leq M$ for $|x - a| \leq d$, then

$$\left|R_n(x)\right| \leq \frac{M}{(n + 1)!}|x - a|^{n+1} \text{ for } |x - a| \leq d.$$

The approximation $T_n(x)$ will be accurate to within any given number ϵ by selecting n such that

$$\frac{M}{(n + 1)!}|x - a|^{n+1} < \epsilon.$$

In the special case where the Taylor series is an alternating series, $\left|R_n(x)\right|$

may be estimated (by the Alternating Series Estimation Theorem) with

$\left|\frac{f^{(n+1)}(x)}{(n + 1)!}(x - a)^{n+1}\right|$, the $(n + 1)^{\text{st}}$ term of the Taylor series for f.

Example: This example puts together the concepts from Section 11.10 and this section.

a) Find the Taylor polynomial of degree 3 about $x = 0$ for $f(x) = xe^{-x}$.

$f(x) = xe^{-x}$ $\qquad\qquad\qquad\qquad\qquad$ $f(0) = 0 \cdot e^{-0} = 0$

$f'(x) = x(-e^{-x}) + e^{-x}(1) = -e^{-x}(x - 1)$ $\qquad$ $f'(0) = -e^{-0}(0 - 1) = 1$

$f''(x) = -(x - 1)(-e^{-x}) + e^{-x}(-1) = e^{-x}(x - 2)$ $\qquad$ $f''(0) = e^{-0}(0 - 2) = -2$

$f'''(x) = (x - 2)(-e^{-x}) + e^{-x}(1) = -e^{-x}(x - 3)$ $\qquad$ $f'''(0) = -e^{-0}(0 - 3) = 3$.

In general, $f^{(n)}(x) = (-1)^n e^{-x}(x - n)$.

Therefore, $T_3(x) = 0 + \dfrac{1}{1!}x + \dfrac{-2}{2!}x^2 + \dfrac{3}{3!}x^3 = x - x^2 + \dfrac{1}{2}x^3$.

Alternately, we could have modified the power series $e^x = \displaystyle\sum_{n=0}^{\infty} \dfrac{x^n}{n!}$ by

replacing x by $-x$: $e^{-x} = \displaystyle\sum_{n=0}^{\infty} \dfrac{(-x)^n}{n!} = \sum_{n=0}^{\infty} \dfrac{(-1)^n x^n}{n!}$,

then multiply by x: $xe^{-x} = \displaystyle\sum_{n=0}^{\infty} \dfrac{(-1)^n x^n}{n!} x = \sum_{n=0}^{\infty} \dfrac{(-1)^n x^{n+1}}{n!}$

and selecting the first 3 terms of the series to obtain $T_3(x) = x - x^2 + \dfrac{1}{2}x^3$.

b) For $-1 < x < 1$, estimate the accuracy of $T_3(x)$ using Taylor's Inequality.

For $n = 3$, the error is $R_3(x) = xe^{-x} - \left(x - x^2 + \frac{1}{2}x^3\right)$.

The graph of $y = |R_3(x)|$, given at the right, shows that for $-1 < x < 1$, $|R_3(x)|$ is small.

Let's see how small.

For any n and $-1 < x < 1$, we have

$0 < e^{-x} < e$ and $|x - n| < n + 1$. Thus

$|f^{(n)}(x)| = |(-1)^n e^{-x}(x - n)| \le e(n + 1)$.

Then $M = e(n + 1)$ is an upper bound for the values of the nth derivative, and by Taylor's Inequality

$$|R_n(x)| \le \frac{e(n + 1)}{(n + 1)!} = \frac{e}{n!}.$$

Therefore, for $-1 < x < 1$, $|R_3(x)| \le \dfrac{e}{3!} \approx 0.4530$.

This means that for any x in $(-1, 1)$, the error from estimating $f(x)$ with $T_3(x)$ is at most 0.4530. Specifically, for $x = 0.8$,

$T_3(0.8) = 0.8 - 0.8^2 + 0.8^3/2 = 0.4160$ is an estimate for

$f(0.8) = 0.8e^{-0.8} \approx 0.3595$. The actual error from this estimate is

$|R_3(0.8)| = |0.3595 - 0.4160| = 0.0565$, which is well within the error estimate of 0.4530.

c) Use the Alternating Series Estimation Theorem to estimate the error for $n = 3$. For any n, the theorem says $|R_n(x)|$ is less than the $(n + 1)$st term. Thus

$$|R_n(x)| \leq \left| \frac{f^{(n+1)}(0)}{(n+1)!} x^{n+1} \right| = \left| \frac{(-1)^{n+1}(0 - (n+1))e^{-0}}{(n+1)!} x^{n+1} \right| = \frac{|n+1|}{(n+1)!} = \frac{1}{n!},$$

which is a smaller estimate for the error than the estimate found in part b).

In particular, for $n = 3, |R_3(x)| \leq \frac{1}{3!} \approx 0.1667$ and for $x = 0.8, |R_3(0.8)| =$

0.0565 is within our error bound of 0.1667.

---

3) a) Use Taylor's Inequality to estimate the accuracy of $f(x) \approx T_3(x)$ for $1 < x < 4$ for question **1**).

From **1 a)**, $f^{(4)}(x) = 24(3 + x)^{-5}$.
If $1 \leq x \leq 4$, then $4 \leq x + 3 \leq 7$.
$4^5 \leq (x + 3)^5 \leq 7^5$

$$\frac{1}{7^5} \leq (x + 3)^{-5} \leq \frac{1}{4^5}$$

$$\frac{24}{7^5} \leq 24(x + 3)^{-5} \leq \frac{24}{4^5}.$$

Thus, $\left| f^{(4)}(x) \right| \leq \frac{24}{4^5}.$

From $1 \leq x \leq 4, -1 \leq x - 2 \leq 2.$
$(x - 2)^4 \leq 2^4 = 16.$
Finally,

$$|R_3(x)| \leq \frac{\frac{24}{4^5}}{4!} (x - 2)^4$$

$$\leq \frac{24}{4^5 \cdot 4!} \cdot 16 = 0.0156.$$

b) Check the accuracy of the estimate when $x = 3$.

$f(3) = \frac{1}{3 + 3} = \frac{1}{6} \approx 0.1667.$

$T_3(3) \approx 0.1664$ by **1 b)**.

$R_3(3) \approx 0.1667 - 0.1664 = 0.0003,$

which is within 0.0156.

4) a) For $f(x) = x^{5/2}$, estimate the error between $f(5)$ and $T_3(5)$. (See question **2**).

From **2 a)**, $f^{(4)}(x) = -\frac{15}{16}x^{-3/2}.$

For $|x - 4| \leq 2, 2 \leq x \leq 6.$

$6^{-3/2} \leq x^{-3/2} \leq 2^{-3/2}.$

$\left| f^{(4)}(x) \right| \leq \frac{15}{16} \cdot 2^{-3/2} = \frac{15}{32\sqrt{2}}.$

$|R_3(x)| \leq \frac{15}{32\sqrt{2}} |x - 4|^4 \leq \frac{\frac{15}{32\sqrt{2}}}{4!}(2)^4$

$\approx 0.221.$

b) Check the accuracy of the estimate $f(5) \approx T_3(5)$.

$f(5) = 5^{5/2} \approx 55.9017.$
$T_3(5) \approx 55.9063$ (question **2b**)).
$R_3(5) \approx 55.9017 - 55.9063 = -0.0046,$
which is within 0.221.

5) a) What degree Maclaurin polynomial is needed to approximate cos 0.5 accurate to within 0.00001?

Rephrased, the question asks:
For what n is $\left|R_n(0.5)\right| < 0.00001$ when
$f(x) = \cos x$ and $a = 0$?
Because $f^{(n)}(x) = \pm\sin x$ or $\pm\cos x$ for all
$n, \left|f^{(n+1)}(x)\right| \le 1.$

Thus, $\left|R_n(x)\right| \le \dfrac{1}{(n+1)!}\left|x\right|^{n+1}$

$= \dfrac{\left|x\right|^{n+1}}{(n+1)!} \le \dfrac{1}{(n+1)!}$ if $\left|x\right| \le 1.$

$\dfrac{1}{(n+1)!} \le 0.00001$ means

$(n+1)! \ge 100{,}000.$
Therefore, $n = 8$, since $8! = 40{,}320$ and
$9! = 362{,}800.$

b) Use your answer to part **a)** to estimate cos 0.5 to within 0.00001.

From the Taylor series for cosine

$$T_8(x) = 1 - \frac{x^2}{2} + \frac{x^4}{24} - \frac{x^6}{720} + \frac{x^8}{40320}.$$

$T_8(0.5)$
$= 1 - \dfrac{(0.5)^2}{2} + \dfrac{(0.5)^4}{24} - \dfrac{(0.5)^6}{720} + \dfrac{(0.5)^8}{40320}$
$\approx 0.877583.$

To 6 decimals, cos 0.5 = 0.877583 so the approximation is within 0.00001 as required by part **a)**.

Chapter 12 — Vectors and the Geometry of Space

Section 12.1   Three-Dimensional Coordinate Systems

This section extends the two-dimensional x-, y-coordinate system to three dimensions (x, y, z) with three axes, each perpendicular to the other two axes. You might visualize the three axes as forming the corner of a room.

Concepts to Master

A. Rectangular three-dimensional coordinates; Distance between two points

B. Planes, spheres, and regions in space

Summary and Focus Questions

Page 810 (ET Page 786)*

A. In three-dimensional space ($\mathbb{R}^3$), draw three axes, labeled x, y, and z, each perpendicular to the other two axes and intersecting at a common point called the **origin**. The most common orientation of the axes is given in the figure. To see the 3-D effect, imagine that the x-axis sticks out toward you. Each triple of real numbers (a, b, c) corresponds to a unique point P in space, which is a units from the x-axis, b units from the y-axis, and c units from the z-axis.

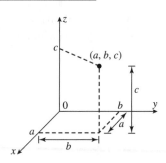

Example: The points $P(2, 3, 5)$ and $Q(3, -2, -3)$ are plotted at the right.

The **distance between points** $P_1(x_1, y_1, z_1)$ and $P_2(x_2, y_2, z_2)$ is

$$|P_1P_2| = \sqrt{(x_2 - x_1)^2 + (y_2 - y_1)^2 + (z_2 - z_1)^2}.$$

Example: The distance between $P(2, 3, 5)$ and $Q(3, -2, -3)$ is

$$\sqrt{(3 - 2)^2 + (-2 - 3)^2 + (-3 - 5)^2} = \sqrt{1 + 25 + 64} = \sqrt{90}.$$

---

*When using the Early Transcendentals text, use the page number in parentheses.

1) Plot the points
$A(0, 5, 0)$, $B(5, 4, 6)$, and $C(1, -1, 3)$.

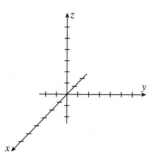

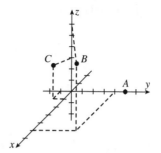

2) The coordinates of the points A and B pictured are _____.

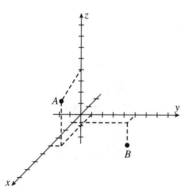

$A(4, 1, 4)$
$B(1, 5, -2)$

3) For the plotted points below, label the other six points.

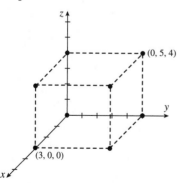

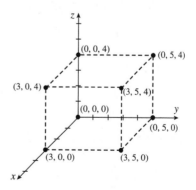

4) The distance between the points $(7, 5, -2)$ and $(4, 6, 1)$ is _____.

$$\sqrt{(7-4)^2 + (5-6)^2 + (-2-1)^2}$$
$$= \sqrt{19}.$$

Page
812
(ET Page
788)

B. Some simple linear equations in x, y, z with one or more of the variables missing represent planes parallel to the axes of the missing variable. Only a portion of each plane is drawn here:

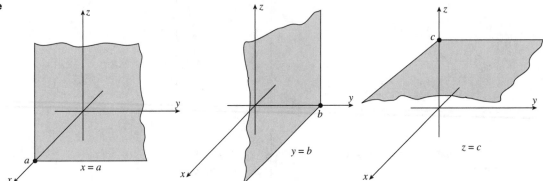

If a variable is missing, the graph will be parallel to the axis of the missing variable. For example, the graph of $3x + 2z = 6$ is a plane parallel to the y-axis.

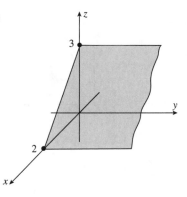

The **equation of the sphere** with center (a, b, c) and radius r is $(x - a)^2 + (y - b)^2 + (z - c)^2 = r^2$.

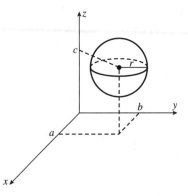

Example: Find an equation of the sphere with center $(5, -6, 3)$ and radius 5, and describe its intersection with each of the coordinate planes (the xy-, xz-, and yz-planes.)

The equation is $(x - 5)^2 + (y + 6)^2 + (z - 3)^2 = 25$.

When $z = 0$ (the xy-plane), the equation becomes $(x - 5)^2 + (y + 6)^2 + (0 - 3)^2 = 25$, so $(x - 5)^2 + (y + 6)^2 = 16$. The sphere intersects the xy-plane in a circle in the xy-plane, centered at $(5, -6)$ with radius 4.

When $y = 0$ (the xz-plane), the equation becomes $(x - 5)^2 + (0 + 6)^2 + (z - 3)^2 = 25$, so $(x - 5)^2 + (z - 3)^2 = -11$. The sphere does not intersect the xz-plane.

When $x = 0$ (the yz-plane), the equation becomes $(0 - 5)^2 + (y + 6)^2 + (z - 3)^2 = 25$ $(z - 3)^2 = 25$, so $(y + 6)^2 + (z - 3)^2 = 0$. The sphere intersects the yz-plane in a single point, $(0, -6, 3)$.

---

5) Describe the coordinates of each set of points:

a)

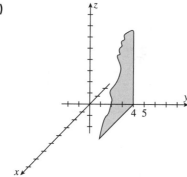

b)

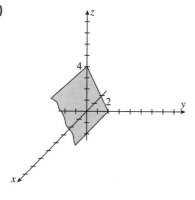

\<Look for which variable, x, y, or z, is missing.\>

A point in the plane in the figure may have any values for the x- and z-coordinates but must have a y-coordinate of 4. The plane is all points (x, y, z) such that $y = 4$.

The plane is parallel to the x-axis, so its equation has no x variable. In the yz-plane it is the line through $(0, 0, 4)$ and $(0, 2, 0)$: $z + 2y = 4$. The plane is $z + 2y = 4$.

6) Find an equation of the sphere with center $(2, 3, -4)$ and radius 5.

$(x - 2)^2 + (y - 3)^2 + (z + 4)^2 = 25.$

7) Find the center and radius of the sphere $x^2 - 6x + y^2 + 4y + z^2 = 1$.

Complete the square for x and y:
$x^2 - 6x + 9 + y^2 + 4y + 4 + z^2$
$\qquad = 1 + 9 + 4$
$(x - 3)^2 + (y + 2)^2 + z^2 = 14.$
The center is $(3, -2, 0)$.
The radius is $\sqrt{14}$.

8) Describe the region given by each:

a) $x^2 + z^2 = 4$

$x^2 + z^2 = 4$ is a circle in the xz-plane with center $(0, 0)$ and radius 2. Therefore, this is a cylinder of radius 2 along the y-axis.

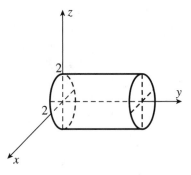

b) $x^2 + y^2 + z^2 \leq 1$

This is a solid ball of radius 1 and center $(0, 0, 0)$.

c) $\{(x, y, z): x = 3 \text{ and } y = 4\}$

This is a line through $(3, 4, 0)$ parallel to the z-axis.

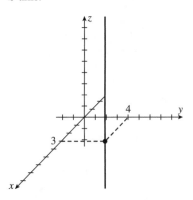

d) $|y| + |z| \leq 1$

$|y| + |z| \leq 1$ is a box in the yz-plane.

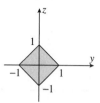

Therefore, this is a solid square box along the x-axis. The diagonals through the box are each 2 units. Each box side is $\sqrt{2}$ units.

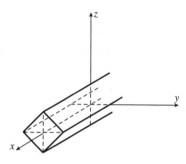

e) $x^2 = 1$

Since $x^2 = 1$ is equivalent to $x = 1$ and $x = -1$ and y and z may have any values, this is two planes, $x = 1$ and $x = -1$, both parallel to the yz-plane.

Section 12.2 Vectors

A vector has both a length and a direction, such as the force of wind at a given place and given time. This section studies two- and three-dimensional vectors and their properties.

Concepts to Master

A. Vectors; Position vector; Length; Unit Vector; Standard basis vectors **i, j, k**

B. Vector arithmetic; Scalars; Vector properties

Summary and Focus Questions

Page 815 (ET Page 791)

A. A **vector** is a quantity having both a magnitude (length) and a direction. A **two-dimensional vector** is an ordered pair of numbers $\mathbf{a} = \langle a_1, a_2 \rangle$. The numbers a_1 and a_2 are called **components** of **a**.

The directed line segment $\overrightarrow{AB}$ from $A(x_1, y_1)$ to $B(x_2, y_2)$ represents the vector $\mathbf{a} = \langle x_2 - x_1, y_2 - y_1 \rangle$.

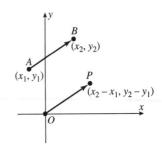

A is the **initial point** (or **tail**), and B is the **terminal point** (**head** or **tip**). This "pointy stick" representation conveys the notions of length and direction.

a is the **position vector** for the line segment from the origin O to $P(x_2 - x_1, y_2 - y_1)$. Both $\overrightarrow{AB}$ and $\overrightarrow{OP}$ represent the same vector.

The **length** (or **magnitude**) of $\mathbf{a} = \langle a_1, a_2 \rangle$ is $|\mathbf{a}| = \sqrt{a_1^2 + a_2^2}$.

Vectors **u** and **v** are **equal** (**equivalent**) if they have the same magnitude and point in the same direction. If $\mathbf{u} = \mathbf{v}$, then their corresponding components are equal.

Example: The directed line segment $\overrightarrow{AB}$ where A is $(-1, 3)$ and B is $(2, 7)$ is the vector $\mathbf{u} = \langle 2 - (-1), 7 - 3 \rangle = \langle 3, 4 \rangle$.

If $\mathbf{v} = \overrightarrow{CD}$, where C is $(2, 5)$ and D is $(5, 9)$, then $\mathbf{v} = \langle 5 - 2, 9 - 5 \rangle = \langle 3, 4 \rangle$. Therefore, $\mathbf{u} = \mathbf{v}$.

The length of **u** is $|\mathbf{u}| = \sqrt{3^2 + 4^2} = 5$.

A **unit vector** has length 1.

The **zero vector** is $\mathbf{0} = \langle 0, 0 \rangle$.

The unit vectors $\mathbf{i} = \langle 1, 0 \rangle$ and $\mathbf{j} = \langle 0, 1 \rangle$ are the **standard basis vectors**.

For all $\mathbf{a} = \langle a_1, a_2 \rangle$, we can write

$$\mathbf{a} = a_1 \mathbf{i} + a_2 \mathbf{j}.$$

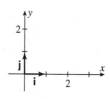

Example: If $\mathbf{a} = \langle 3, 2 \rangle$ then $\mathbf{a} = 3\mathbf{i} + 2\mathbf{j}$.

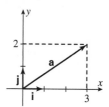

We will have more to say about vector addition in part B.

All the preceding definitions may be modified to fit three dimensions. For example, the **length** of the three dimensional vector $\mathbf{a} = \langle a_1, a_2, a_3 \rangle$ is $|\mathbf{a}| = \sqrt{a_1^2 + a_2^2 + a_3^2}$.
The zero vector is $\mathbf{0} = \langle 0, 0, 0 \rangle$.

The **standard basis vectors** in three dimensions are:

$$\mathbf{i} = \langle 1, 0, 0 \rangle,$$
$$\mathbf{j} = \langle 0, 1, 0 \rangle, \text{ and}$$
$$\mathbf{k} = \langle 0, 0, 1 \rangle.$$

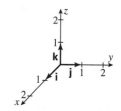

They may be used to write

$$\mathbf{a} = a_1 \mathbf{i} + a_2 \mathbf{j} + a_3 \mathbf{k}.$$

Example: The three-dimensional vector $\mathbf{a} = \langle 2, 4, 3 \rangle$ is pictured at the right. Its length is

$$|\mathbf{a}| = \sqrt{2^2 + 4^2 + 3^2} = \sqrt{29}.$$

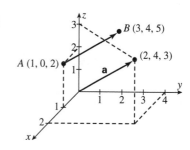

We can write $\mathbf{a}$ as $\mathbf{a} = 2\mathbf{i} + 4\mathbf{j} + 3\mathbf{k}$. If $A(1, 0, 2)$ and $B(3, 4, 5)$, then the directed line segment $\overrightarrow{AB}$ is a representation of the vector $\mathbf{a}$.

You should think of vectors in four different, yet equivalent, ways. For instance, given the vector $\mathbf{a} = \langle 2, 4, 3 \rangle$, we may imagine $\mathbf{a}$ as:

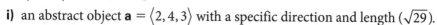

i) an abstract object $\mathbf{a} = \langle 2, 4, 3 \rangle$ with a specific direction and length ($\sqrt{29}$).

ii) the point in three-dimensional space with coordinates $(2, 4, 3)$.

iii) a pointy stick whose tail is at the origin and head at $(2, 4, 3)$.

iv) a pointy stick whose tail is at A and head is at B (or any other pair of points A and B, with $\overrightarrow{AB}$ representing the same direction and length).

The ability to shift back and forth between such interpretations is very useful for understanding the calculus of vectors and its applications.

All the previous ideas may be extended to higher dimensions as well. Although we lose our ability to draw pictures in other dimensions, we can imagine similar figures.

For example, in four dimensions with the four unit basis vectors $\mathbf{u}_1, \mathbf{u}_2, \mathbf{u}_3, \mathbf{u}_4$, a vector $\mathbf{a} = \langle a_1, a_2, a_3, a_4 \rangle$ may be written $\mathbf{a} = a_1\mathbf{u}_1 + a_2\mathbf{u}_2 + a_3\mathbf{u}_3 + a_4\mathbf{u}_4$ and $|\mathbf{a}| = \sqrt{a_1^2 + a_2^2 + a_3^2 + a_4^2}$. The set of all n-dimensional vectors is denoted V_n.

1) The vector represented by the directed line segment from $(5, 6)$ to $(3, 7)$ is _____.

$\mathbf{a} = \langle 3 - 5, 7 - 6 \rangle = \langle -2, 1 \rangle.$

2) The vector represented by the directed line segment from $(10, 4, 3)$ to $(5, 4, 6)$ is _____.

$\mathbf{b} = \langle 5 - 10, 4 - 4, 6 - 3 \rangle = \langle -5, 0, 3 \rangle.$

3) What are the lengths of the vectors in questions 1 and 2?

$|\mathbf{a}| = \sqrt{(-2)^2 + 1^2} = \sqrt{5}.$

$|\mathbf{b}| = \sqrt{(-5)^2 + 0^2 + 3^2} = \sqrt{34}.$

4) Let A be $(-1, 2)$, B be $(3, 8)$, C be $(7, -11)$, and D be $(13, -2)$. Are the vectors determined by $\overrightarrow{AB}$ and $\overrightarrow{CD}$ equal?

No, $\overrightarrow{AB}$ is the vector $\langle 3 - (-1), 8 - 2 \rangle = \langle 4, 6 \rangle$ and $\overrightarrow{CD}$ is $\langle 13 - 7, -2 - (-11) \rangle = \langle 6, 9 \rangle$. The vectors do point in the same direction but are not equal because they have different lengths.

5) Sketch the position vector for each:
  a) $\mathbf{a} = \langle 2, -6 \rangle$.

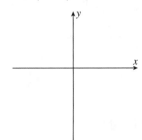

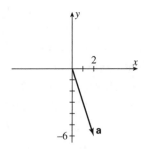

b) $\mathbf{a} = \langle 3, 6, 4 \rangle.$

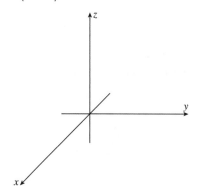

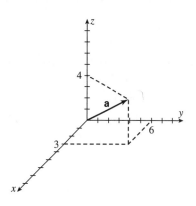

6) Is $\mathbf{a} = \left\langle \dfrac{-1}{4}, \dfrac{\sqrt{3}}{2}, \dfrac{\sqrt{3}}{4} \right\rangle$ a unit vector?

Yes, since

$$|\mathbf{a}| = \sqrt{\left(\dfrac{-1}{4}\right)^2 + \left(\dfrac{\sqrt{3}}{2}\right)^2 + \left(\dfrac{\sqrt{3}}{4}\right)^2} = 1.$$

7) Write each in terms of standard basis vectors:

   a) $\mathbf{a} = \langle 4, -3 \rangle.$

                          $\mathbf{a} = 4\mathbf{i} - 3\mathbf{j}$

   b) $\mathbf{a} = \langle 0, 4, 5 \rangle.$

                          $\mathbf{a} = 4\mathbf{j} + 5\mathbf{k}$

   c) $\mathbf{a}$ is represented by $\overrightarrow{AB}$, where
       A is $(2, 5)$ and B is $(6, 1)$.

                          $\mathbf{a} = \langle 6 - 2, \, 1 - 5 \rangle = \langle 4, \, -4 \rangle = 4\mathbf{i} - 4\mathbf{j}$

Page
816
(ET Page
792)

B. We can combine two vectors using vector addition:

For $\mathbf{a} = \langle a_1, a_2 \rangle$ and $\mathbf{b} = \langle b_1, b_2 \rangle$,

$$\mathbf{a} + \mathbf{b} = \langle a_1 + b_1, a_2 + b_2 \rangle.$$

In the diagram, $\mathbf{a} + \mathbf{b}$ is the vector obtained by positioning the initial point of $\mathbf{b}$ at the terminal point of $\mathbf{a}$ and drawing the directed line segment from the initial point of $\mathbf{a}$ to the terminal point of $\mathbf{b}$.

Another way to visualize $\mathbf{a} + \mathbf{b}$ is to place the tails of $\mathbf{a}$ and $\mathbf{b}$ together and draw the resulting parallelogram. $\mathbf{a} + \mathbf{b}$ is the diagonal of the parallelogram.

For real number c (called a **scalar**), $c\mathbf{a}$ is the **scalar multiple** of $\mathbf{a}$:

$$c\mathbf{a} = \langle ca_1, ca_2 \rangle.$$

For $c > 0$, $c\mathbf{a}$ is the vector in the same direction as $\mathbf{a}$ with length c times the length of $\mathbf{a}$. For $c < 0$, $c\mathbf{a}$ is the vector in the *opposite* direction of $\mathbf{a}$ with length $-c$ times the length of $\mathbf{a}$.

$(c > 0)$ $\qquad\qquad\qquad\qquad\qquad$ $(c < 0)$

Vector subtraction is $\mathbf{a} - \mathbf{b} = \mathbf{a} + (-1)\mathbf{b} = \langle a_1 - b_1, a_2 - b_2 \rangle$.

Examples: Let $\mathbf{a} = \langle 3, 5 \rangle$ and $\mathbf{b} = \langle 2, -1 \rangle$. Then

$$\mathbf{a} + \mathbf{b} = \langle 3 + 2, 5 + (-1) \rangle = \langle 5, 4 \rangle.$$
$$3\mathbf{a} = \langle 3(3), 3(5) \rangle = \langle 9, 15 \rangle.$$
$$\mathbf{b} - \mathbf{a} = \langle 2 - 3, -1 - 5 \rangle = \langle -1, -6 \rangle.$$

Vector addition and scalar multiplication for three or more dimensions are defined similarly—coordinate-wise.

Vector operations allow us to write $\mathbf{a} = \langle a_1, a_2 \rangle$ as $\mathbf{a} = a_1\mathbf{i} + a_2\mathbf{j}$ because
$\langle a_1, a_2 \rangle = \langle a_1, 0 \rangle + \langle 0, a_2 \rangle = a_1 \langle 1, 0 \rangle + a_2 \langle 0, 1 \rangle = a_1\mathbf{i} + a_2\mathbf{j}$.

For any nonzero vector $\mathbf{a}$, $\dfrac{\mathbf{a}}{|\mathbf{a}|}$ is a unit vector.

Nonzero vectors $\mathbf{a}$ and $\mathbf{b}$ are **parallel** means $\mathbf{b} = c\mathbf{a}$ for some c.

Properties of vector operations:

$\mathbf{a} + \mathbf{b} = \mathbf{b} + \mathbf{a}$
$\mathbf{a} + (\mathbf{b} + \mathbf{c}) = (\mathbf{a} + \mathbf{b}) + \mathbf{c}$
$\mathbf{a} + \mathbf{0} = \mathbf{a}$ ($\mathbf{0}$ is the zero vector.)
$\mathbf{a} + (-\mathbf{a}) = \mathbf{0}$
$c(\mathbf{a} + \mathbf{b}) = c\mathbf{a} + c\mathbf{b}$
$(c + d)\mathbf{a} = c\mathbf{a} + d\mathbf{a}$
$(cd)\mathbf{a} = c(d\mathbf{a})$
$1\mathbf{a} = \mathbf{a}$.

8) For $\mathbf{a} = \langle 4, 3 \rangle$ and $\mathbf{b} = \langle -5, 1 \rangle$:

   a) $\mathbf{a} + \mathbf{b} =$ _____ .

$\langle 4, 3 \rangle + \langle -5, 1 \rangle = \langle -1, 4 \rangle.$

   b) $6\mathbf{a} =$ _____ .

$\langle 6(4), 6(3) \rangle = \langle 24, 18 \rangle.$

   c) Sketch the position vector of $\mathbf{b} - \mathbf{a}$.

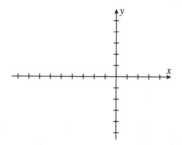

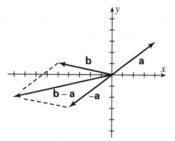

9) For $\mathbf{a} = \langle 1, 3, 2 \rangle$ and
$\mathbf{b} = \langle 2, 1, 6 \rangle$, $2\mathbf{a} + 3\mathbf{b} =$ _____ .

$$2\mathbf{a} = \langle 2, 6, 4 \rangle$$
$$3\mathbf{b} = \langle 6, 3, 18 \rangle$$
$$2\mathbf{a} + 3\mathbf{b} = \langle 8, 9, 22 \rangle.$$

10) True or False:
$c(\mathbf{a} - \mathbf{b}) = c\mathbf{a} - c\mathbf{b}.$

True.

11) Find the unit vector in the direction of
$\mathbf{a} = \langle 4, 3, -1 \rangle$.

$|\mathbf{a}| = \sqrt{4^2 + 3^2 + (-1)^2} = \sqrt{26}.$

The unit vector is

$$\frac{\mathbf{a}}{|\mathbf{a}|} = \left\langle \frac{4}{\sqrt{26}}, \frac{3}{\sqrt{26}}, \frac{-1}{\sqrt{26}} \right\rangle.$$

12) For $\mathbf{a} = 7\mathbf{i} + 4\mathbf{j} - 3\mathbf{k}$ and $\mathbf{b} = 2\mathbf{i} + \mathbf{j} + 5\mathbf{k}$:

   a) $\mathbf{a} + \mathbf{b} =$ _____ .

$9\mathbf{i} + 5\mathbf{j} + 2\mathbf{k}$

   b) $-2\mathbf{b} =$ _____ .

$-4\mathbf{i} - 2\mathbf{j} - 10\mathbf{k}$

13) Is $\mathbf{a} = \langle 4, -8, 6 \rangle$ parallel to $\mathbf{b} = \langle -2, 3, -4 \rangle$?

No. Comparing first components, $\mathbf{a} = -2\mathbf{b}$ if $\mathbf{a}$ and $\mathbf{b}$ are parallel. But comparing second components, $\mathbf{a} = -\dfrac{8}{3}\mathbf{b}$.

14) A plane is heading north at 250 km/hr, but there is a west-to-east crosswind pushing the plane eastward at 40 km/hr. What is the true direction the plane is heading?

Let $\mathbf{u}$ be the vector representing due north at 250 km/hr. Let $\mathbf{v}$ be the wind from the west at 40 km/hr. The true heading is
$\mathbf{u} + \mathbf{v} = \langle 0, 250 \rangle + \langle 40, 0 \rangle = \langle 40, 250 \rangle$.

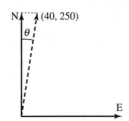

For the angle θ from north,

$$\tan \theta = \frac{40}{250} = 0.16.$$
$$\theta = \tan^{-1} 0.16 \approx 0.159.$$
$$0.159 \cdot \frac{180°}{\pi} \approx 9.1°.$$

The true course is approximately 9° east of due north.

Section 12.3   The Dot Product

The previous section showed you how to add vectors and stretch them (scalar multiplication). This section gives one way to multiply vectors, with the result being a scalar. The dot product of two vectors defined in this section provides a convenient way to determine the angle between the vectors.

Concepts to Master

A. Dot product; Properties of $\mathbf{a} \cdot \mathbf{b}$

B. Angle between two vectors; Orthogonal vectors

C. Scalar projection; Work; Direction cosines

Summary and Focus Questions

Page 824 (ET Page 800)

A. For $\mathbf{a} = \langle a_1, a_2 \rangle$ and $\mathbf{b} = \langle b_1, b_2 \rangle$, the **dot product** is
$$\mathbf{a} \cdot \mathbf{b} = a_1 b_1 + a_2 b_2.$$
For $\mathbf{a} = \langle a_1, a_2, a_3 \rangle$ and $\mathbf{b} = \langle b_1, b_2, b_3 \rangle$, the **dot product** is
$$\mathbf{a} \cdot \mathbf{b} = a_1 b_1 + a_2 b_2 + a_3 b_3.$$

Examples: $\langle 3, 5 \rangle \cdot \langle 2, 4 \rangle = 3(2) + 5(4) = 26.$
$\langle 1, 4, 3 \rangle \cdot \langle 2, 2, 3 \rangle = 1(2) + 4(2) + 3(3) = 19.$

Properties of the dot product include:

$\mathbf{a} \cdot \mathbf{b} = \mathbf{b} \cdot \mathbf{a}$

$\mathbf{a} \cdot (\mathbf{b} + \mathbf{c}) = \mathbf{a} \cdot \mathbf{b} + \mathbf{a} \cdot \mathbf{c}$

$c(\mathbf{a} \cdot \mathbf{b}) = (c\mathbf{a}) \cdot \mathbf{b}$

$\mathbf{a} \cdot \mathbf{a} = |\mathbf{a}|^2$ (the square of the length of $\mathbf{a}$).

$\mathbf{0} \cdot \mathbf{a} = 0.$

1) Find $\mathbf{a} \cdot \mathbf{b}$ for

   a) $\mathbf{a} = \langle 6, 3 \rangle, \mathbf{b} = \langle 2, -1 \rangle.$ $\mathbf{a} \cdot \mathbf{b} = 6(2) + 3(-1) = 9.$

   b) $\mathbf{a} = 4\mathbf{i} - 3\mathbf{j} + \mathbf{k}, \mathbf{b} = 5\mathbf{j} + 10\mathbf{k}.$ $\mathbf{a} \cdot \mathbf{b} = 4(0) + (-3)5 + 1(10) = -5.$

   c) $\mathbf{a} = \langle 8, 1, 4 \rangle, \mathbf{b} = \langle 3, 0, -6 \rangle.$ $\mathbf{a} \cdot \mathbf{b} = 8(3) + 1(0) + 4(-6) = 0.$

2) Why is $\mathbf{a} \cdot \mathbf{b} \cdot \mathbf{c}$ not defined although $(\mathbf{a} \cdot \mathbf{b})\mathbf{c}$ is defined?

$\mathbf{a} \cdot \mathbf{b} \cdot \mathbf{c}$ makes no sense because the dot product $\mathbf{a} \cdot \mathbf{b}$ is scalar and the dot product of a scalar with a vector $\mathbf{c}$ is not defined. But $(\mathbf{a} \cdot \mathbf{b})\mathbf{c}$ (with no dot between the parenthesis and $\mathbf{c}$) is a scalar multiple of $\mathbf{c}$.

3) True or False:
$\mathbf{a} \cdot \mathbf{a} = 1$ if and only if $\mathbf{a}$ is a unit vector.

True

4) Let $\mathbf{a} = \langle 1, 3, 1 \rangle$, $\mathbf{b} = \langle 4, 1, -2 \rangle$, and $\mathbf{c} = \langle 0, 1, 5 \rangle$. Compute $\mathbf{a} \cdot (\mathbf{b} + \mathbf{c})$ two ways.

(1) $\mathbf{b} + \mathbf{c} = \langle 4, 2, 3 \rangle$.
$\langle 1, 3, 1 \rangle \cdot \langle 4, 2, 3 \rangle$
$= 1(4) + 3(2) + 1(3) = 13$.

(2) $\mathbf{a} \cdot \mathbf{b} = 1(4) + 3(1) + 1(-2) = 5$
$\mathbf{a} \cdot \mathbf{c} = 1(0) + 3(1) + 1(5) = 8$
$\mathbf{a} \cdot \mathbf{b} + \mathbf{a} \cdot \mathbf{c} = 5 + 8 = 13$.

Page 825 (ET Page 801)

B. Placing the tails of two vectors $\mathbf{a}$ and $\mathbf{b}$ together forms and angle θ **between a and b**. The angle θ satisfies the equation

$$\mathbf{a} \cdot \mathbf{b} = |\mathbf{a}|\,|\mathbf{b}| \cos \theta, \text{ or equivalently,}$$

$$\cos \theta = \frac{\mathbf{a} \cdot \mathbf{b}}{|\mathbf{a}|\,|\mathbf{b}|}.$$

Example: Find the angle between the vectors $\mathbf{a} = \langle 1, 3, 1 \rangle$ and $\mathbf{b} = \langle 4, 1, -2 \rangle$.
$|\mathbf{a}| = \sqrt{1^2 + 3^2 + 1^2} = \sqrt{11}$, $|\mathbf{b}| = \sqrt{4^2 + 1^2 + (-2)^2} = \sqrt{21}$, and
$\mathbf{a} \cdot \mathbf{b} = 1(4) + 3(1) + 1(-2) = 5$.
Then, $\cos \theta = \dfrac{5}{\sqrt{11}\sqrt{21}} \approx 0.3290$, so $\theta = \cos^{-1}(0.3290) \approx 1.24$ radians $\approx 71°$.

When $\mathbf{a}$ and $\mathbf{b}$ are nonzero and $\mathbf{a} \cdot \mathbf{b} = 0$, $\cos \theta = 0$, so $\theta = 90°$. Therefore, we say vectors $\mathbf{a}$ and $\mathbf{b}$ are **orthogonal** (or **perpendicular**) if $\mathbf{a} \cdot \mathbf{b} = 0$.

Examples: Let $\mathbf{a} = \langle 5, 3 \rangle$, $\mathbf{b} = \langle 2, -1 \rangle$ and $\mathbf{c} = \langle 6, 12 \rangle$. Then:
$\mathbf{a}$ and $\mathbf{b}$ are not orthogonal since $\mathbf{a} \cdot \mathbf{b} = 5(2) + 3(-1) = 7$.
$\mathbf{b}$ and $\mathbf{c}$ are orthogonal since $\mathbf{b} \cdot \mathbf{c} = 2(6) + (-1)(12) = 0$.

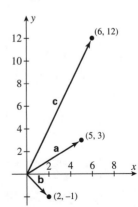

4) Find the angle θ between the vectors
$\mathbf{a} = \langle 1, -1, -1 \rangle$ and $\mathbf{b} = \langle -1, 5, 1 \rangle$.

$$\cos \theta = \frac{\mathbf{a} \cdot \mathbf{b}}{|\mathbf{a}||\mathbf{b}|}$$

$$= \frac{1(-1) + (-1)5 + (-1)1}{\sqrt{1^2 + (-1)^2 + (-1)^2} \sqrt{(-1)^2 + 5^2 + 1^2}}$$

$$= \frac{-7}{9}.$$

$$\theta = \cos^{-1}\left(\frac{-7}{9}\right) \approx 2.46 \text{ rad}$$

$$\approx 141°.$$

5) Is $6\mathbf{i} - 3\mathbf{j} + 2\mathbf{k}$ orthogonal to
$2\mathbf{i} + 10\mathbf{j} + 9\mathbf{k}$?

Yes. The dot product is
$6(2) + (-3)10 + 2(9) = 0$.

6) Find any nonzero vector $\mathbf{x}$ orthogonal to
$\mathbf{y} = \langle 3, 4 \rangle$.

If $\mathbf{x} = \langle a, b \rangle$ is orthogonal to y, then
$\mathbf{x} \cdot \mathbf{y} = 3a + 4b = 0$. Any solution other than
$(0,0)$ to this equation will do. Let $a = 4$ and
$b = -3$. Then $\mathbf{x} = \langle 4, -3 \rangle$ is orthogonal to $\mathbf{y}$.

7) Find a unit vector orthogonal to
$\mathbf{a} = \langle 4, -3, 6 \rangle$.

First, find a vector $\mathbf{b} = \langle x, y, z \rangle$ where
$\mathbf{a} \cdot \mathbf{b} = 0$.
$\mathbf{a} \cdot \mathbf{b} = 4x - 3y + 6z = 0$.
One of infinitely many solutions is
$x = 3$, $y = 4$, $z = 0$. Thus $\mathbf{b} = \langle 3, 4, 0 \rangle$
$|\mathbf{b}| = \sqrt{3^2 + 4^2 + 0^2} = 5$.
The unit vector is $\dfrac{\mathbf{b}}{|\mathbf{b}|} = \left\langle \dfrac{3}{5}, \dfrac{4}{5}, 0 \right\rangle$.

**Page 828
(ET Page 804)**

C. The **scalar projection** of $\mathbf{b}$ onto $\mathbf{a}$ is the number $|\mathbf{b}| \cos \theta$.
Intuitively, the scalar projection is the length of the shadow
of $\mathbf{b}$ cast upon $\mathbf{a}$ by a light source directly over $\mathbf{a}$.

The scalar projection is denoted $\text{comp}_{\mathbf{a}}\mathbf{b}$, (for **component** of
$\mathbf{b}$ along $\mathbf{a}$.) Also,

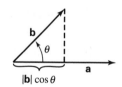

$$\text{comp}_{\mathbf{a}}\mathbf{b} = |\mathbf{b}|\cos \theta = \frac{\mathbf{a} \cdot \mathbf{b}}{|\mathbf{a}|}.$$

The **vector projection** of **b** along **a**, $\text{proj}_\mathbf{a}\mathbf{b}$, is the vector along **a** with length $\text{comp}_\mathbf{a}\mathbf{b}$.

Thus,

$$\text{proj}_\mathbf{a}\mathbf{b} = \left(\frac{|\mathbf{b}|\cos\theta}{|\mathbf{a}|}\right)\mathbf{a} = \left(\frac{\mathbf{a}\cdot\mathbf{b}}{|\mathbf{a}|^2}\right)\mathbf{a}.$$

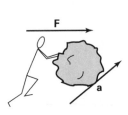

The **work** done by a constant force **F** moving an object along a directed line segment represented by a vector **a** is $\mathbf{F}\cdot\mathbf{a}$. In SI units, force is measured in newtons (kilogram meters per second^2) denoted N. Work is measured in joules (meter-kilograms), denoted J.

If the displacement is measured in feet and the force in pounds, then the work is measured in foot-pounds (ft-lb).

Example: Suppose a force of $\mathbf{F} = 3\mathbf{i} + 4\mathbf{j}$ N moves an object from $A(1, 3, 1)$ to $B(2, 5, 3)$ (in meters). How much work is done?

The displacement vector describing the movement is $\mathbf{a} = \overrightarrow{AB} = \langle 2-1, 5-3, 3-1\rangle = \langle 1, 2, 2\rangle$. The amount of work is $\mathbf{F}\cdot\mathbf{a} = 3(1) + 4(2) + 0(2) = 11$ J.

For nonzero $\mathbf{a} = \langle a_1, a_2, a_3\rangle$ let

α = angle between **a** and positive x-axis,
β = angle between **a** and positive y-axis,
γ = angle between **a** and positive z-axis.

The angles α, β and γ are **direction angles**.

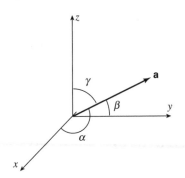

$\cos\alpha = \dfrac{a_1}{|\mathbf{a}|}, \cos\beta = \dfrac{a_2}{|\mathbf{a}|}, \cos\gamma = \dfrac{a_3}{|\mathbf{a}|}$ are the

direction cosines for the direction angles α, β and γ.

Direction angles and cosines provide ways to specify the direction, but not the length, of a three-dimensional vector.

8) Find the scalar projection of $\mathbf{x} = 3\mathbf{i} + \mathbf{j} + \mathbf{k}$ onto $\mathbf{y} = 2\mathbf{i} + 3\mathbf{j} - 4\mathbf{k}$.

$|\mathbf{x}| = \sqrt{11}, |\mathbf{y}| = \sqrt{29}.$
For θ, the angle between **x** and **y**,

$$\cos\theta = \frac{\mathbf{x}\cdot\mathbf{y}}{|\mathbf{x}||\mathbf{y}|} = \frac{5}{\sqrt{11}\sqrt{29}}.$$

The projection is

$$|\mathbf{x}|\cos\theta = \sqrt{11}\frac{5}{\sqrt{11}\sqrt{29}} = \frac{5}{\sqrt{29}}.$$

9) Find the amount of work done by a horizontal force of 4 N moving an object 3 meters up a hill that makes a 30° angle with the horizontal ground. (Disregard gravity.)

<*Find the force vector **F** and the direction vector **a**.*>

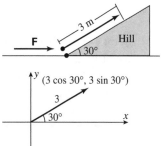

The vector representing the distance traveled is

$$\mathbf{d} = \langle 3\cos 30°, 3\sin 30° \rangle = \left\langle \frac{3\sqrt{3}}{2}, \frac{3}{2} \right\rangle.$$

The vector representing the force is $\mathbf{F} = \langle 4,0 \rangle$. The work done is

$$\mathbf{F} \cdot \mathbf{d} = 4\left(\frac{3\sqrt{3}}{2} \right) + 0\left(\frac{3}{2} \right) = 6\sqrt{3} \text{ J}.$$

10) Let $\mathbf{a} = 3\mathbf{i} - \mathbf{j} + 4\mathbf{k}$.

    a) Find the direction cosines of $\mathbf{a}$.

$$|\mathbf{a}| = \sqrt{3^2 + (-1)^2 + 4^2} = \sqrt{26}.$$
$$\cos\alpha = \frac{3}{\sqrt{26}}, \cos\beta = \frac{-1}{\sqrt{26}}, \cos\gamma = \frac{4}{\sqrt{26}}.$$

    b) Find the direction angles.

$$\alpha = \cos^{-1}\frac{3}{\sqrt{26}} \approx 0.942 \text{ rad} \approx 54°.$$
$$\beta = \cos^{-1}\frac{-1}{\sqrt{26}} \approx 1.768 \text{ rad} \approx 101°.$$
$$\gamma = \cos^{-1}\frac{4}{\sqrt{26}} \approx 0.669 \text{ rad} \approx 38°.$$

11) What may be concluded about a vector $\mathbf{v}$ whose direction cosines are $\cos\alpha = 0$, $\cos\beta = -1$, and $\cos\gamma = 0$?

The only nonzero component of $\mathbf{v}$ is in the $\mathbf{y}$ direction. The vector $\mathbf{v}$ points in the direction of the negative y-axis.

12) What may be concluded if a force $\mathbf{F}$ tries to move an object along a direction $\mathbf{a}$ orthogonal to $\mathbf{F}$?

Since $\mathbf{F}$ and $\mathbf{a}$ are orthogonal, $\mathbf{F} \cdot \mathbf{a} = 0$. No work is done. The object does not move.

Section 12.4 The Cross Product

This section provides a way to multiply two three-dimensional vectors that results in a third vector perpendicular to both given vectors. Cross product is defined only for three-dimensional vectors. It has many uses. For example, in the next section the cross product of two vectors will used to determine the equation of the plane formed by those vectors.

Concepts to Master

A. Cross product; Properties of $\mathbf{a} \times \mathbf{b}$

B. Area of parallelogram; Scalar triple product; Volume of parallelepiped

C. Technology Plus

Summary and Focus Questions

Page
832
(ET Page
808)

A. Let $\mathbf{a} = \langle a_1, a_2, a_3 \rangle$ and $\mathbf{b} = \langle b_1, b_2, b_3 \rangle$. The **cross product** of $\mathbf{a}$ and $\mathbf{b}$ is

$$\mathbf{a} \times \mathbf{b} = \langle a_2 b_3 - a_3 b_2, a_3 b_1 - a_1 b_3, a_1 b_2 - a_2 b_1 \rangle.$$

Example: For $\mathbf{a} = \langle 2, 1, 5 \rangle$ and $\mathbf{b} = \langle 4, 6, 3 \rangle$,
$\mathbf{a} \times \mathbf{b} = \langle 1(3) - 6(5), 5(4) - 2(3), 2(6) - 4(1) \rangle = \langle -27, 14, 8 \rangle.$

Rather than remembering the formula, calculating $\mathbf{a} \times \mathbf{b}$ may be done using determinants.

For a two-by-two array (a square of four numbers), the **determinant** is:

$$\begin{vmatrix} x & y \\ z & w \end{vmatrix} = xw - zy.$$

This calculation may be remembered as the product of the numbers on the upper-left to lower-right diagonal (xw) minus the product from the lower-left to the upper-right diagonal (zy).

For example, $\begin{vmatrix} 6 & 2 \\ 3 & 5 \end{vmatrix} = 6(5) - 3(2) = 30 - 6 = 24.$

The determinant of a three-by-three square array of nine numbers may be used to compute $\mathbf{a} \times \mathbf{b}$: use vectors $\mathbf{i}, \mathbf{j}, \mathbf{k}$ as symbols in the first row of the determinant, the components of $\mathbf{a}$ in the second row, and the components of $\mathbf{b}$ in the third row.

For example, using $\mathbf{a} = \langle 2, 1, 5 \rangle$ and $\mathbf{b} = \langle 4, 6, 3 \rangle$ from the example above,

$$\mathbf{a} \times \mathbf{b} = \begin{vmatrix} \mathbf{i} & \mathbf{j} & \mathbf{k} \\ 2 & 1 & 5 \\ 4 & 6 & 3 \end{vmatrix}.$$

This is evaluated as follows:

the determinant obtained by deleting the row and column of **i** times **i**

$$\begin{vmatrix} 1 & 5 \\ 6 & 3 \end{vmatrix} \mathbf{i} = [1(3) - 6(5)]\mathbf{i} = -27\mathbf{i}$$

minus

the determinant obtained by deleting the row and column of **j** times **j**

$$-\begin{vmatrix} 2 & 5 \\ 4 & 3 \end{vmatrix} \mathbf{j} = -[2(3) - 4(5)]\mathbf{j} = -(-14\mathbf{j})$$

plus

the determinant obtained by deleting the row and column of **k** times **k**

$$+\begin{vmatrix} 2 & 1 \\ 4 & 6 \end{vmatrix} \mathbf{k} = [2(6) - 4(1)]\mathbf{k} = 8\mathbf{k}.$$

Thus $\mathbf{a} \times \mathbf{b} = -27\mathbf{i} - (-14\mathbf{j}) + 8\mathbf{k} = -27\mathbf{i} + 14\mathbf{j} + 8\mathbf{k}$.

Right-Hand Rule The direction of $\mathbf{a} \times \mathbf{b}$ follows the right-hand rule—if the fingers of your right hand curve in the direction of the angle from **a** to **b**, then your thumb points in the direction of $\mathbf{a} \times \mathbf{b}$.

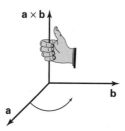

Properties of cross products include:

$\mathbf{a} \times \mathbf{b}$ is orthogonal to both **a** and **b**.
$|\mathbf{a} \times \mathbf{b}| = |\mathbf{a}||\mathbf{b}| \sin\theta$ (θ = the angle between **a** and **b**, $0 \leq \theta \leq \pi$)
$\mathbf{a} \times \mathbf{b} = -\mathbf{b} \times \mathbf{a}$
$c(\mathbf{a} \times \mathbf{b}) = c\mathbf{a} \times \mathbf{b} = \mathbf{a} \times c\mathbf{b}$ (c is a scalar)
$\mathbf{a} \times (\mathbf{b} + \mathbf{c}) = \mathbf{a} \times \mathbf{b} + \mathbf{a} \times \mathbf{c}$
$(\mathbf{a} + \mathbf{b}) \times \mathbf{c} = \mathbf{a} \times \mathbf{c} + \mathbf{b} \times \mathbf{c}$
$\mathbf{a} \times (\mathbf{b} \times \mathbf{c}) = (\mathbf{a} \cdot \mathbf{c})\mathbf{b} - (\mathbf{a} \cdot \mathbf{b})\mathbf{c}.$

For the standard basis vectors:

$\mathbf{i} \times \mathbf{j} = \mathbf{k}$ and $\mathbf{j} \times \mathbf{i} = -\mathbf{k}$
$\mathbf{j} \times \mathbf{k} = \mathbf{i}$ and $\mathbf{k} \times \mathbf{j} = -\mathbf{i}$
$\mathbf{i} \times \mathbf{k} = -\mathbf{j}$ and $\mathbf{k} \times \mathbf{i} = \mathbf{j}.$

1) Find $\mathbf{a} \times \mathbf{b}$, where $\mathbf{a} = \langle 4, -1, 3 \rangle$ and $\mathbf{b} = \langle 1, 6, 2 \rangle$.

$$\mathbf{a} \times \mathbf{b} = \begin{vmatrix} \mathbf{i} & \mathbf{j} & \mathbf{k} \\ 4 & -1 & 3 \\ 1 & 6 & 2 \end{vmatrix}$$

$$= \begin{vmatrix} -1 & 3 \\ 6 & 2 \end{vmatrix} \mathbf{i} - \begin{vmatrix} 4 & 3 \\ 1 & 2 \end{vmatrix} \mathbf{j} + \begin{vmatrix} 4 & -1 \\ 1 & 6 \end{vmatrix} \mathbf{k}$$

$$= -20\mathbf{i} - 5\mathbf{j} + 25\mathbf{k}.$$

2) If $\mathbf{a} = \langle 2,5,0 \rangle$ and $\mathbf{b} = \langle 4,3,0 \rangle$, does $\mathbf{a} \times \mathbf{b}$ point upward or downward?

Downward. Since the third components are 0 for both $\mathbf{a}$ and $\mathbf{b}$, both vectors lie in the xy-plane. Using the right-hand rule in the figure, $\mathbf{a} \times \mathbf{b}$ points downward.

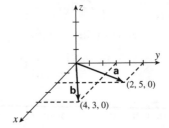

Another way to determine that $\mathbf{a} \times \mathbf{b}$ points downward is to calculate $\mathbf{a} \times \mathbf{b}$:

$$\mathbf{a} \times \mathbf{b} = \begin{vmatrix} \mathbf{i} & \mathbf{j} & \mathbf{k} \\ 2 & 5 & 0 \\ 4 & 3 & 0 \end{vmatrix}$$

$$= \begin{vmatrix} 5 & 0 \\ 3 & 0 \end{vmatrix} \mathbf{i} - \begin{vmatrix} 2 & 0 \\ 4 & 0 \end{vmatrix} \mathbf{j} + \begin{vmatrix} 2 & 5 \\ 4 & 3 \end{vmatrix} \mathbf{k}$$

$$= 0\mathbf{i} - 0\mathbf{j} - 14\mathbf{k}.$$

The vector $-14\mathbf{k}$ points downward.

3) For any vectors $\mathbf{a}, \mathbf{b}$ what is $\mathbf{a} \times \mathbf{b} + \mathbf{b} \times \mathbf{a}$?

Since $\mathbf{a} \times \mathbf{b} = -\mathbf{b} \times \mathbf{a}, \mathbf{a} \times \mathbf{b} + \mathbf{b} \times \mathbf{a} = \mathbf{0}$, the zero vector.

4) For any three-dimensional vector $\mathbf{a}$, what is $\mathbf{a} \times \mathbf{a}$?

$\mathbf{a} \times \mathbf{a} = \mathbf{0}$, the zero vector. The angle θ between $\mathbf{a}$ and $\mathbf{a}$ is 0. Therefore, $|\mathbf{a} \times \mathbf{a}| = |\mathbf{a}|\,|\mathbf{a}|\sin\theta = 0$.

5) Evaluate $2\mathbf{i} \times (4\mathbf{j} + 3\mathbf{k})$.

This may be done using determinants, but we use cross product properties:

$$2\mathbf{i} \times (4\mathbf{j} + 3\mathbf{k}) = 2\mathbf{i} \times 4\mathbf{j} + 2\mathbf{i} \times 3\mathbf{k}$$
$$= 8(\mathbf{i} \times \mathbf{j}) + 6(\mathbf{i} \times \mathbf{k})$$
$$= 8\mathbf{k} + 6(-\mathbf{j})$$
$$= 8\mathbf{k} - 6\mathbf{j}.$$

6) Find a vector orthogonal to both $2\mathbf{i} - 3\mathbf{j} + \mathbf{k}$ and $\mathbf{i} + \mathbf{j} + 2\mathbf{k}$.

The cross product will do:

$$\begin{vmatrix} \mathbf{i} & \mathbf{j} & \mathbf{k} \\ 2 & -3 & 1 \\ 1 & 1 & 2 \end{vmatrix} = \begin{vmatrix} -3 & 1 \\ 1 & 2 \end{vmatrix}\mathbf{i} - \begin{vmatrix} 2 & 1 \\ 1 & 2 \end{vmatrix}\mathbf{j} + \begin{vmatrix} 2 & -3 \\ 1 & 1 \end{vmatrix}\mathbf{k}$$

$$= -7\mathbf{i} - 3\mathbf{j} + 5\mathbf{k}.$$

7) True or False:
$\mathbf{a} \times (\mathbf{b} \times \mathbf{c}) = (\mathbf{a} \times \mathbf{b}) \times \mathbf{c}$.

False. For example, let $\mathbf{a} = \langle 1, 2, 3 \rangle$, $\mathbf{b} = \langle 4, 5, 6 \rangle$ and $\mathbf{c} = \langle 7, 8, 9 \rangle$. Then $\mathbf{a} \times (\mathbf{b} \times \mathbf{c}) = \langle -24, -6, 12 \rangle$ while $(\mathbf{a} \times \mathbf{b}) \times \mathbf{c} = \langle 78, 6, -66 \rangle$.

Page 385 (ET Page 811)

B. For two nonzero vectors $\mathbf{a}$ and $\mathbf{b}$, $|\mathbf{a} \times \mathbf{b}|$ is the area of the parallelogram formed by $\mathbf{a}$ and $\mathbf{b}$.

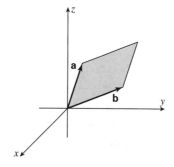

Example: Find the area of the parallelogram formed by the vectors $\mathbf{a} = \langle 2, 1, 5 \rangle$ and $\mathbf{b} = \langle 4, 6, 3 \rangle$.

In an earlier example we saw that $\mathbf{a} \times \mathbf{b} = -27\mathbf{i} + 14\mathbf{j} + 8\mathbf{k}$. Therefore, the area is $|\mathbf{a} \times \mathbf{b}| = \sqrt{(-27)^2 + 14^2 + 8^2} = \sqrt{989} \approx 31.4$.

The **triple scalar product** of vectors $\mathbf{a} = \langle a_1, a_2, a_3 \rangle, \mathbf{b} = \langle b_1, b_2, b_3 \rangle$, and $\mathbf{c} = \langle c_1, c_2, c_3 \rangle$ is the number $\mathbf{a} \cdot (\mathbf{b} \times \mathbf{c})$. It may be evaluated directly or by using

$$\mathbf{a} \cdot (\mathbf{b} \times \mathbf{c}) = \begin{vmatrix} a_1 & a_2 & a_3 \\ b_1 & b_2 & b_3 \\ c_1 & c_2 & c_3 \end{vmatrix}.$$

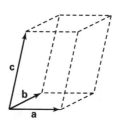

The absolute value of the triple product, $|\mathbf{a} \cdot (\mathbf{b} \times \mathbf{c})|$, is the volume of the parallelepiped (slanted box) formed by $\mathbf{a}, \mathbf{b},$ and $\mathbf{c}$.

Example: For $\mathbf{a} = \langle 2, -1, 2 \rangle$, $\mathbf{b} = \langle 0, 3, 2 \rangle$, and $\mathbf{c} = \langle 4, -3, 1 \rangle$,

$$\mathbf{a} \cdot (\mathbf{b} \times \mathbf{c}) = \begin{vmatrix} 2 & -1 & 2 \\ 0 & 3 & 2 \\ 4 & -3 & 1 \end{vmatrix} = 2 \begin{vmatrix} 3 & 2 \\ -3 & 1 \end{vmatrix} - (-1) \begin{vmatrix} 0 & 2 \\ 4 & 1 \end{vmatrix} + 2 \begin{vmatrix} 0 & 3 \\ 4 & -3 \end{vmatrix}$$

$$= 2(9) + 1(-8) + 2(-12) = -14.$$

8) Find the area of the parallelogram with vertices P (4, 1, 3), Q (7, 5, 3), R (6, 4, 2), and S (9, 8, 2).

Let $\mathbf{a} = \overrightarrow{PQ} = \langle 7 - 4, 5 - 1, 3 - 3 \rangle$
$$= \langle 3, 4, 0 \rangle.$$

Let $\mathbf{b} = \overrightarrow{PR} = \langle 6 - 4, 4 - 1, 2 - 3 \rangle$
$$= \langle 2, 3, -1 \rangle.$$

The area is $|\mathbf{a} \times \mathbf{b}|$.

$$\mathbf{a} \times \mathbf{b} = \begin{vmatrix} \mathbf{i} & \mathbf{j} & \mathbf{k} \\ 3 & 4 & 0 \\ 2 & 3 & -1 \end{vmatrix} = -4\mathbf{i} + 3\mathbf{j} + \mathbf{k}.$$

$$|\mathbf{a} \times \mathbf{b}| = \sqrt{(-4)^2 + 3^2 + (-1)^2} = \sqrt{26}.$$

9) Let $\mathbf{a} = \langle 1, 3, 1 \rangle$, $\mathbf{b} = \langle 0, 4, 2 \rangle$ and $\mathbf{c} = \langle -2, 2, 3 \rangle$. Find $\mathbf{a} \cdot (\mathbf{b} \times \mathbf{c})$

   a) from the definitions.

$$\mathbf{b} \times \mathbf{c} = \begin{vmatrix} \mathbf{i} & \mathbf{j} & \mathbf{k} \\ 0 & 4 & 2 \\ -2 & 2 & 3 \end{vmatrix}$$

$$= \begin{vmatrix} 4 & 2 \\ 2 & 3 \end{vmatrix} \mathbf{i} - \begin{vmatrix} 0 & 2 \\ -2 & 3 \end{vmatrix} \mathbf{j} + \begin{vmatrix} 0 & 4 \\ -2 & 2 \end{vmatrix} \mathbf{k}$$

$$= 8\mathbf{i} - 4\mathbf{j} + 8\mathbf{k}.$$

$$\mathbf{a} \cdot (\mathbf{b} \times \mathbf{c}) = 1(8) + 3(-4) + 1(8)$$
$$= 4.$$

   b) as a triple scalar product.

$$\mathbf{a} \cdot (\mathbf{b} \times \mathbf{c}) = \begin{vmatrix} 1 & 3 & 1 \\ 0 & 4 & 2 \\ -2 & 2 & 3 \end{vmatrix}$$

$$= 1 \begin{vmatrix} 4 & 2 \\ 2 & 3 \end{vmatrix} - 3 \begin{vmatrix} 0 & 2 \\ -2 & 3 \end{vmatrix} + 1 \begin{vmatrix} 0 & 4 \\ -2 & 2 \end{vmatrix}$$

$$= 1(8) - 3(4) + 1(8) = 4.$$

10) Find the volume of the parallelepiped formed by $\mathbf{a} = -2\mathbf{i} + 3\mathbf{j}$, $\mathbf{b} = 4\mathbf{i} + \mathbf{j} - \mathbf{k}$, $\mathbf{c} = 6\mathbf{j} + \mathbf{k}$.

$$\mathbf{a} \cdot (\mathbf{b} \times \mathbf{c}) = \begin{vmatrix} -2 & 3 & 0 \\ 4 & 1 & -1 \\ 0 & 6 & 1 \end{vmatrix}$$

$$= -2 \begin{vmatrix} 1 & -1 \\ 6 & 1 \end{vmatrix} - 3 \begin{vmatrix} 4 & -1 \\ 0 & 1 \end{vmatrix} + 0 \begin{vmatrix} 4 & 1 \\ 0 & 6 \end{vmatrix}$$

$$= -2(7) - 3(4) + 0 = -26.$$

$$|\mathbf{a} \cdot (\mathbf{b} \times \mathbf{c})| = 26.$$

C. Technology Plus. Use a computer algebra system or a graphing calculator to solve.

T-1) Some calculators and CAS systems have a built-in vector cross product function (for example, "crossP" on some Texas Instruments calculators and "crossproduct" for Maple). Find $\mathbf{a} \times \mathbf{b}$ for

a) $\mathbf{a} = \langle 3, 7, 2 \rangle$
$\mathbf{b} = \langle 1, 2, 1 \rangle$

$\mathbf{a} \times \mathbf{b} = \langle 3, -1, -1 \rangle$

b) $\mathbf{a} = \langle -3.51, 2.62, 8.01 \rangle$
$\mathbf{b} = \langle 4.41, -1.02, 6.32 \rangle$

$\mathbf{a} \times \mathbf{b} = \langle 24.73, 57.51, -7.97 \rangle$

Section 12.5 Equations of Lines and Planes

This section gives three different, but equivalent, "equations" for a line in three-dimensional space—a vector form, a parametric form, and a form similar to that of a line in two dimensions. A plane in three-dimensional space, which may be thought of as all vectors perpendicular to a given vector, has an equation form very similar to that of a line in two dimensions. We will use equations of lines and planes in later chapters when we discuss tangent planes to surfaces—a higher-dimensional version of tangent lines to curves in two dimensions.

Concepts to Master

A. Equations of lines in space (vector, parametric, and symmetric); Parallel lines; Skew lines

B. Normal vector; Equations of planes (vector and scalar); Parallel planes; Angle between intersecting planes; Distance from a point to a plane and distance between two planes

C. Technology Plus

Summary and Focus Questions

Page 840 (ET Page 816)

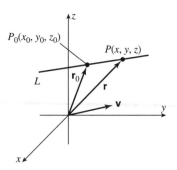

A. Let L be a line in three dimensions passing through $P_0(x_0, y_0, z_0)$ and parallel to vector $\mathbf{v} = \langle a, b, c \rangle$. Let $\mathbf{r}_0 = \langle x_0, y_0, z_0 \rangle$. If $P(x, y, z)$ is also on the line and $\mathbf{r} = \langle x, y, z \rangle$, then $\mathbf{r} - \mathbf{r}_0 = \langle x - x_0, y - y_0, z - z_0 \rangle$ is parallel to $\mathbf{v}$. Therefore, the three forms of the equations for L are

Vector Form	Parametric Form	Symmetric Form
$\mathbf{r} = \mathbf{r}_0 + t\mathbf{v}$	$x = x_0 + at$	$\dfrac{x - x_0}{a} = \dfrac{y - y_0}{b} = \dfrac{z - z_0}{c}$
	$y = y_0 + bt$	$(a, b, c \text{ nonzero})$
	$z = z_0 + ct$	

In case one of the components of $\mathbf{v}$ is 0, that component for points on L is a constant. For example, if $b = 0$, the symmetric form is modified to $y = y_0$ and $\dfrac{x - x_0}{a} = \dfrac{z - z_0}{c}$.

Note: The vector **v** is *not* on the L unless the line passes through the origin. Also, even though **r**, $\mathbf{r}_0$, and **v** are vectors, when we interpret the equation $\mathbf{r} = \mathbf{r}_0 + t\mathbf{v}$, think of **r** and $\mathbf{r}_0$ as points and **v** as the pointy stick in the same direction as L.

For lines $L_1: \mathbf{r} = \mathbf{r}_0 + t_1\mathbf{v}_1$ and $L_2: \mathbf{r} = \mathbf{s}_0 + t_2\mathbf{v}_2$:

1) L_1 and L_2 are parallel iff $\mathbf{v}_1$ is a scalar multiple of $\mathbf{v}_2$. ($\mathbf{v}_1$ and $\mathbf{v}_2$ are parallel.)

2) L_1 and L_2 intersect iff $\mathbf{r}_0 + t_1\mathbf{v}_1 = \mathbf{s}_0 + t_2\mathbf{v}_2$ for some t_1, t_2.

Two lines are **skew** if they are not parallel and do not intersect.

Parallel and intersecting lines will lie within a plane. Skew lines never lie in a common plane.

Example: Are the lines $L_1: \begin{matrix} x = 3 + 5t \\ y = 8 - 2t \\ z = 1 + 3t \end{matrix}$ and $L_2: \begin{matrix} x = 4 + 3s \\ y = 1 - s \\ z = 14 + 4s \end{matrix}$ parallel, skew, or intersecting?

The direction vector for L_1, $\langle 5, -2, 3 \rangle$, is not a multiple of the direction vector for L_2, $\langle 3, -1, 4 \rangle$. Therefore, the lines are not parallel.

If L_1 and L_2 intersect, there must be numbers t and s so that:

$$3 + 5t = 4 + 3s$$
$$8 - 2t = 1 - s$$
$$1 + 3t = 14 + 4s.$$

If we multiply the second equation by $3 (24 - 6t = 3 - 3s)$ and the third equation by $2 (2 + 6t = 28 + 8s)$ and add the results, we eliminate the variable t:

$$26 = 31 + 5s$$
$$5s = -5$$
$$s = -1$$

Substituting $s = -1$ in the second equation, $8 - 2t = 1 - (-1)$, so $2t = 6$ and $t = 3$. Therefore, the lines, if they do intersect, will intersect when $t = 3$ and $s = -1$. These values must also satisfy the first equation. If $t = 3$, $x = 3 + 5(3) = 18$. If $s = -1$, $x = 4 + 3(-1) = 1$. Since $t = 3$ and $s = -1$ do not satisfy the first equation, the lines do not intersect. Therefore, L_1 and L_2 are skew.

The equation of the **line segment** between the points $\mathbf{r}_0$ and $\mathbf{r}_1$ is:

$$\mathbf{r} = (1 - t)\mathbf{r}_0 + t\mathbf{r}_1 \text{ for } 0 \le t \le 1.$$

1) Find the three forms for the equation of the line L through $(4, 1, 6)$ parallel to $\langle 7, 3, 0 \rangle$.

Vector form: $\mathbf{r} = \langle 4, 1, 6 \rangle + \mathbf{t} \langle 7, 3, 0 \rangle$,
or $\mathbf{r} = (4 + 7t)\mathbf{i} + (1 + 3t)\mathbf{j} + 6\mathbf{k}$.

Parametric form: $x = 4 + 7t$
$$y = 1 + 3t$$
$$z = 6 + 0t \text{ (or } z = 6).$$

Symmetric form: $z = 6$ and $\dfrac{x-4}{7} = \dfrac{y-1}{3}$.

2) Find the vector form of the equation of the line that passes through $A(5, 9, 7)$ and $B(12, 6, 10)$.

$\overline{AB}$ is a directed line segment along the line;
$\overline{AB} = \langle 12 - 5, \ 6 - 9, \ 10 - 7 \rangle = \langle 7, -3, 3 \rangle$.
The line is $\mathbf{r} = \langle 5, 9, 7 \rangle + t \langle 7, -3, 3 \rangle$.

3) Determine whether the lines L_1 and L_2 are parallel, intersect, or are skew:

$L_1: x = 1 + 5t$ $L_2: x = 10 + 3s$
    $y = 3 + 4t$ $y = 17 - s$
    $z = -2 + t$ $z = -3 + 2s$

The direction $\langle 5, 4, 1 \rangle$ is not a multiple of the direction $\langle 3, -1, 2 \rangle$ so L_1 and L_2 are not parallel.

If L_1 and L_2 intersect, then

$$1 + 5t = 10 + 3s$$
$$3 + 4t = 17 - s$$
$$-2 + t = -3 + 2s$$

must have a solution. Multiplying the second equation by 2 and adding to the third gives

$$4 + 9t = 31,$$
$$9t = 27, t = 3.$$

Using $t = 3$ in the third equation, $-2 + 3 = -3 + 2s$. Thus $2s = 4$, $s = 2$.

The solution to the second and third equations is $t = 3$, $s = 2$. Since this also satisfies the first $[1 + 5(3) = 10 + 3(2)]$ the lines intersect.

At $t = 3$ and $s = 2$, $x = 16$, $y = 15$, $z = 1$ so the point of intersection is $(16, 15, 1)$.

4) Find the equation of the line parallel to the line L through the point P. Use the same form for your answer as that used to describe L.

a) $L: \dfrac{x-1}{8} = \dfrac{y+2}{5} = \dfrac{z+3}{-2}$
    $P(4, 5, 1)$

$$\dfrac{x-4}{8} = \dfrac{y-5}{5} = \dfrac{z-1}{-2}.$$

b) $L: \mathbf{r} = (3 + 4t)\mathbf{i} - t\mathbf{j} + (8 + t)\mathbf{k}$
$P(-1, 3, 2)$

$\mathbf{r} = (-1 + 4t)\mathbf{i} + (3 - t)\mathbf{j} + (2 + t)\mathbf{k}.$

c) $L: x = 9 - 9t$
$\quad y = 8 + 3t$
$\quad z = 2 + 4t$
$P(7, -3, 5)$

$x = 7 - 9t.$
$y = -3 + 3t.$
$z = 5 + 4t.$

5) Can three line L_1, L_2, and L_3 be drawn so that any two of them are skew?

Yes. In three dimensions there is lots of room for any number of skew lines. For example, let L_1 be the line through $(0, 0, 4)$ parallel to the x-axis. L_2 be the line through $(3, 0, 0)$ parallel to the y-axis, and L_3 be the line through $(1, 2, 0)$ parallel to the z-axis.

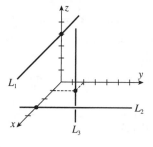

B. A vector orthogonal to a plane is called a **normal vector** for the plane.

Page 833 (ET Page 797)

Suppose a plane has a normal vector $\mathbf{n} = \langle a, b, c \rangle$. Let $\mathbf{r}_0(x_0, y_0, z_0)$ be a point in the plane and $\mathbf{r}(x, y, z)$ be any other point in the plane. We note that, in general, neither $\mathbf{r}$ nor $\mathbf{r}_0$ is a vector that lies in the plane. However, $\mathbf{r} - \mathbf{r}_0$ is in the plane, and hence $\mathbf{r} - \mathbf{r}_0$ is orthogonal to $\mathbf{n}$. Therefore, $\mathbf{n} \cdot (\mathbf{r} - \mathbf{r}_0) = 0$. The two forms of the equation of the plane are:

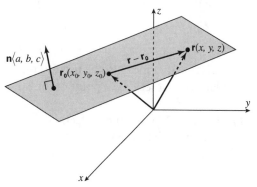

Vector Form	Scalar Form
$\mathbf{n} \cdot \mathbf{r} = \mathbf{n} \cdot \mathbf{r}_0$	$a(x - x_0) + b(y - y_0) + c(z - z_0) = 0$

Example: The equation of the plane through the point $r_0(2, -3, 5)$ with normal vector $n = \langle 8, 4, 7 \rangle$ is

$$\langle 8, 4, 7 \rangle \cdot \langle x, y, z \rangle = \langle 8, 4, 7 \rangle \cdot \langle 2, -3, 5 \rangle$$
$$8x + 4y + 7z = 8(2) + 4(-3) + 7(5)$$
$$8x + 4y + 7z = 39.$$

Every linear equation of the form $ax + by + cz = d$ is an equation of a plane in three-dimensional space with normal vector $n = \langle a, b, c \rangle$.

If a and b are nonparallel vectors in a plane, then $a \times b$ may be used as a normal vector for the plane.

Two planes are **parallel** if their normal vectors are parallel.

The **angle θ between two intersecting planes** with normal vectors n_1 and n_2 satisfies

$$\cos \theta = \frac{n_1 \cdot n_2}{|n_1| \, |n_2|}.$$

Example: The planes $8x + 4y + z = 42$ and $x + y + 6z = 7$ are not parallel because their normal vectors, $n_1 = \langle 8, 4, 1 \rangle$ and $n_2 = \langle 1, 1, 6 \rangle$, are not parallel (neither vector is a scalar multiple of the other.)

The angle of intersection of the planes, θ, satisfies:

$$\cos \theta = \frac{n_1 \cdot n_2}{|n_1| \, |n_2|} = \frac{8(1) + 4(1) + 1(6)}{\sqrt{8^2 + 4^2 + 1^2} \sqrt{1^2 + 1^2 + 6^2}} = \frac{18}{\sqrt{81}\sqrt{38}} = \frac{2}{\sqrt{38}} \approx 0.3244.$$

Thus, $\theta = \cos^{-1}(0.3244) \approx 1.2404$ rad $\approx 71°$.

The **distance from a point $P(x_0, y_0, z_0)$ to a plane** $ax + by + cz = d$ is:

$$\frac{|ax_0 + by_0 + cz_0 - d|}{\sqrt{a^2 + b^2 + c^2}}.$$

This formula may also be used to find the distance between two parallel planes—just let P be a point on one of the planes.

---

6) Find the two forms of the equation of the plane through $(8, 2, 3)$ with normal vector $\langle 6, 2, 5 \rangle$.

Vector form:
$$\langle 6, 2, 5 \rangle \cdot r = \langle 6, 2, 5 \rangle \cdot \langle 8, 2, 3 \rangle$$
$$\langle 6, 2, 5 \rangle \cdot r = 67.$$
Scalar form:
$$6(x - 8) + 2(y - 2) + 5(z - 3) = 0$$
$$6x + 2y + 5z = 67.$$

7) Find the scalar form for the equation of the plane through $P(1, 3, 1)$, $Q(3, 0, 4)$, $R(4, -1, 2)$.

Let $a = \overrightarrow{PQ} = \langle 2, -3, 3 \rangle$ and $b = \overrightarrow{PR} = \langle 3, -4, 1 \rangle$.
A normal vector to the plane is

$$a \times b = \begin{vmatrix} i & j & k \\ 2 & -3 & 3 \\ 3 & -4 & 1 \end{vmatrix} = 9i + 7j + k.$$

Use P as a point on the plane.
The equation of the plane is
$9(x - 1) + 7(y - 3) + (z - 1) = 0$ or
$9x + 7y + z = 31$.

8) Find the equation of the plane through the point $(1, 2, 3)$ that is parallel to the plane $2x + 5y + 2z = 9$.

Since the planes are parallel, they have the same normal vector $\langle 2, 5, 2 \rangle$:
$2(1) + 5(2) + 2(3) = 18$.
The equation is $2x + 5y + 2z = 18$.

9) Find θ, the angle of intersection of the planes:
$$3x + y + 3z = 5$$
$$6x - 3y - 2z = 14.$$

Let $\mathbf{n}_1 = \langle 3, 1, 3 \rangle$, $\mathbf{n}_2 = \langle 6, -3, -2 \rangle$.
$$\cos\theta = \frac{\mathbf{n}_1 \cdot \mathbf{n}_2}{|\mathbf{n}_1||\mathbf{n}_2|} = \frac{9}{\sqrt{19}\sqrt{49}} = \frac{9}{7\sqrt{19}}$$
$$\approx 0.29496.$$
$$\theta = \cos^{-1}(0.29496) \approx 1.2714 \approx 72.8°.$$

10) Find the line of intersection of the planes:
$$2x + y - z = -4$$
$$3x + y + 2z = 5.$$

<*The line of intersection will be orthogonal to the normal vectors of both planes.*>

Let $\mathbf{n}_1 = \langle 2, 1, -1 \rangle$, $\mathbf{n}_2 = \langle 3, 1, 2 \rangle$.

The line of intersection has the same direction as
$$\mathbf{n}_1 \times \mathbf{n}_2 = \begin{vmatrix} \mathbf{i} & \mathbf{j} & \mathbf{k} \\ 2 & 1 & -1 \\ 3 & 1 & 2 \end{vmatrix} = \langle 3, -7, -1 \rangle.$$

To find a point of intersection, set $x = 0$ ($y = 0$ or $z = 0$ also could be used):
$$y - z = -4$$
$$y + 2z = 5$$

Subtracting: $3z = 9$, $z = 3$.

$y - 3 = -4$, $y = -1$.

Therefore, $(0, -1, 3)$ is a point on the line of intersection. The equation of the line is
$$\frac{x}{3} = \frac{y + 1}{-7} = \frac{z - 3}{-1}.$$

11) Find the distance from $(1, 2, -2)$ to the plane $x + 3y + 4z = 12$.

$$\frac{|1(1) + 3(2) + 4(-2) - 12|}{\sqrt{1^2 + 3^2 + 4^2}} = \frac{|-13|}{\sqrt{26}} = \frac{\sqrt{26}}{2}.$$

12) Find the distance between the parallel planes:

$$2x + 3y + 7z = 10 \text{ and}$$
$$2x + 3y + 7z = 134.$$

<Pick a point on one plane and use the equation of the other plane.>

$(5, 0, 0)$ is a point on the first plane. The distance between the planes is

$$\frac{|2(5) + 3(0) + 7(0) - 134|}{\sqrt{2^2 + 3^2 + 7^2}} = \frac{|-124|}{\sqrt{62}}$$

$$= \frac{124}{\sqrt{62}} = 2\sqrt{62}.$$

13) Where does the line $\dfrac{x-1}{1} = \dfrac{y+2}{3} = \dfrac{z-1}{2}$ intersect the plane $2x - 3y + z = -6$?

First rewrite the line in parametric form:

$$x = 1 + t$$
$$y = -2 + 3t$$
$$z = 1 + 2t.$$

Now substitute into the equation of the plane:

$$2(1 + t) - 3(-2 + 3t) + (1 + 2t) = -6$$
$$9 - 5t = -6$$
$$-5t = -15$$
$$t = 3.$$

Thus, $x = 1 + 3 = 4$, $y = -2 + 3(3) = 7$, and $z = 1 + 2(3) = 7$. The point of intersection is $(4, 7, 7)$.

C. Technology Plus. Use a computer algebra system or a graphing calculator to solve.

T-1) Sketch the graphs of these two lines:

$$L_1: x = 1 - 3t$$
$$y = -3 - 7t$$
$$z = 2 + 4t$$

$$L_2: x = -5 - 4t$$
$$y = 6 + 2t$$
$$z = -1 + 3t$$

Are the lines skew or do they intersect?

With viewing angles of 110° for x and 70° for z, the graph is

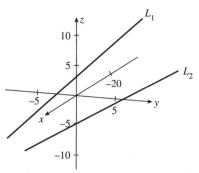

The lines are skew.

Section 12.6 Cylinders and Quadric Surfaces

This section describes the equations and graphs of several types of surfaces in three dimensions; they are reminiscent of the conic sections (ellipses, parabolas, and hyperbolas) in two dimensions. We begin with cylinders, which are collections of parallel lines.

Concepts to Master

A. Cylinders in three-dimensional space

B. Nondegenerate quadric surfaces

C. Technology Plus

Summary and Focus Questions

Page 840 (ET Page 804)

A. A **cylinder** is a surface composed of all lines parallel to a given line that pass through a given curve. A cylinder need not be a "round tube," as the common use of the term cylinder suggests. The plane $z = 2$, at the right, fits the definition of a cylinder—all lines parallel to the y-axis through the line L: $x = t, y = 0, z = 2$.

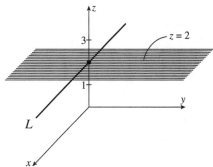

Some of the cylinders that are easiest to visualize are defined by equations where one or more of the variables (x, y, or z) is missing. Help in identifying a surface can come from examining the surface's **traces**—the curves that are the intersections of the surface and planes parallel to one of the xy-, xz-, or yz-planes.

Example: The equation $(x - 3)^2 + (y - 4)^2 = 4$ has no z variable, so it is a cylinder along the z-axis. In the xy-plane, the trace is the circle $(x - 3)^2 + (y - 4)^2 = 4$ with center $(3, 4)$ and radius 2. The graph of the trace is given at the right. The graph of the cylinder is obtained from this graph by extending the trace in the z direction.

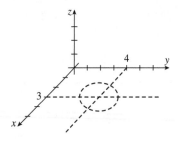 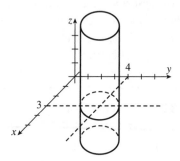

Here are some cylinders with conic sections as traces.

Surface	Typical Equation	Traces			Typical Graph
		$xy\,(z = 0)$	$xz\,(y = 0)$	$yz\,(x = 0)$	
Parabolic Cylinder	$x^2 = 4ay$	parabola	z-axis	z-axis	
Elliptic Cylinder	$\dfrac{x^2}{a^2} + \dfrac{y^2}{b^2} = 1$	ellipse	two parallel lines	two parallel lines	
Hyperbolic Cylinder	$\dfrac{x^2}{a^2} - \dfrac{y^2}{b^2} = 1$	hyperbola	two parallel lines	none	

1) True or False:

If one of x, y, z is missing from an equation, the surface is a cylinder.

True.

2) Identify each:

a) $x^2 - 5z^2 = 0$

Hyperbolic cylinder along the y-axis.

b)

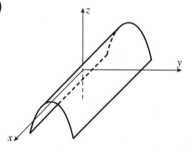

Parabolic cylinder along the x-axis.

3) Sketch a graph of each:

a) $x + 2y = 6$

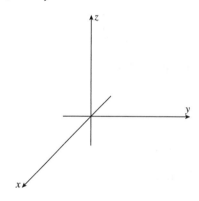

<Determine what variable is missing.>

The variable z is not present. The trace in the xy-plane is a line intercepting the x-axis at $x = 6$ and the y-axis at $y = 3$. This is a plane parallel to the z-axis.

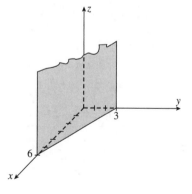

b) $\dfrac{x^2}{9} + \dfrac{z^2}{25} = 1$

The *y* variable is missing. This is an elliptic cylinder along the *y*-axis.

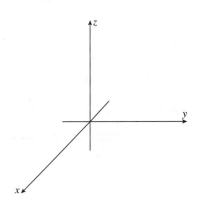

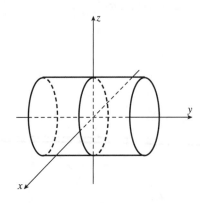

B. The three-dimensional analogues of the ellipse, parabola, and hyperbola are called **quadric surfaces.** To visualize these surfaces, it will help to first determine traces.

Page 851 (ET Page 827)

The following table summarizes the equations and their traces for six types of quadric surfaces. Interchanging *x*, *y*, and *z* will produce surfaces of the same type but oriented along a different axis. For example $z^2 = \dfrac{x^2}{a^2} + \dfrac{y^2}{b^2}$, $y^2 = \dfrac{x^2}{a^2} + \dfrac{z^2}{c^2}$, and $x^2 = \dfrac{y^2}{b^2} + \dfrac{z^2}{c^2}$ are elliptic cones along the *z*-, *y*-, and *x*-axis, respectively.

		Traces			
Surface	**Typical Equation**	**xy (z = 0)**	**xz (y = 0)**	**yz (x = 0)**	**Typical Graph**
Ellipsoid	$\dfrac{x^2}{a^2} + \dfrac{y^2}{b^2} + \dfrac{z^2}{c^2} = 1$	ellipse	ellipse	ellipse	
Hyperboloid of One Sheet (along z-axis)	$\dfrac{x^2}{a^2} + \dfrac{y^2}{b^2} - \dfrac{z^2}{c^2} = 1$	ellipse	hyberbola	hyberbola	

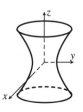

(continued)

Surface	Typical Equation	Traces			Typical Graph
		$xy\,(z=0)$	$xz\,(y=0)$	$yz\,(x=0)$	
Hyperboloid of Two Sheets (along *z*-axis)	$-\dfrac{x^2}{a^2}-\dfrac{y^2}{b^2}+\dfrac{z^2}{c^2}=1$	None	hyperbola	hyperbola	
Elliptic cone (along *z*-axis)	$\dfrac{z^2}{c^2}=\dfrac{x^2}{a^2}+\dfrac{y^2}{b^2}$	(0, 0, 0)	two interesting lines	two interesting lines	
Elliptic Paraboloid (along *z*-axis)	$\dfrac{z}{c}=\dfrac{x^2}{a^2}+\dfrac{y^2}{b^2}$	(0, 0, 0)	parabola upward	parabola upward	
Hyperbolic Paraboloid (along *z*-axis)	$\dfrac{z}{c}=\dfrac{y^2}{b^2}-\dfrac{x^2}{a^2}$	two interesting lines	parabola downward	parabola upward	

In ordinary life, an American football is an ellipsoid, some nuclear cooling towers are shaped like a hyperboloid of one sheet, a satellite TV antenna is an elliptic paraboloid, and a Pringle's® potato chip is a hyperbolic paraboloid.

Some equations may need to be rewritten by completing the square before trying to identify the surface.

Example: The equation $(x-4)^2+(y-2)^2+(z-3)^2=13$ is a sphere (ellipsoid) with center (4, 2, 3) and radius $\sqrt{13}$.

The traces in the coordinate planes are:

xy-plane ($z=0$): The equation becomes $(x-4)^2+(y-2)^2+(0-3)^2=13$, so $(x-4)^2+(y-2)^2=4$, a circle of radius 2.

xz-plane ($y = 0$): The equation becomes
$(x - 4)^2 + (0 - 2)^2 + (z - 3)^2 = 13$,
so $(x - 4)^2 + (z - 3)^2 = 9$, a circle of
radius 3.

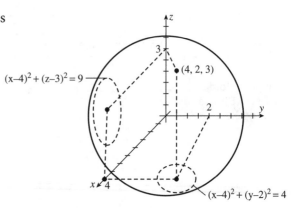

yz-plane ($x = 0$): The equation becomes
$(0 - 4)^2 + (y - 2)^2 + (z - 3)^2 = 13$,
so $(y - 2)^2 + (z - 3)^2 = -3$. The
sphere does not intersect the *yz*-plane.

The graph of the sphere and the two
traces are at the right.

4) True or False: A hyperbolic paraboloid must
have an x^2 term, a y^2 term, and a z term.

False. Two variables must be squared and
the third one must be linear, but they need
not be x^2, y^2, and z. For example,

$y = \dfrac{x^2}{4} - \dfrac{z^2}{9}$ is a hyperbolic

paraboloid along the *y*-axis.

5) Identify each:

   a) $\dfrac{x^2}{25} - \dfrac{y^2}{4} - \dfrac{z^2}{16} = 1$

Hyperboloid of two sheets.

   b) $y - 4x^2 = 16z^2$

Elliptic paraboloid.

   c) $y^2 = x^2 + z^2$

Elliptic cone.

   d) $x^2 - 8x - y^2 - 2y + z^2 + 15 = 0$

Complete the square:
$x^2 - 8x + 16 - (y^2 + 2y + 1) + z^2 = 0$
$(x - 4)^2 - (y + 1)^2 + z^2 = 0$
$(y + 1)^2 = (x - 4)^2 + z^2$

This is an elliptic cone with center at
$(4, -1, 0)$.

e)

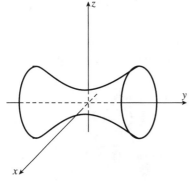

Hyperboloid of one sheet.

f)

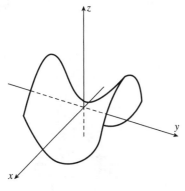

Hyperbolic paraboloid.

g)

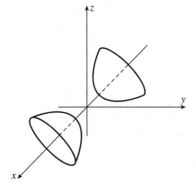

Hyperboloid of two sheets.

6) Sometimes, Always, or Never:
The linear terms are important in
determining the type of quadric surface.

Sometimes. If the equation contains both a
first and a second power of a variable, the
presence of the first power does not affect
the *type* of surface it is.

C. Technology Plus. Use a computer algebra system or a graphing calculator to solve.

T-1) Graph $z = x^2 + 2xy$ for
$-2 \leq x \leq 2, -2 \leq y \leq 2$.

Try different viewpoints.

With viewing angles of 15° for the x-axis and 70° for the z-axis, the graph is:

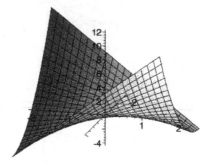

With viewing angles of 60° for the x-axis and 50° for the z-axis, the graph is:

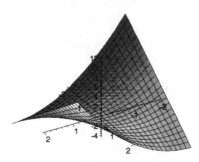

T-2) Graph $x^2 - y^2 + z^2 = 0$ for
$-3 \leq x \leq 3$,
$-3 \leq y \leq 3$,
$-3 \leq z \leq 3$.

This is a cone along the y-axis. For many systems you will need to either plot it parametrically or use an implicit plotting command. In Maple, for example, the command is

implicitplot3d $((x\verb|^|2 - y\verb|^|2 + z\verb|^|2 = 0,$
$x = -3..3, y = -3..3, z = -3..3);)$

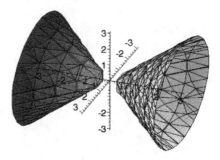

Chapter 13 — Vector Functions

Section 13.1   Vector Functions and Space Curves

Until now if $y = f(x)$ is a function, then x is a real number and $f(x)$ is a real number. This section discusses functions f where x is still a real number but $f(x)$ is a vector. We will extend the concepts of limits and continuity to vector functions. The "graphs" of vector functions are curves in two- or three-dimensional space. Vector functions are another way to look at curves defined by parametric equations.

Concepts to Master

A. Vector functions; Limits; Continuity; Curves in space

B. Technology Plus

Summary and Focus Questions

Page 864 (ET Page 840)*

A. A **vector function** (or **vector-valued function**) **r** associates to a real number t a vector $\mathbf{r}(t)$. The **domain** of a vector function **r** is the set of all real numbers t for which the vector $\mathbf{r}(t)$ is defined.

Most of this section is concerned with three-dimensional vectors. We write:

$$\mathbf{r}(t) = f(t)\mathbf{i} + g(t)\mathbf{j} + h(t)\mathbf{k}$$

The real-valued functions f, g, and h are **component functions**.

Example: The function $\mathbf{r}(t) = \sqrt{t}\,\mathbf{i} + t^2\mathbf{j} + 2t\mathbf{k}$ has component functions $f(t) = \sqrt{t}$, $g(t) = t^2$, and $h(t) = 2t$. $\mathbf{r}(t)$ may also be written $\mathbf{r}(t) = \langle \sqrt{t}, t^2, 2t \rangle$. $\mathbf{r}(t)$ associates to $t = 4$ the vector $\mathbf{r}(4) = 2\mathbf{i} + 16\mathbf{j} + 8\mathbf{k} = \langle 2, 16, 8 \rangle$. $\mathbf{r}(0) = \mathbf{0}$, the zero vector. The domain $\mathbf{r}(t)$ is $[0, \infty)$.

Limits of vector functions are defined component-wise:

$$\lim_{t \to t_0} \mathbf{r}(t) = (\lim_{t \to t_0} (f(t)))\mathbf{i} + (\lim_{t \to t_0} (g(t)))\mathbf{j} + (\lim_{t \to t_0} (h(t)))\mathbf{k}.$$

$\mathbf{r}(t)$ **is continuous at a** means $\lim_{t \to a} \mathbf{r}(t) = \mathbf{r}(a)$.

The **range** of a vector function **r** is the set C of all vectors $\mathbf{r}(t)$, where t is in the domain of **r**. Those

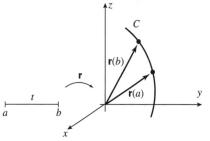

*When using the Early Transcendentals text, use the page number in parentheses.

133

vectors in C, when thought of as points, trace out a **space curve**. We say that the curve C is defined parametrically by

$$x = f(t),\ y = g(t),\ z = h(t).$$

Visualizing the curve for a given vector function can be difficult. It may help to draw it in a box coordinate instead of simply with x-, y-, and z-axes or to determine what solid or surface the curve wraps itself around.

Example: Sketch the graph of the curve
$\mathbf{r}(t) = \langle \sin t, \cos t, 1 + \sin 8t \rangle$ for $0 \le t \le 2\pi$.
Let $x = \sin t,\ y = \cos t,$ and $z = 1 + \sin 8t$.
Notice that z oscillates between 0 and 2 and that
$x^2 + y^2 = 1$. The curve will be on the cylinder
$x^2 + y^2 = 1$.
A computer generated sketch is at the right.

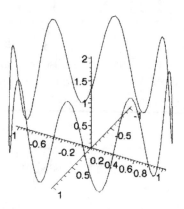

1) Let $\mathbf{r}(t) = 2t\mathbf{i} - \dfrac{1}{t}\mathbf{j} + \sqrt{t+1}\,\mathbf{k}$.

  a) What is the domain of $\mathbf{r}(t)$?

$2t$ is defined for all t. $\dfrac{1}{t}$ is defined for all $t \ne 0$. $\sqrt{t+1}$ is defined for $t \ge -1$. The domain of $\mathbf{r}(t)$ is $[-1,\ 0) \cup (0,\ \infty)$.

  b) Find $\mathbf{r}(2)$ and $\mathbf{r}(-1)$.

$\mathbf{r}(2) = 2(2)\mathbf{i} - \tfrac{1}{2}\mathbf{j} + \sqrt{2+1}\,\mathbf{k}$
$\qquad = 4\mathbf{i} - \tfrac{1}{2}\mathbf{j} + \sqrt{3}\,\mathbf{k}.$
$\mathbf{r}(-1) = -2\mathbf{i} + \mathbf{j}.$

2) For $\mathbf{r}(t) = \langle \sin t,\ t^2 + 4,\ e^t \rangle$
$\lim\limits_{t \to 0} \mathbf{r}(t) = $ _____ .

$\langle \sin 0,\ 0^2 + 4,\ e^0 \rangle = \langle 0,\ 4,\ 1 \rangle.$

3) Sketch a graph of
$\mathbf{r}(t) = (1 + \sin t)\mathbf{i} + (1 + \sin t)\mathbf{j} + t\mathbf{k}$.

We note that $x = y$ so the curve is in the plane $x = y$ parallel to the z-axis.

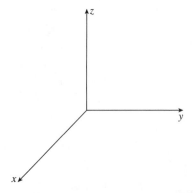

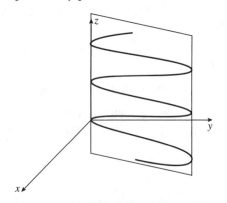

4) Find a vector function whose graph is the line segment from $(1, 2, 1)$ to $(4, 6, 5)$.

Use parametric equations to describe the line determined by the two points.

$$x = 1 + (4 - 1)t = 1 + 3t$$
$$y = 2 + (6 - 2)t = 2 + 4t$$
$$z = 1 + (5 - 1)t = 1 + 4t.$$

Therefore, $\mathbf{r}(t) = \langle 1 + 3t, \ 2 + 4t, \ 1 + 4t \rangle$ for $0 \leq t \leq 1$ has a graph that is the line segment from $(1, 2, 1)$ to $(4, 6, 5)$.

5) Do the curves $\mathbf{r}(t) = \langle t + 3, \ t^2, \ 2t + 1 \rangle$ and $\mathbf{s}(t) = \langle 10 - t, \ 3t - 11, \ t \rangle$ intersect?

<Note the question does not ask if there is a t such that $\mathbf{r}(t) = \mathbf{s}(t)$ because the curves might not intersect at a common value of t.>

The question asks whether there are numbers t and k such that $\mathbf{r}(t) = \mathbf{s}(k)$. If so, then

$$t + 3 = 10 - k$$
$$t^2 = 3k - 11$$
$$2t + 1 = k$$

must all be true for some t and k.

Using the third equation we may substitute for k in the first equation to solve for t:

$$t + 3 = 10 - (2t + 1)$$
$$3t = 6$$
$$t = 2$$

Then $k = 2t + 1 = 2(2) + 1 = 5$.
The pair $t = 2$ and $k = 5$ does satisfy the second equation, so the curves do intersect at $\mathbf{r}(2) = \mathbf{s}(5) = \langle 5, \ 4, \ 5 \rangle$.

B. Technology Plus. Use a computer algebra system or a graphing calculator to solve.

T-1) Graph $\mathbf{r}(t)$, where

$$x = \cos 5t \cos 6t$$
$$y = \sin 5t \cos 6t$$
$$z = \sin 6t, \ 0 \leq t \leq 2\pi.$$

On what surface does the curve lie?

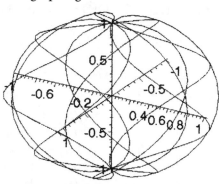

We note that $x^2 + y^2 + z^2 = 1$, so $\mathbf{r}(t)$ lies on the sphere with center $(0, 0, 0)$ and radius 1.

Section 13.2 Derivatives and Integrals of Vector Functions

This section extends the concepts of derivatives and integrals to vector functions. As you might expect, differentiation and integration is performed component-wise.

Concepts to Master

A. Derivatives of vector functions; Tangent vector; Properties of derivatives

B. Integrals of vector functions

C. Technology Plus

Summary and Focus Questions

Page 871 (ET Page 847)

A. The **derivative** of a vector function $\mathbf{r}(t)$ is defined to be

$$\frac{d\mathbf{r}}{dt} = \mathbf{r}'(t) = \lim_{h \to 0} \frac{\mathbf{r}(t+h) - \mathbf{r}(t)}{h}.$$

$\mathbf{r}'(t)$ is the **tangent vector** to the curve defined by the function $\mathbf{r}(t)$.

$\mathbf{T}(t) = \dfrac{\mathbf{r}'(t)}{|\mathbf{r}'(t)|}$ is the **unit tangent vector.** $\mathbf{T}(t)$ points in the same direction as $\mathbf{r}'(t)$ but has length 1.

$\mathbf{r}'(t)$ may be found by finding the derivatives of the components.
If $\mathbf{r}(t) = f(t)\mathbf{i} + g(t)\mathbf{j} + h(t)\mathbf{k}$, then

$$\mathbf{r}'(t) = f'(t)\mathbf{i} + g'(t)\mathbf{j} + h'(t)\mathbf{k}.$$

Example: Find the unit tangent vector at $t = 1$ for $\mathbf{r}(t) = (t^2 + t)\mathbf{i} + 4t^3\mathbf{j} - 10t\mathbf{k}$.
$\mathbf{r}'(t) = (2t + 1)\mathbf{i} + 12t^2\mathbf{j} - 10\mathbf{k}$, so $\mathbf{r}'(1) = (2(1) + 1)\mathbf{i} + 12(1)^2\mathbf{j} - 10\mathbf{k} = 3\mathbf{i} + 12\mathbf{j} - 10\mathbf{k}$.

Then, $|\mathbf{r}'(t)| = \sqrt{3^2 + 12^2 + 10^2} = \sqrt{253}$ and $\mathbf{T}(1) = \dfrac{2}{\sqrt{253}}\mathbf{i} + \dfrac{12}{\sqrt{253}}\mathbf{j} - \dfrac{10}{\sqrt{253}}\mathbf{k}$.

For differentiable vector functions $\mathbf{r}_1(t)$ and $\mathbf{r}_2(t)$ and scalar (real-valued) function $a(t)$:

$$[\mathbf{r}_1(t) + \mathbf{r}_2(t)]' = \mathbf{r}_1'(t) + \mathbf{r}_2'(t)$$
$$[c\mathbf{r}_1(t)]' = c[\mathbf{r}_1'(t)] \quad (c, \text{ a constant})$$
$$[a(t)\mathbf{r}_1(t)]' = a'(t)\mathbf{r}_1(t) + a(t)\mathbf{r}_1'(t)$$
$$[\mathbf{r}_1(t) \cdot \mathbf{r}_2(t)]' = \mathbf{r}_1'(t) \cdot \mathbf{r}_2(t) + \mathbf{r}_1(t) \cdot \mathbf{r}_2'(t)$$
$$[\mathbf{r}_1(t) \times \mathbf{r}_2(t)]' = \mathbf{r}_1'(t) \times \mathbf{r}_2(t) + \mathbf{r}_1(t) \times \mathbf{r}_2'(t)$$
$$[\mathbf{r}_1(a(t))]' = a'(t)\mathbf{r}_1'(a(t)) \quad \text{(Chain Rule)}$$

1) Find $\mathbf{r}'(t)$ for $\mathbf{r}(t) = \left\langle e^{2t}, \tan t, t^3 \right\rangle$.

$\mathbf{r}'(t) = \left\langle 2e^{2t}, \sec^2 t, 3t^2 \right\rangle$.

2) Find the unit tangent vector to the curve $\mathbf{r}(t) = t^2\mathbf{i} - t^3\mathbf{j} + t^4\mathbf{k}$ at $t = 1$.

$\mathbf{r}'(t) = 2t\mathbf{i} - 3t^2\mathbf{j} + 4t^3\mathbf{k}$
$\mathbf{r}'(1) = 2\mathbf{i} - 3\mathbf{j} + 4\mathbf{k}$

$$\mathbf{T}(1) = \frac{2\mathbf{i} - 3\mathbf{j} + 4\mathbf{k}}{\sqrt{2^2 + (-3)^2 + 4^2}}$$

$$= \frac{2}{\sqrt{29}}\mathbf{i} - \frac{3}{\sqrt{29}}\mathbf{j} + \frac{4}{\sqrt{29}}\mathbf{k}.$$

3) Find $\mathbf{r}''(t)$ for $\mathbf{r}(t) = \left\langle \cos t, t^3 \right\rangle$.

$\mathbf{r}'(t) = \left\langle -\sin t, 3t^2 \right\rangle$.
$\mathbf{r}''(t) = \left\langle -\cos t, 6t \right\rangle$.

4) Let $\mathbf{f}(t) = t^2\mathbf{i} + t^3\mathbf{j}$ and $\mathbf{g}(t) = 7t\mathbf{i} + 3t^2\mathbf{j}$. Find $[\mathbf{f}(t) \cdot \mathbf{g}(t)]'$ two ways.

a) $\mathbf{f}(t) \cdot \mathbf{g}(t) = t^2(7t) + t^3(3t^2)$
$\qquad = 7t^3 + 3t^5$.
Therefore, $[\mathbf{f}(t) \cdot \mathbf{g}(t)]' = 21t^2 + 15t^4$.

b) $[\mathbf{f}(t) \cdot \mathbf{g}(t)]' = \mathbf{f}'(t) \cdot \mathbf{g}(t) + \mathbf{f}(t) \cdot \mathbf{g}'(t)$
$\qquad = [2t\mathbf{i} + 3t^2\mathbf{j}] \cdot [7t\mathbf{i} + 3t^2\mathbf{j}]$
$\qquad\quad + [t^2\mathbf{i} + t^3\mathbf{j}] \cdot [7\mathbf{i} + 6t\mathbf{j}]$
$\qquad = (14t^2 + 9t^4) + (7t^2 + 6t^4)$
$\qquad = 21t^2 + 15t^4$.

5) The curves
$\mathbf{r}_1(t) = (t^3 + t)\mathbf{i} + e^{2t}\mathbf{j} + (\sin t)\mathbf{k}$ and
$\mathbf{r}_2(s) = (2s - 4)\mathbf{i} + (s^2 - 3)\mathbf{j} + (s - 2)^2\mathbf{k}$
intersect at $(0, 1, 0)$, where $t = 0$ and $s = 2$. Find the angle of intersection of the curves.

<*The angle of intersection of the curves, θ, is the same as the angle of intersection of the tangent vectors to the curves.*>
$\mathbf{r}_1'(t) = (3t^3 + 1)\mathbf{i} + 2e^{2t}\mathbf{j} + (\cos t)\mathbf{k}$
$\mathbf{r}_1'(0) = (3(0)^2 + 1)\mathbf{i} + 2e^0\mathbf{j} + (\cos 0)\mathbf{k}$
$\qquad = \mathbf{i} + 2\mathbf{j} + \mathbf{k}$.
$\mathbf{r}_2'(s) = 2\mathbf{i} + 2s\mathbf{j} + 2(s - 2)\mathbf{k}$
$\mathbf{r}_2'(2) = 2\mathbf{i} + 2(2)\mathbf{j} + 2(2 - 2)\mathbf{k}$
$\qquad = 2\mathbf{i} + 4\mathbf{j}$.

$$\cos\theta = \frac{\langle 1, 2, 1 \rangle \cdot \langle 2, 4, 0 \rangle}{\sqrt{1^2 + 2^2 + 1^2}\sqrt{2^2 + 4^2 + 0^2}}$$

$$= \frac{2 + 8 + 0}{\sqrt{6}\sqrt{20}} = \frac{\sqrt{30}}{6}.$$

$$\theta = \cos^{-1}\left(\frac{\sqrt{30}}{6}\right) \approx 0.4205 \text{ rad} \approx 24°.$$

Page
874
(ET Page
850)

B. A form of the Fundamental Theorem of Calculus holds for vector functions: If $\mathbf{r}(t) = f(t)\mathbf{i} + g(t)\mathbf{j} + h(t)\mathbf{k}$ is continuous on $[a, b]$, then

$$\int_a^b \mathbf{r}(t)\,dt = \left(\int_a^b f(t)\,dt\right)\mathbf{i} + \left(\int_a^b g(t)\,dt\right)\mathbf{j} + \left(\int_a^b h(t)\,dt\right)\mathbf{k}.$$

In vector form the theorem says

$$\int_a^b \mathbf{r}(t)\,dt = \mathbf{R}(b) - \mathbf{R}(a), \text{ where } \mathbf{R}'(t) = \mathbf{r}(t).$$

6) Evaluate $\displaystyle\int_1^3 \left(\mathbf{i} + 2t\mathbf{j} + \frac{1}{t}\mathbf{k}\right) dt.$

The integral equals

$$\left(\int_1^3 dt\right)\mathbf{i} + \left(\int_1^3 2t\,dt\right)\mathbf{j} + \left(\int_1^3 \frac{1}{t}\,dt\right)\mathbf{k}$$

$$= t\Big]_1^3 \mathbf{i} + t^2 \Big]_1^3 \mathbf{j} + \ln t\Big]_1^3 \mathbf{k} = 2\mathbf{i} + 8\mathbf{j} + (\ln 3)\mathbf{k}.$$

7) Find $\mathbf{r}(t)$ if $\mathbf{r}'(t) = e^{2t}\mathbf{i} - \mathbf{j} + t^2\mathbf{k}$ and $\mathbf{r}(0) = \mathbf{i} - \mathbf{j}.$

$$\int(e^{2t}\mathbf{i} - \mathbf{j} + t^2\mathbf{k})\,dt = \frac{e^{2t}}{2}\mathbf{i} - t\mathbf{j} + \frac{t^3}{3}\mathbf{k} + \mathbf{C}$$

At $t = 0$, $\mathbf{r}(0) = \dfrac{e^0}{2}\mathbf{i} - 0\mathbf{j} + \dfrac{0^3}{3}\mathbf{k} + \mathbf{C}$

$= \frac{1}{2}\mathbf{i} + \mathbf{C}$. But we are given $\mathbf{r}(0) = \mathbf{i} - \mathbf{j}$.

Therefore, $\frac{1}{2}\mathbf{i} + \mathbf{C} = \mathbf{i} - \mathbf{j}$.

$\mathbf{C} = \frac{1}{2}\mathbf{i} - \mathbf{j}.$

$$\mathbf{r}(t) = \frac{e^{2t}}{2}\mathbf{i} - t\mathbf{j} + \frac{t^3}{3}\mathbf{k} + \left(\frac{1}{2}\mathbf{i} - \mathbf{j}\right)$$

$$= \frac{e^{2t} + 1}{2}\mathbf{i} - (t + 1)\mathbf{j} + \frac{t^3}{3}\mathbf{k}.$$

C. Technology Plus. Use a computer algebra system or a graphing calculator to solve.

T-1) Find the parametric equations for the tangent line to the curve $\mathbf{r}(t) = \left\langle 4t, 2t^2, t^3 \right\rangle$ at $t = 1.5$. Graph the curve and unit tangent vector on the same screen.

$\mathbf{r}(t) = \left\langle 4t, 2t^2, t^3 \right\rangle$
$\mathbf{r}'(t) = \left\langle 4, 4t, 3t^2 \right\rangle$
At $t = 1.5$, $\mathbf{r}(1.5) = \left\langle 6, 4.5, 3.375 \right\rangle$ and
$\mathbf{r}'(1.5) = \left\langle 4, 6, 6.75 \right\rangle.$

The tangent line is

$x = 6 + 4t$
$y = 4.5 + 6t$
$z = 3.375 + 6.75t$

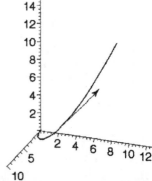

Section 13.3   Arc Length and Curvature

This section describes how to determine the arc length of a continuous space curve. We will see that the process is the same as that for two-dimensional curves defined parametrically. This section also contains several concepts for describing the shape of a curve. Curvature, for example, is a measure of how fast the curve is bending at a given point.

Concepts to Master

A. Length of a curve in space; Arc length function; Parameterization

B. Smooth curve; Curvature

C. Tangent vector; Unit normal vector; Normal plane

D. Technology Plus

Summary and Focus Questions

Page 877 (ET Page 853)

A. If $\mathbf{r}(t) = f(t)\mathbf{i} + g(t)\mathbf{j} + h(t)\mathbf{k}$ and $\mathbf{r}'(t)$ is a continuous function for $a \le t \le b$, the **length of the space curve** determined by $\mathbf{r}$ is

$$L = \int_a^b |\mathbf{r}'(t)|\, dt = \int_a^b \sqrt{[f'(t)]^2 + [g'(t)]^2 + [h'(t)]^2}\, dt.$$

Example: The function $\mathbf{r}(t) = t^3\mathbf{i} + t^2\mathbf{j} + 2t\mathbf{k}$ for $0 \le t \le 1$ defines a curve C from $(0, 0, 0)$ to $(1, 1, 2)$. Since $\mathbf{r}'(t) = 3t^2\mathbf{i} + 2t\mathbf{j} + 2\mathbf{k}$ the length of C is

$$\int_0^1 \sqrt{(3t^2)^2 + (2t)^2 + (2)^2}\, dt = \int_0^1 \sqrt{9t^4 + 4t^2 + 4}\, dt \approx 2.60.$$

For a given curve, there are many different vector functions that describe the curve. Each of these is a different **parameterization** of the curve.

Example: The unit circle C in the first quadrant of the plane may be parameterized many ways. Here are two:

$$\mathbf{r}(t) = \left\langle t, \sqrt{1 - t^2} \right\rangle,\ 0 \le t \le 1$$

$$\mathbf{q}(t) = \langle \sin t,\ \cos t \rangle,\ 0 \le t \le \frac{\pi}{2}.$$

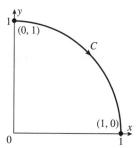

Both traverse the curve C from $(0, 1)$ to $(1, 0)$ but at different velocities along the way. Regardless, the arc length is the same:

$$\int_0^1 |\mathbf{r}'(t)|\, dt = \int_0^1 \sqrt{1^2 + \left(\frac{-t}{\sqrt{1-t^2}}\right)^2}\, dt = \int_0^1 \frac{1}{\sqrt{1-t^2}}\, dt = \sin^{-1} t\Big]_0^1 = \frac{\pi}{2}.$$

$$\int_0^{\frac{\pi}{2}} |\mathbf{q}'(t)|\, dt = \int_0^{\frac{\pi}{2}} \sqrt{\cos^2 t + \sin^2 t}\, dt = \int_0^{\frac{\pi}{2}} 1\, dt = \frac{\pi}{2}.$$

The **arc length function** $s(t) = \int_a^t |\mathbf{r}'(u)|\, du$ measures the length of that part of the curve from a to t. By definition of $s(t)$, $s'(t)$.

If we can solve for t in terms of s (so that we can write $t = t(s)$), we can "parameterize" the curve in the form $\mathbf{r} = \mathbf{r}(t(s))$. For this particular parameterization, $\mathbf{r}(t(1))$ is 1 unit along the curve, $\mathbf{r}(t(2))$ is 2 units along the curve, and so on. Therefore, $\mathbf{r}(t(s))$ represents moving a point from the beginning of the curve to the end with uniform speed (with respect to distance traveled along the curve.) As an example, think of a car travelling with constant velocity along a winding road.

Example: In the preceding example, $\mathbf{r}(t) = \left\langle t, \sqrt{1-t^2}\right\rangle, 0 \le t \le 1$, was used to parameterize the unit circle in the first quadrant from $(0, 1)$ to $(1, 0)$. Using $\mathbf{r}(t)$, reparameterize the circle with respect to arc length at $t = 0$ in the direction of increasing t.

The arc length function is:

$$s(t) = \int_0^t |\mathbf{r}'(u)|\, du = \int_0^t \sqrt{1^2 + \left(\frac{-u}{\sqrt{1-u^2}}\right)^2}\, du$$

$$= \int_0^t \frac{1}{\sqrt{1-u^2}}\, du = \sin^{-1} u\Big]_0^t = \sin^{-1} t.$$

Solving $s = \sin^{-1} t$ for t, we have $t = \sin s$. Then:

$$\mathbf{r}(t(s)) = \left\langle t(s), \sqrt{1-(t(s))^2}\right\rangle = \left\langle \sin s, \sqrt{1-(\sin s)^2}\right\rangle = \left\langle \sin s,\ \cos s\right\rangle.$$

This is the parameterization $\mathbf{q}$ in the previous example. Therefore, as a particle moves along the unit circle with parameterization $\mathbf{q}$, the particle moves at uniform velocity.

Since t varies from 0 to $\frac{\pi}{2}$, let's choose, for example, the value of t that is a third of the way from 0 to $\frac{\pi}{2}$ (that is, $t = \frac{1}{3}\left(\frac{\pi}{2} - 0\right) = \frac{\pi}{6}$). The particle will be at $\mathbf{q}\left(\frac{\pi}{6}\right) = \left\langle \sin\frac{\pi}{6},\ \cos\frac{\pi}{6}\right\rangle = \left\langle \frac{1}{2},\ \frac{\sqrt{3}}{2}\right\rangle$, which is the point on the circle C that is third of the way along the circle from $(0, 1)$ toward $(1, 0)$.

1) Find the length of the curve given by
$\mathbf{r}(t) = \ln t\mathbf{i} - t^2\mathbf{j} + 2t\mathbf{k}$ for $t \in [1, 4]$.

$\mathbf{r}'(t) = \frac{1}{t}\mathbf{i} - 2t\mathbf{j} + 2\mathbf{k}$.

$$|\mathbf{r}'(t)| = \sqrt{\frac{1}{t^2} + 4t^2 + 4} = \sqrt{\frac{1 + 4t^2 + 4t^4}{t^2}}$$

$$= \sqrt{\frac{(1 + 2t^2)^2}{t^2}} = \frac{1 + 2t^2}{t} = \frac{1}{t} + 2t.$$

The length of the curve is

$$L = \int_1^4 \left(\frac{1}{t} + 2t\right) dt = (\ln t + t^2)\Big]_1^4$$

$$= 15 + \ln 4.$$

2) Use methods of this section to find the length of the curve $y = x^2$ for $0 \le x \le 2$.

A simple parameterization of $y = x^2$ for $0 \le x \le 2$ is $\mathbf{r}(t) = \langle t, t^2 \rangle$, $0 \le t \le 2$. $\mathbf{r}'(t) = \langle 1, 2t \rangle$ and the arc length is

$$\int_0^2 \sqrt{(1)^2 + (2t)^2}\, dt = \int_0^2 \sqrt{1 + 4t^2}\, dt \approx 4.65.$$

3) a) Find the arc length function for the curve C given by
$\mathbf{r}(t) = (1 - t^2)\mathbf{i} + t^2\mathbf{j} + \sqrt{2}t^2\mathbf{k}$
for $0 \le t \le 4$.

$\mathbf{r}'(t) = -2t\mathbf{i} + 2t\mathbf{j} + 2\sqrt{2}t\mathbf{k}$.

$|\mathbf{r}'(t)| = \sqrt{(-2t)^2 + (2t)^2 + (2\sqrt{2}t)^2} = 4t$.

$s(t) = \int_0^t |\mathbf{r}'(u)|\, du = \int_0^t 4u\, du = 2t^2$.

b) Re parameterize the curve C using the arc length function.

From part a), $s = 2t^2$, $t^2 = \frac{s}{2}$ and $t = \sqrt{\frac{s}{2}}$.

At $t = 0$, $s = 0$ and at $t = 4$, $s = 32$. The reparameterization is

$$\mathbf{r}(t(s)) = \left(1 - \left(\sqrt{\frac{s}{2}}\right)^2\right)\mathbf{i} + \left(\sqrt{\frac{s}{2}}\right)^2\mathbf{j} + \sqrt{2}\left(\sqrt{\frac{s}{2}}\right)^2\mathbf{k}$$

$$= \left(1 - \frac{s}{2}\right)\mathbf{i} + \frac{s}{2}\mathbf{j} + \frac{s}{\sqrt{2}}\mathbf{k},$$

for $0 \le s \le 32$.

B. A curve C given by a vector function $\mathbf{r}(t)$ for t in an interval I is **smooth** if $\mathbf{r}'(t)$ is continuous and $\mathbf{r}'(t) \neq 0$ for all t in I. Informally, smooth means the curve has no breaks or sharp corners.

Example: The curve determined by $\mathbf{r}(t) = \langle \cos t, \ \sin t, t^2 \rangle$ is smooth for any interval $[a, b]$ because $\mathbf{r}'(t) = \langle -\sin t, \ \cos t, 2t \rangle$ is never the zero vector $\left(|\mathbf{r}'(t)| = \sqrt{(-\sin t)^2 + (\cos t)^2 + (2t)^2} = \sqrt{1 + 4t^2} \text{ is never zero} \right)$.

The **curvature** κ for C parameterized by a twice differentiable $\mathbf{r}(t)$ is

$$\kappa(t) = \left| \frac{d\mathbf{T}}{ds} \right| = \left| \frac{\mathbf{T}'(t)}{\mathbf{r}'(t)} \right| = \frac{|\mathbf{r}'(t) \times \mathbf{r}''(t)|}{|\mathbf{r}'(t)|^3}.$$

In the special case of $y = f(x)$ in the plane the curvature κ may be rewritten as

$$\kappa(x) = \frac{|y''|}{[1 + (y')^2]^{3/2}}.$$

Curvature is a measure of how quickly a curve C changes direction. For example, if y' is large, then $y = f(x)$ is increasing rapidly and the graph of f has a small curvature.

4) For each curve C defined by $\mathbf{r}(t)$, determine whether C is smooth.

  a) $\mathbf{r}(t) = \langle e^t, t^2, t^3 \rangle$, $t \in [0, 1]$.

$\mathbf{r}'(t) = \langle e^t, 2t, 3t^2 \rangle$. Since the x-coordinate of $\mathbf{r}'(t)$ is never 0, $\mathbf{r}'(t) \neq \mathbf{0}$. Therefore, C is smooth everywhere and, in particular, for $t \in [0, 1]$.

  b) $\mathbf{r}(t) = \langle t \ln t - t, \ t^2 - 2t, \ e^t - t \rangle$ for $\frac{1}{2} \leq t \leq 2$.

$\mathbf{r}'(t) = \langle \ln t, \ 2t - 2, \ e^t - 1 \rangle$. For $\frac{1}{2} \leq t \leq 2$, $e^t - 1 > 0$. Therefore $\mathbf{r}'(t) \neq \mathbf{0}$ for $\frac{1}{2} \leq t \leq 2$ and C is smooth.

5) Which curve would have greater curvature at $t = 0$?

  $C_1 : x = t \qquad C_2 : x = t$
  $\quad\ \ y = t^2 \qquad\quad\ y = t^4$

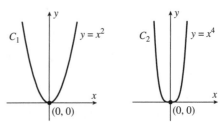

At $(0, 0)$, C_1 has a greater curvature. (Curve C_2 is flatter near $(0, 0)$.)

6) Find the curvature at $t = 1$ for the curve
$\mathbf{r}(t) = t^2\mathbf{i} + \dfrac{2}{3}t^3\mathbf{j} + 2t\mathbf{k}$.

$\mathbf{r}'(t) = 2t\mathbf{i} + 2t^2\mathbf{j} + 2\mathbf{k}$.

$|\mathbf{r}'(t)| = \sqrt{4t^2 + 4t^4 + 4} = 2\sqrt{t^2 + t^4 + 1}$.

$\mathbf{r}''(t) = 2\mathbf{i} + 4t\mathbf{j}$.

$$\mathbf{r}'(t) \times \mathbf{r}''(t) = \begin{vmatrix} \mathbf{i} & \mathbf{j} & \mathbf{k} \\ 2t & 2t^2 & 2 \\ 2 & 4t & 0 \end{vmatrix}$$

$$= -8t\mathbf{i} + 4\mathbf{j} + 4t^2\mathbf{k}.$$

$|\mathbf{r}' \times \mathbf{r}''| = \sqrt{64t^2 + 16 + 16t^4}$

$\qquad = 4\sqrt{4t^2 + 1 + t^4}$.

$$\kappa(t) = \frac{|\mathbf{r}' \times \mathbf{r}''|}{|\mathbf{r}'|^3} = \frac{4\sqrt{4t^2 + 1 + t^4}}{\left(2\sqrt{t^2 + t^4 + 1}\right)^3}$$

$$= \frac{4\sqrt{4t^2 + 1 + t^4}}{8(t^2 + t^4 + 1)^{3/2}}.$$

At $t = 1$, $\kappa = \dfrac{\sqrt{4 + 1 + 1}}{2(1 + 1 + 1)^{3/2}} = \dfrac{\sqrt{6}}{2(3)^{3/2}}$

$$= \frac{\sqrt{6}}{2(3)\sqrt{3}} = \frac{\sqrt{2}}{6}.$$

7) Find the curvature of $y = x^2$ at $x = \sqrt{6}$.

For $y = x^2$, $y' = 2x$, $y'' = 2$ and

$$\kappa(x) = \frac{|2|}{(1 + (2x)^2)^{3/2}}.$$

At $x = \sqrt{6}$, $\kappa = \dfrac{2}{(25)^{3/2}} = \dfrac{2}{125}$.

Page 882 (ET Page 858)

C. For a smooth curve C given by the function $\mathbf{r}(t)$, the
unit tangent vector T is $\mathbf{T}(t) = \dfrac{\mathbf{r}'(t)}{|\mathbf{r}'(t)|}$ and points in the
same direction as the curve.

The **principal unit normal vector N** is $\mathbf{N}(t) = \dfrac{\mathbf{T}'(t)}{|\mathbf{T}'(t)|}$.

N is orthogonal to **T**. **N** is the direction in which the
curve is turning and thus always points "inward" on
the curve C.

The **binormal vector B** is $\mathbf{B}(t) = \mathbf{T}(t) \times \mathbf{N}(t)$.

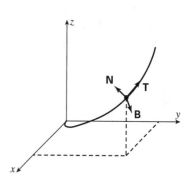

Example: Find $\mathbf{T}(t)$ and $\mathbf{N}(t)$ for $\mathbf{r}(t) = \left\langle \dfrac{t^2}{2}, \dfrac{t^2}{2}, t \right\rangle$, $t \geq 0$.

a) First, $\mathbf{T}(t)$: $\mathbf{r}'(t) = \langle t, t, 1 \rangle$ and $|\mathbf{r}'(t)| = \sqrt{t^2 + t^2 + 1} = \sqrt{2t^2 + 1}$.

Therefore, $\mathbf{T}(t) = \left\langle \dfrac{t}{\sqrt{2t^2 + 1}}, \dfrac{t}{\sqrt{2t^2 + 1}}, \dfrac{1}{\sqrt{2t^2 + 1}} \right\rangle$.

b) Next, $\mathbf{T}'(t) = \left\langle \dfrac{1}{(2t^2 + 1)^{3/2}}, \dfrac{1}{(2t^2 + 1)^{3/2}}, \dfrac{-2t}{(2t^2 + 1)^{3/2}} \right\rangle$ and

$$|\mathbf{T}'(t)| = \sqrt{\frac{1}{(2t^2 + 1)^3} + \frac{1}{(2t^2 + 1)^3} + \frac{4t^2}{(2t^2 + 1)^3}} = \frac{1}{2t^2 + 1}\sqrt{\frac{4t^2 + 2}{2t^2 + 1}} = \frac{\sqrt{2}}{2t^2 + 1}.$$

Therefore, $\mathbf{N}(t) = \dfrac{2t^2 + 1}{\sqrt{2}} \left\langle \dfrac{1}{(2t^2 + 1)^{3/2}}, \dfrac{1}{(2t^2 + 1)^{3/2}}, \dfrac{-2t}{(2t^2 + 1)^{3/2}} \right\rangle$

$= \left\langle \dfrac{1}{\sqrt{2}\sqrt{2t^2 + 1}}, \dfrac{1}{\sqrt{2}\sqrt{2t^2 + 1}}, \dfrac{-2t}{\sqrt{2}\sqrt{2t^2 + 1}} \right\rangle$.

The plane formed by $\mathbf{N}$ and $\mathbf{B}$ is the **normal plane**. While $\mathbf{T}$ points in the direction of the curve C, $\mathbf{N}$ and $\mathbf{B}$ are both orthogonal to $\mathbf{T}(t)$. Therefore, all vectors in the normal plane are orthogonal to $\mathbf{T}$, and the curve C pierces the normal plane at a right angle.

The plane formed by $\mathbf{T}$ and $\mathbf{N}$ is called the **osculating plane**. It is the plane that best approximates the direction of the curve. We may draw a circle in that plane with radius $\frac{1}{\kappa}$ and through the point on the curve C. This **osculating circle** has the same tangent vector, normal vector and curvature as C (and, therefore, is a good approximation to the curve). If κ is small (not much curvature), then $\frac{1}{\kappa}$ will be large and the osculating circle will have a large radius (and be rather flat at the point of tangency.)

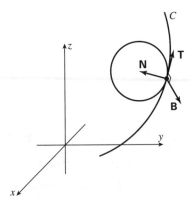

8) Find the principal unit normal vector for
$\mathbf{r}(t) = \sin t\mathbf{i} + 2t\mathbf{j} - \cos t\mathbf{k}$ at $t = \frac{\pi}{4}$.

$\mathbf{r}'(t) = \cos t\mathbf{i} + 2\mathbf{j} + \sin t\mathbf{k}.$

$|\mathbf{r}'(t)| = \sqrt{\cos^2 t + 4 + \sin^2 t} = \sqrt{5}.$

$\mathbf{T} = \dfrac{1}{\sqrt{5}}(\cos t\mathbf{i} + 2\mathbf{j} + \sin t\mathbf{k}).$

$\mathbf{T}' = \dfrac{1}{\sqrt{5}}(-\sin t\mathbf{i} + \cos t\mathbf{k}).$

$|\mathbf{T}'| = \dfrac{1}{\sqrt{5}}\sqrt{\sin^2 t + \cos^2 t} = \dfrac{1}{\sqrt{5}}.$

$\mathbf{N} = \dfrac{\mathbf{T}'}{|\mathbf{T}'|} = -\sin t\mathbf{i} + \cos t\mathbf{k}.$

At $t = \frac{\pi}{4}$, $\mathbf{N} = -\dfrac{\mathbf{i}}{\sqrt{2}} + \dfrac{\mathbf{k}}{\sqrt{2}}.$

9) For $\mathbf{r}(t) = \left\langle 2t, \dfrac{t^2}{2}, t \right\rangle$ find

   a) the unit tangent vector $\mathbf{T}$ at $t = 2$.

The graph of the curve $\mathbf{r}(t)$ is plotted here along with the point $(4, 2, 2)$ corresponding to $t = 2$.

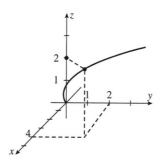

$\mathbf{r}'(t) = \langle 2,\, t,\, 1 \rangle.$

$|\mathbf{r}'(t)| = \sqrt{4 + t^2 + 1} = \sqrt{t^2 + 5}.$

$\mathbf{T}(t) = \left\langle \dfrac{2}{\sqrt{t^2 + 5}}, \dfrac{t}{\sqrt{t^2 + 5}}, \dfrac{1}{\sqrt{t^2 + 5}} \right\rangle.$

$\mathbf{T}(2) = \left\langle \dfrac{2}{3}, \dfrac{2}{3}, \dfrac{1}{3} \right\rangle.$

   b) the unit normal $\mathbf{N}$ at $t = 2$.

$\mathbf{T}'(t) = \left\langle \dfrac{-2t}{(t^2 + 5)^{3/2}}, \dfrac{5}{(t^2 + 5)^{3/2}}, \dfrac{-t}{(t^2 + 5)^{3/2}} \right\rangle.$

$\mathbf{T}'(2) = \left\langle \dfrac{-4}{27}, \dfrac{5}{27}, \dfrac{-2}{27} \right\rangle.$

$|\mathbf{T}'(2)| = \sqrt{\dfrac{16 + 25 + 4}{27^2}} = \dfrac{\sqrt{5}}{9}.$

Therefore, $\mathbf{N}(2) = \dfrac{\mathbf{T}'(2)}{|\mathbf{T}'(2)|}$

$= \dfrac{9}{\sqrt{5}} \left\langle \dfrac{-4}{27}, \dfrac{5}{27}, \dfrac{-2}{27} \right\rangle$

$= \left\langle \dfrac{-4}{3\sqrt{5}}, \dfrac{5}{3\sqrt{5}}, \dfrac{-2}{3\sqrt{5}} \right\rangle.$

c) the binormal $\mathbf{B}$ at $t = 2$.

$\mathbf{B}(2) = \mathbf{T}(2) \times \mathbf{N}(2) = \begin{vmatrix} \mathbf{i} & \mathbf{j} & \mathbf{k} \\ \frac{2}{3} & \frac{2}{3} & \frac{1}{3} \\ \frac{-4}{3\sqrt{5}} & \frac{5}{3\sqrt{5}} & \frac{-2}{3\sqrt{5}} \end{vmatrix}$

$= \left(\dfrac{-4}{9\sqrt{5}} - \dfrac{5}{9\sqrt{5}} \right) \mathbf{i} - \left(\dfrac{-4}{9\sqrt{5}} + \dfrac{4}{9\sqrt{5}} \right) \mathbf{j}$

$\qquad + \left(\dfrac{10}{9\sqrt{5}} - \dfrac{-8}{9\sqrt{5}} \right) \mathbf{k}$

$= \left\langle \dfrac{-1}{\sqrt{5}}, 0, \dfrac{2}{\sqrt{5}} \right\rangle.$

d) the normal plane at $t = 2$.

The normal plane contains vectors
orthogonal to $\mathbf{T}(2) = \left\langle \frac{2}{3}, \frac{2}{3}, \frac{1}{3} \right\rangle$.
$\mathbf{r}(2) = \langle 4, 2, 2 \rangle$ is a point on the plane.
The equation of the normal plane is
$\dfrac{2}{3}(x - 4) + \dfrac{2}{3}(y - 2) + \dfrac{1}{3}(z - 2) = 0$
$2x + 2y + z = 14.$
The curve and the plane are graphed
here, with the curve piercing the plane at a
right angle.

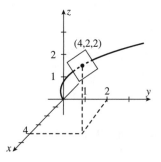

e) the osculating plane at $t = 2$.

This plane contains vectors orthogonal to

$$\mathbf{B}(2) = \left\langle \frac{-1}{\sqrt{5}}, 0, \frac{2}{\sqrt{5}} \right\rangle.$$

The equation of the plane is

$$\frac{-1}{\sqrt{5}}(x - 4) + 0(y - 2) + \frac{2}{\sqrt{5}}(z - 2) = 0$$

$$x - 2z = 0.$$

D. Technology Plus. Use a computer algebra system or a graphing calculator to solve.

T-1) Graph $y = e^x$ and the osculating circle at $(0, 1)$.

For $y = e^x$, $\kappa = \dfrac{y''}{(1 + y'^2)^{3/2}}$

$$= \frac{e^t}{(1 + (e^t)^2)^{3/2}} = \frac{e^t}{(1 + e^{2t})^{3/2}}.$$

At $t = 0$, $\kappa = \dfrac{1}{(1 + 1)^{3/2}} = \dfrac{1}{2\sqrt{2}}$.

$\dfrac{1}{\kappa} = 2\sqrt{2}$ is the radius.

The tangent at $(0, 1)$ has slope 1, so the normal has slope -1. The center of the osculating circle is on the normal:

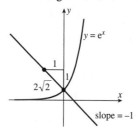

$a = 2$, so the center of the circle is $(-2, 3)$. The equation is $(x + 2)^2 + (y - 3)^2 = 8$. The graph of $y = e^x$ and the osculating circle are given below:

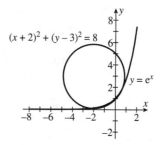

Section 13.4   Motion in Space: Velocity and Acceleration

In this section, all previous concepts in this chapter, such as the tangent vector and curvature, are used to describe motion of an object along a curve in three-dimensional space.

Concepts to Master

A. Velocity; Speed; Acceleration; Newton's Second Law of Motion

B. Tangent and normal components of acceleration

Summary and Focus Questions

Page 866 (ET Page 862)

A. A vector function $\mathbf{r}(t)$ may be thought of as the position of a particle in space at time t. As t varies from a to b, the particle moves along the curve determined by $\mathbf{r}(t)$.

$\mathbf{v}(t) = \mathbf{r}'(t)$ is the **velocity vector** and $|\mathbf{r}'(t)|$ is the **speed**.

$\mathbf{a}(t) = \mathbf{r}''(t)$ is the **acceleration** vector.

The unit tangent $\mathbf{T}(t)$ points in the direction the particle is moving and the unit normal vector $\mathbf{N}(t)$ points in the direction the curve is turning and is orthogonal to the motion. At every point along the curve, the particle is moving directly toward the normal plane of the curve.

Vector integrals may be used to determine velocity:

$$\mathbf{v}(t) = \int \mathbf{a}(t)\,dt$$

and position:

$$\mathbf{r}(t) = \int \mathbf{v}(t)\,dt.$$

Newton's Second Law of Motion is $\mathbf{F}(t) = m\mathbf{a}(t)$, where $\mathbf{F}(t)$ is the force acting on an object of mass m whose acceleration is $\mathbf{a}(t)$.

In the particular case of a projectile of mass m launched at an angle α with initial velocity $\mathbf{v}_0 = (v_0 \cos\alpha)\mathbf{i} + (v_0 \sin\alpha)\mathbf{j}$, assume gravity is the only force acting on the projectile after it is fired. We can use Newton's Second Law to find the acceleration, velocity, and position of the projectile:

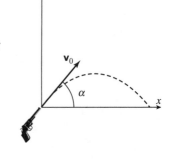

$\mathbf{F}(t) = -mg\mathbf{j}$ $(g = 9.8 \text{ m/s}^2)$

$\mathbf{a}(t) = -g\mathbf{j}$

$\mathbf{v}(t) = -gt\mathbf{j} + \mathbf{v}_0$

$\mathbf{r}(t) = -\dfrac{1}{2}gt^2\mathbf{j} + t\mathbf{v}_0$

1) Find the velocity, speed, and acceleration of a particle whose position at time t is $\mathbf{r}(t) = \langle 3t^2, t^3 + 1, e^{-t} \rangle$.

$\mathbf{v}(t) = \mathbf{r}'(t) = \langle 6t, 3t^2, -e^{-t} \rangle$.

Speed is $|\mathbf{v}(t)| = \sqrt{36t^2 + 9t^4 + e^{-2t}}$.

$\mathbf{a}(t) = \langle 6, 6t, e^{-t} \rangle$.

2) Find the velocity and position at time t of a particle whose initial position is $\mathbf{i} + \mathbf{j}$, initial velocity is $\mathbf{j} + \mathbf{k}$, and acceleration is $\mathbf{a}(t) = 12t\mathbf{i} + 2\mathbf{k}$.

$\mathbf{v}(t) = \int \mathbf{a}(t)\,dt = 6t^2\mathbf{i} + 2t\mathbf{k} + \mathbf{C}$.

Initially $(t = 0)\ \mathbf{v}_0 = \mathbf{j} + \mathbf{k}$.

$0\mathbf{i} + 0\mathbf{k} + \mathbf{C} = \mathbf{j} + \mathbf{k}$

$\mathbf{C} = \mathbf{j} + \mathbf{k}$.

$\mathbf{v}(t) = 6t^2\mathbf{i} + 2t\mathbf{k} + (\mathbf{j} + \mathbf{k})$
$\qquad = 6t^2\mathbf{i} + \mathbf{j} + (2t + 1)\mathbf{k}$.

$\mathbf{r}(t) = \int \mathbf{v}(t)\,dt = 2t^3\mathbf{i} + t\mathbf{j} + (t^2 + t)\mathbf{k} + \mathbf{D}$

Initially $\mathbf{r}(0) = \mathbf{i} + \mathbf{j}$.

$0\mathbf{i} + 0\mathbf{j} + 0\mathbf{k} + \mathbf{D} = \mathbf{i} + \mathbf{j}$.

$\mathbf{D} = \mathbf{i} + \mathbf{j}$.

$\mathbf{r}(t) = 2t^3\mathbf{i} + t\mathbf{j} + (t^2 + t)\mathbf{k} + \mathbf{i} + \mathbf{j}$
$\qquad = (2t^3 + 1)\mathbf{i} + (t + 1)\mathbf{j} + (t^2 + t)\mathbf{k}$.

3) What force is required to push a 5 kg mass in such a way that its position is $\mathbf{r}(t) = 3t^2\mathbf{i} + t^4\mathbf{j} + 2t^3\mathbf{k}$ at time t?

$\mathbf{v}(t) = \mathbf{r}'(t) = 6t\mathbf{i} + 4t^3\mathbf{j} + 6t^2\mathbf{k}$.

$\mathbf{a}(t) = \mathbf{v}'(t) = 6\mathbf{i} + 12t^2\mathbf{j} + 12t\mathbf{k}$.

$\mathbf{F}(t) = 5\mathbf{a}(t) = 30\mathbf{i} + 60t^2\mathbf{j} + 60t\mathbf{k}$.

4) A projectile is fired with a velocity of 200 m/s at an inclination of 30° from a point 10 m above the ground. Find the vector function that describes the path of motion.

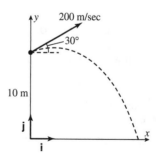

Choose a coordinate system with origin at ground level so that from Newton's Second Law, $\mathbf{a}(t) = -g\mathbf{j}$.

$\mathbf{v}(t) = \int -g\mathbf{j}\,dt = -gt\mathbf{j} + \mathbf{C}$.

At $t = 0$, $\mathbf{v}_0 = \mathbf{C}$. Since the velocity is
200 m/s at a 30° inclination,
$$\mathbf{C} = (200 \cos 30°)\mathbf{i} + (200 \sin 30°)\mathbf{j}$$
$$= 100\sqrt{3}\mathbf{i} + 100\mathbf{j}.$$
$$\mathbf{v}(t) = 100\sqrt{3}\mathbf{i} + (100 - gt)\mathbf{j}.$$
$$\mathbf{r}(t) = \int \mathbf{v}(t)\,dt$$
$$= 100\sqrt{3}t\mathbf{i} + \left(100t - \frac{gt^2}{2}\right)\mathbf{j} + \mathbf{D}$$

At $t = 0$, $\mathbf{r}(0) = \mathbf{D}$. Since the projectile is 10 m
above the ground at $t = 0$, $\mathbf{D} = 10\mathbf{j}$. Thus,
$$\mathbf{r}(t) = 100\sqrt{3}t\mathbf{i} + \left(100t - \frac{gt^2}{2} + 10\right)\mathbf{j}.$$

**Page
890
(ET Page
866)**

B. For a curve given by $\mathbf{r}(t)$, the unit tangent vector $\mathbf{T}(t)$ points in the direction
the curve is heading and the unit normal vector $\mathbf{N}(t)$ points in the direction
the curve is turning. Therefore, the acceleration vector $\mathbf{a}(t)$ lies in the plane
determined by $\mathbf{T}$ and $\mathbf{N}$, and may be written as a combination of $\mathbf{T}$ and $\mathbf{N}$:

$$\mathbf{a} = a_T\mathbf{T} + a_N\mathbf{N},$$

where the **tangential component** is $a_T = v'$ (v is the speed, $v = |\mathbf{v}|$)
and the **normal component** is $a_N = \kappa v^2$ (κ is the curvature).

In terms of the position function $\mathbf{r}(t)$ the components are:

$$a_T = \frac{\mathbf{r}' \times \mathbf{r}''}{|\mathbf{r}'|} \text{ and } a_N = \frac{|\mathbf{r}' \times \mathbf{r}''|}{|\mathbf{r}'|}.$$

Example: Find the acceleration vector for $\mathbf{r}(t) = \left\langle \frac{t^2}{2}, \frac{t^2}{2}, t \right\rangle$. Then

find the tangent and normal components of acceleration and verify that
$\mathbf{a} = a_T\mathbf{T} + a_N\mathbf{N}$.

a) $\mathbf{r}'(t) = \langle t, t, 1 \rangle$ and $\mathbf{a}(t) = \mathbf{r}''(t) = \langle 1, 1, 0 \rangle$. Acceleration is constant along
the curve $\mathbf{r}$.

b) To find a_T, we first calculate $\mathbf{r}' \cdot \mathbf{r}'' = \langle t, t, 1 \rangle \cdot \langle 1, 1, 0 \rangle = t + t + 0 = 2t$.

Next, $|\mathbf{r}'(t)| = \sqrt{t^2 + t^2 + 1} = \sqrt{2t^2 + 1}$. Therefore, $a_T = \dfrac{2t}{\sqrt{2t^2 + 1}}$.

c) To find a_N, first calculate $\mathbf{r}' \times \mathbf{r}'' = \begin{vmatrix} \mathbf{i} & \mathbf{j} & \mathbf{k} \\ t & t & 1 \\ 1 & 1 & 0 \end{vmatrix} = (-1)\mathbf{i} - (-1)\mathbf{j} + (0)\mathbf{k} = -\mathbf{i} + \mathbf{j}.$

Next, $|\mathbf{r}' \times \mathbf{r}''| = \sqrt{(-1)^2 + 1^2 + 0^2} = \sqrt{2}$. Therefore, $a_N = \dfrac{\sqrt{2}}{\sqrt{2t^2 + 1}}$.

d) To verify that $\mathbf{a} = a_T\mathbf{T} + a_N\mathbf{N}$, recall from the example in part C of Section 13.3 that we found

$$\mathbf{T}(t) = \left\langle \frac{t}{\sqrt{2t^2 + 1}}, \frac{t}{\sqrt{2t^2 + 1}}, \frac{1}{\sqrt{2t^2 + 1}} \right\rangle \text{ and}$$

$$\mathbf{N}(t) = \left\langle \frac{1}{\sqrt{2}\sqrt{2t^2 + 1}}, \frac{1}{\sqrt{2}\sqrt{2t^2 + 1}}, \frac{-2t}{\sqrt{2}\sqrt{2t^2 + 1}} \right\rangle.$$

Therefore,

$$a_T\mathbf{T} = \frac{2t}{\sqrt{2t^2 + 1}} \left\langle \frac{t}{2t^2 + 1}, \frac{t}{\sqrt{2t^2 + 1}}, \frac{1}{\sqrt{2t^2 + 1}} \right\rangle = \left\langle \frac{2t^2}{2t^2 + 1}, \frac{2t^2}{2t^2 + 1}, \frac{2t}{2t^2 + 1} \right\rangle$$

and

$$a_N\mathbf{N} = \frac{\sqrt{2}}{\sqrt{2t^2 + 1}} = \left\langle \frac{1}{\sqrt{2}\sqrt{2t^2 + 1}}, \frac{1}{\sqrt{2}\sqrt{2t^2 + 1}}, \frac{-2t}{\sqrt{2}\sqrt{2t^2 + 1}} \right\rangle$$

$$= \left\langle \frac{1}{2t^2 + 1}, \frac{1}{2t^2 + 1}, \frac{-2t}{2t^2 + 1} \right\rangle.$$

When we add the components, the result is the acceleration vector **a** we found in part a):

$$a_T\mathbf{T} + a_N\mathbf{N} = \left\langle \frac{2t^2}{2t^2 + 1}, \frac{2t^2}{2t^2 + 1}, \frac{2t}{2t^2 + 1} \right\rangle + \left\langle \frac{1}{2t^2 + 1}, \frac{1}{2t^2 + 1}, \frac{-2t}{2t^2 + 1} \right\rangle = \langle 1, 1, 0 \rangle.$$

5) Find the tangential and normal components of acceleration for $\mathbf{r}(t) = t^2\mathbf{i} + t^4\mathbf{j} + t^3\mathbf{k}$.

$\mathbf{r}'(t) = 2t\mathbf{i} + 4t^3\mathbf{j} + 3t^2\mathbf{k}$.

$|\mathbf{r}'(t)| = \sqrt{4t^2 + 16t^6 + 9t^4}$

$\qquad = t\sqrt{4 + 16t^4 + 9t^2}$.

$\mathbf{r}''(t) = 2\mathbf{i} + 12t^2\mathbf{j} + 6t\mathbf{k}$.

$\mathbf{r}' \cdot \mathbf{r}'' = 4t + 48t^5 + 18t^3$.

$$\mathbf{r}' \times \mathbf{r}'' = \begin{vmatrix} \mathbf{i} & \mathbf{j} & \mathbf{k} \\ 2t & 4t^3 & 3t^2 \\ 2 & 12t^2 & 6t \end{vmatrix}$$

$\qquad = -12t^4\mathbf{i} - 6t^2\mathbf{j} + 16t^3\mathbf{k}$.

$|\mathbf{r}' \times \mathbf{r}''| = \sqrt{144t^8 + 36t^4 + 256t^6}$

$\qquad = 2t^2\sqrt{36t^4 + 9 + 64t^2}$.

$$a_T = \frac{4t + 48t^5 + 18t^3}{t\sqrt{4 + 16t^4 + 9t^2}}$$

$$= \frac{4 + 48t^4 + 18t^2}{\sqrt{4 + 16t^4 + 9t^2}} \text{ and}$$

$$a_N = \frac{2t^2\sqrt{36t^4 + 9 + 64t^2}}{t\sqrt{4 + 16t^4 + 9t^2}}$$

$$= \frac{2t\sqrt{36t^4 + 9 + 64t^2}}{\sqrt{4 + 16t^4 + 9t^2}}.$$

Chapter 14 — Partial Derivatives

Section 14.1 Functions of Several Variables

In the last chapter we studied vector functions whose domain elements are real numbers and whose range elements are vectors. In this section we introduce functions "of several variables," for which the domains are sets of vectors and the values are real numbers. A function of two variables, $z = f(x, y)$, will have a graph that is a surface in three dimensions. Since these graphs are sometimes difficult to visualize, we introduce the concept of a level curve—a subset of the xy-plane that corresponds to a specific function value.

Concepts to Master

A. Functions of two variables; Domain; Range; Graphs

B. Level curves

C. Functions of more than two variables

D. Technology Plus

Summary and Focus Questions

Page 902 (ET Page 878)*

A. Let D be a subset of $\mathbb{R}^2$. A **function of two variables** assigns to each ordered pair (x, y) in D a unique real number $f(x, y)$. D is the **domain** of f. The **range** of f is the set of all $f(x, y)$ values.

Example: Let $f(x, y) = \dfrac{x + 1}{y - 2}$. Then $f(9, 4) = 5, f(0, 5) = \dfrac{1}{3}$, and $f(7, 2)$ is not defined. The domain of f is $\{(x, y) \big| y \neq 2\}$. The range consists of all real numbers.

As is the case with a function of one variable, a function of several variables may be defined verbally, by a table of values, with a graph, or by an explicit formula.

The **graph** of a function of two variables is a surface in three dimensions. Similar to the single-variable case where $y = f(x)$ is the height of a point on a curve above the point x on the x-axis, $z = f(x, y)$ will be the height of a point on a surface above the point (x, y) in the xy-plane. The point (x, y, z) will be on the graph of f if and only if $z = f(x, y)$.

---

*When using the Early Transcendentals text, use the page number in parentheses.

A terrain with hills and valleys is a good example of the graph of a function of two variables. At any point with coordinates (a, b), let $f(a, b)$ be the elevation of the point above sea level. A point on the surface, such as $(25, 30, 200)$, would be the point on the ground with coordinates $x = 25$ and $y = 30$ that is 200 meters above sea level. As you move around the terrain (changing the values of x and y), the elevation $f(x, y)$ changes.

Example: Let $D = \mathbb{R}^2$ and $f(x, y) = x^2 + y^2$. Then for $x = 2$ and $y = 1$, $f(2, 1) = 2^2 + 1^2 = 5$. Also, $f(3, -1) = 3^2 + (-1)^2 = 10$. The graph of $z = x^2 + y^2$ is a circular paraboloid with vertex at $(0, 0, 0)$. The two points $(2, 1, 5)$ and $(3, -1, 10)$ are plotted on the graph of f at the right.

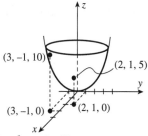

There are three different ways to view a function $f(x, y)$ with the domain D:

i) f is a function of two independent variables x and y, where $(x, y) \in D$.

ii) f is a function whose domain is all points (x, y) in D.

iii) f is a function whose domain is all vectors $\langle x, y \rangle$ in D.

In the preceding example, if $\mathbf{u} = 2\mathbf{i} + 1\mathbf{j}$, then $f(\mathbf{u}) = f(2, 1) = 2^2 + 1^2 = 5$. It will be helpful to be able to switch among thinking of a domain element as a pair of independent variables, as a point in the xy-plane, or as a vector.

1) What is the domain of $f(x, y) = \dfrac{1}{5x - 10y}$?

$f(x, y)$ is defined for all (x, y) except when $5x - 10y = 0$ or $x = 2y$. The domain is $\{(x, y) \mid x \neq 2y\}$.

2) Find the domain and range of
$$f(x, y) = \frac{1}{x^2 + y^2}.$$

$x^2 + y^2 = 0$ only for $(x, y) = (0, 0)$. The domain is $\{(x, y) \mid x \neq 0 \text{ and } y \neq 0\}$. $x^2 + y^2 > 0$ for all $(x, y) \neq (0, 0)$. The range is $(0, \infty)$.

3) Let $f(x, y)$ be defined by the table:

		y	
		4	5	6
	1	8	7	−1
x	2	4	−10	7
	3	3	−3	2

a) What is the domain?

The domain consists of the nine pairs: $(1, 4), (1, 5), (1, 6), (2, 4), (2, 5), (2, 6),$ $(3, 4), (3, 5), (3, 6)$.

b) What is the range?

$\{-10, -3, -1, 2, 3, 4, 7, 8\}$.

c) What is $f(2, 5)$?

$f(2, 5) = -10$.

d) What is $f(4, 3)$?

$f(4, 3)$ does not exist.

4) Describe the graph of $f(x, y) = x^2$.

$z = x^2$ is a parabolic cylinder along the y-axis.

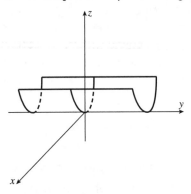

5) Sketch the graph of $z = e^{-x^2 - y^2}$.

<First determine traces in the yz- and xz-planes.>

In the yz-plane, $x = 0$ and $z = e^{-x^2 - y^2}$ becomes $z = e^{-y^2}$. The graph is:

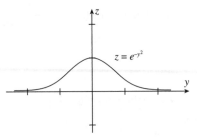

In the xz-plane where $y = 0$, $z = e^{-x^2}$ has a similar graph. We draw it with a view of the x-axis coming toward us.

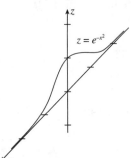

Finally, the graph of $z = e^{-x^2-y^2}$ is:

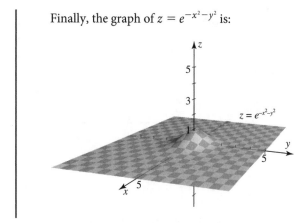

B. For a constant k, the **level curve determined by k** for a function $z = f(x, y)$ is the set of all points (x, y) in the domain of f such that $f(x, y) = k$.

It is important to remember that:

i) Level curves are subsets of the domain of f; level curves are not part of the graph of f.

ii) All points on a given level curve have the same f value.

iii) Two different level curves determined by k_1 and k_2 with $k_1 \neq k_2$ will never intersect.

An elevation map of a hilly terrain is a good example of a set of level curves. At each point (x, y) in a region, let $H(x, y)$ be the elevation above sea level. On a paper map of the terrain, for each elevation a curve is drawn through all points that have given height. The level curve determined by 240 m would be the curve traced out on the elevation map of all points that are 240 m above sea level. If you were to walk along the terrain using the level curve for your x- and y-coordinates, you would always stay at 240 m above sea level.

Example: Let $f(x, y) = x^2 + y^2$. For each $k > 0$, the level curve $x^2 + y^2 = k$ is a circle in the xy-plane with center $(0, 0)$ and radius $\sqrt{k}$.

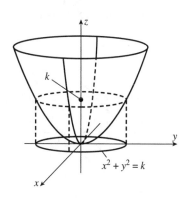

Page 907 (ET Page 883)

6) Sketch the level curves for $f(x,y) = x - y^2$ for $k = 0, 2, -2, 4$.

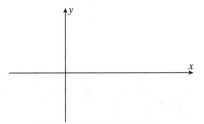

Each level curve has the form $x - y^2 = k$, which is a parabola opening to the right with vertex $(k, 0)$.

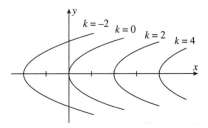

7) What are the level curves for $f(x,y) = xy$?

For any real number $k \neq 0$, $xy = k$ is $y = \dfrac{k}{x}$, a hyperbola with the x- and y-axes as asymptotes and passing through $(k, 1)$ and $(-k, -1)$.

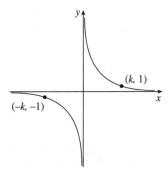

For $k = 0$, the level curve consists of the x- and y-axes.

8) Find a function $z = f(x, y)$ that has the following set of level curves.

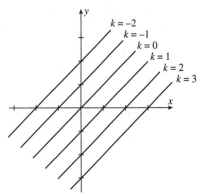

The lines are of form $y = x - k$ for each k. Therefore, $x - y = k$. Let f be the function $f(x, y) = x - y$.

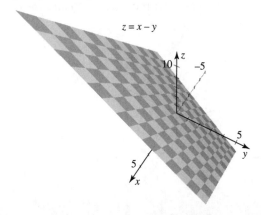

C. A function of three variables $f(x, y, z)$ has a graph in four-dimensional space. We cannot draw the graph but we can visualize the **level surfaces** since they are subsets of $\mathbb{R}^3$. Functions of more than three variables and their level surfaces may be defined similarly.

Page 910 (ET Page 887)

9) For $f(x, y, z) = x^2 - xy + z$, what is $f(2, 2, -1)$?

$f(2, 2, -1) = 2^2 - (2)(2) + (-1) = -1.$

10) Describe the level surfaces of $f(x, y, z) = x^2 + 4y^2 + 9z^2$.

For any $k \geq 0$, the level surface is all (x, y, z) such that $x^2 + 4y^2 + 9z^2 = k$, which is an ellipsoid with center $(0, 0, 0)$.

11) What is the domain of $f(x, y, z) = \dfrac{\sqrt{x}}{y + z}$?

$f(x, y, z)$ is defined for all $x \geq 0$ and whenever $y + z \neq 0$. The domain is $\{(x, y, z) \mid x \geq 0 \text{ and } y \neq -z\}$.

D. Technology Plus. Use a computer algebra system or a graphing calculator to solve.

T-1) Let $f(x, y) = 2xy$.

    a) Draw level curves for $c = 0.25, 0.5,$ and 1.

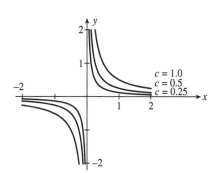

    b) Draw level curves for $c = -0.25, -0.5,$ and -1.

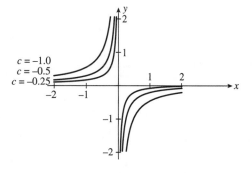

c) Graph $f(x, y)$ for $-1 \leq x \leq 1$, $-1 \leq y \leq 1$.

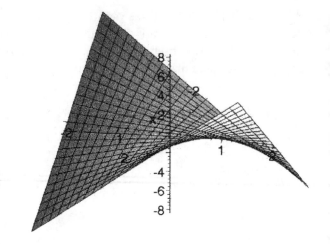

Section 14.2 Limits and Continuity

This section extends the concepts of limits and continuity to functions of two or more variables. Be careful to note that a function can be "one-dimensionally" continuous along every line approaching a point and still fail to be continuous at that point. Drawing sketches of graphs may be difficult or impossible, so rely on generalizing what you know about single-variable functions.

Concepts to Master

A. Limits of functions of two or more variables; Limits along curves

B. Continuity of functions of two or more variables

Summary and Focus Questions

Page 916 (ET Page 89)

A. Let f be a function of two variables whose domain D includes points arbitrarily close to (a, b).

$$\lim_{(x,y)\to(a,b)} f(x,y) = L \text{ means that if } (x,y) \text{ is near } (a, b) \text{ then } f(x,y) \text{ is near } L.$$

More precisely, $\displaystyle\lim_{(x,y)\to(a,b)} f(x,y) = L$ means

for all $\varepsilon > 0$, there exists $\delta > 0$ such that if $(x, y) \in D$ and
$0 < \sqrt{(x-a)^2 + (y-b)^2} < \delta$, then $|f(x,y) - L| < \varepsilon$.

Using $\mathbf{x} = \langle x, y \rangle$ and $\mathbf{a} = \langle a, b \rangle$, the definition may be rewritten in a familiar way:

For all $\varepsilon > 0$, there exists $\delta > 0$ such that $|f(\mathbf{x}) - L| < \varepsilon$ whenever
$0 < |\mathbf{x} - \mathbf{a}| < \delta$.

This vector version of the definition may be used for functions of three or more variables without modification.

All the usual limit theorems for functions of one variable extend to multivariable functions. For example, the limit of a sum is the sum of the limits:

$$\text{if } \lim_{(x,y)\to(a,b)} f(x,y) = L \text{ and } \lim_{(x,y)\to(a,b)} g(x,y) = M,$$

$$\text{then } \lim_{(x,y)\to(a,b)} [f(x,y) + g(x,y)] = L + M.$$

Examples: a) $\displaystyle\lim_{(x,y)\to(1,2)} (x^2 + xy) = \lim_{(x,y)\to(1,2)} x^2 + \lim_{(x,y)\to(1,2)} xy = 1 + 2 = 3.$

b) $\displaystyle\lim_{(x,y)\to(1,2)} 3x^2 + 5xy - 2y = 3(1)^2 + 5(1)(2) - 2(2) = 9.$

c) $\lim\limits_{(x,y)\to(0,0)} \ln(1+x^2+y^2) = 0$ because $\ln 1 = 0$.

d) For $f(x, y, z) = x^2 - y^2 + z$, $\lim\limits_{\mathbf{u}\to(3,1,2)} f(\mathbf{u}) = 10$.

If $\lim\limits_{(x,y)\to(a,b)} f(x,y) = L$, then $f(x,y)$ approaches L as (x,y) approaches (a, b) along any curve containing (a, b). If two different curves containing (a, b) yield different limits as (x, y) approaches (a, b) along each, then the limit of f does not exist.

Example: To evaluate $\lim\limits_{(x,y)\to(0,0)} \dfrac{\sin x}{y}$ for $y \neq 0$, we find the limit along

various lines. Along the line $y = x$, $\dfrac{\sin x}{y} = \dfrac{\sin x}{x} \to 1$ as (x, x) nears $(0, 0)$.

However, for the line $x = 0$, $\lim\limits_{(0,y)\to(0,0)} \dfrac{\sin 0}{y} = \lim\limits_{(0,y)\to(0,0)} \dfrac{0}{y} = 0$. Since the

limits along these lines are different, $\lim\limits_{(x,y)\to(0,0)} \dfrac{\sin x}{y}$ does not exist.

1) Evaluate each:

a) $\lim\limits_{(x,y)\to(4,2)} (6x + 3y)$

$6(4) + 3(2) = 30.$

b) $\lim\limits_{(x,y)\to(0,0)} e^{1/(x^2+y^2)}$

$\infty.$

c) $\lim\limits_{(x,y)\to(3,2)} (y^3 - y)$

$2^3 - 2 = 6.$

d) $\lim\limits_{(x,y)\to(1,-1)} \dfrac{x^2 - 2xy + y^2 - 4}{x - y - 2}$

At $(1, -1)$ we have $\dfrac{0}{0}$. We can factor and cancel:

$$\dfrac{x^2 - 2xy + y^2 - 4}{x - y - 2} = \dfrac{(x - y)^2 - 4}{x - y - 2}$$

$$= \dfrac{(x - y + 2)(x - y - 2)}{x - y - 2} = x - y + 2.$$

$$\lim\limits_{(x,y)\to(1,-1)} (x - y + 2) = 1 + 1 + 2 = 4.$$

e) $\lim\limits_{(x,y,z)\to(0,1,2)} x^3 + y^2 + z$

$0^3 + 1^2 + 2 = 3.$

2) Let $f(x,y) = \dfrac{x^3 y}{2x^6 + y^2}$ for $(x,y) \neq (0,0)$.

Evaluate $\displaystyle\lim_{(x,y)\to(0,0)} f(x,y)$:

a) along the curve $y = 0$.

$f(x,0) = 0$ for all x so f has limit 0 along $y = 0$.

b) along the curve $y = x^2$.

$f(x,x^2) = \dfrac{x^3 x^2}{2x^6 + x^4} = \dfrac{x}{2x^2 + 1}$.

As $x \to 0$, $\dfrac{x}{2x^2 + 1} \to 0$, so f has limit 0 along $y = x^2$.

c) along the curve $y = x^3$.

$f(x,x^3) = \dfrac{x^3 x^3}{2x^6 + x^6} = \dfrac{1}{3}$ for $x \neq 0$.

f has limit $\dfrac{1}{3}$ along $y = x^3$.

3) Find $\displaystyle\lim_{(x,y)\to(0,0)} f(x,y)$ for the function in question 2.

By parts b) and c), f approaches different limits along different curves. The limit does not exist.

4) True, False:

If $\displaystyle\lim_{(x,y)\to(0,0)} f(x,y) = L$ and

$\displaystyle\lim_{(x,y)\to(0,0)} g(x,y) = M$, then

$\displaystyle\lim_{(x,y)\to(0,0)} [f(x,y)g(x,y)] = LM$.

True.

B. The function $z = f(x,y)$ is **continuous at (a, b)** if $\displaystyle\lim_{(x,y)\to(a,b)} f(x,y) = f(a,b)$.

f is **continuous on a region D** if f is continuous at every point (a, b) in D.

Page 920 (ET Page 896)

Polynomials in x and y are continuous everywhere. Rational functions are continuous for all points in their domains.

Example: Let $f(x,y) = \dfrac{xy+1}{x-y^2}$. Then f is a rational function and is, therefore, continuous everywhere it is defined. Since $x - y^2 = 0$ when $x = y^2$, f is continuous at (x, y) when $x \neq y^2$.

5) Where is $f(x,y) = \ln(x-y)$ continuous?

$\ln(x-y)$ is continuous on its domain. The domain is $\{(x,y) \mid x > y\}$.

6) Where is $f(x,y) = \dfrac{2x+y}{x^2+xy}$ continuous?

Since f is a rational function, f is continuous everywhere on its domain. The domain is all points (x, y) such that
$x^2 + xy \neq 0$
$x(x+y) \neq 0$, so $x \neq 0$ and $x \neq -y$.

7) Is $f(x,y) = \begin{cases} \dfrac{x^3 y}{2x^6 + y^2} & (x,y) \neq (0,0) \\ 0 & (x,y) = (0,0) \end{cases}$

continuous at $(0,0)$?

No. By problem 3 $\displaystyle\lim_{(x,y)\to(0,0)} f(x,y)$ does not exist.

8) Where is $f(x,y) = e^{\frac{1}{x^2 + y^2}}$ continuous? Graph f for $0.5 \leq x \leq 2$, $0.5 \leq y \leq 2$.

$f(x,y)$ is continuous everywhere except at $(0,0)$.

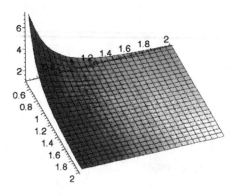

9) Write the definition of f is continuous at $\mathbf{u} = \langle a, b \rangle$ in terms of vectors.

f is continuous at $\mathbf{u}$ in D if $\displaystyle\lim_{\mathbf{x}\to\mathbf{u}} f(\mathbf{x}) = f(\mathbf{u})$.

Section 14.3   Partial Derivatives

This section begins the process of finding derivatives for functions of two or more variables. For two variables, we shall see that there is a derivative in the *x*-direction and another in the *y*-direction, and these may be obtained by a method similar to that for functions of one variable. Each of these two "partial" derivatives has, in turn, two partial derivatives; so the original function has four second partial derivatives. Clairaut's Theorem says that sometimes two of these four second partial derivatives are the same.

Concepts to Master

A. Partial derivatives; Slopes of tangent lines; Instantaneous rate of change; Implicit partial differentiation

B. Higher-order partial derivatives; Clairaut's Theorem

C. Technology Plus

Summary and Focus Questions

A. For $z = f(x, y)$, **the partial derivative of f with respect to x is**

$$f_x(x, y) = \lim_{h \to 0} \frac{f(x + h, y) - f(x, y)}{h}.$$

Since y is unchanging in this definition, $f_x(x, y)$ is computed by treating y as a constant.

Examples: a) If $f(x, y) = x^3 y^2$, then $f_x(x, y) = (3x^2)y^2 = 3x^2 y^2$.

b) If $f(x, y) = 3x^4 + 7y^2$, then $f_x(x, y) = 12x^3 + 0 = 12x^3$.

f_x is the instantaneous rate of change of f with respect to x.

$f_x(a, b)$ may be interpreted as the slope of the tangent line to the surface $z = f(x, y)$ determined by the trace $y = b$. It is the slope at (a, b) in the *x*-direction. Other notations for f_x include:

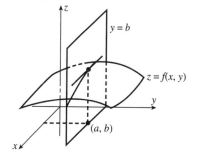

$$\frac{\partial f}{\partial x}, \quad \frac{\partial}{\partial x} f(x, y), \quad \frac{\partial z}{\partial x}, \quad D_x f, \quad D_x f(x, y), \quad D_1 f, \quad f_1.$$

Likewise, $f_y(x, y) = \lim_{h \to 0} \dfrac{f(x, y + h) - f(x, y)}{h}$, the **partial derivative of f with respect to y,** is computed by holding x constant. f_y is the slope of the tangent line to $z = f(x,y)$ in the y-direction and is the instantaneous rate of change of f with respect to y.

Suppose $H(x,y)$ is the elevation above sea level of a point (x,y) in a hilly terrain and you are standing on a hillside at (x,y) holding a compass for which north corresponds to the positive y-axis and east corresponds to the positive x-axis. The partial derivative $H_x(x,y)$ is the slope of the hill as you look eastward, and $H_y(x,y)$ is the slope of the hill as you look north.

Partial derivatives for functions of more than two variables are defined and calculated in ways similar to those for functions of two variables.

Examples: Let $f(x,y,z,w) = 2x^2y + 3xz^3w + z^2y^2w$.

a) Treating y, z, and w as constants, we compute:

$$\frac{\partial f}{\partial x} = 4xy + 3z^3w.$$

b) Holding x, y, and w constant:

$$\frac{\partial f}{\partial z} = 0 + 3x(3z^2)w + (2z)(y^2w) = 9xz^2w + 2zy^2w.$$

Implicit partial differentiation is performed in the same manner as was done for functions of one variable, but you must remember to treat all other variables as constants.

Example: Compute $\dfrac{\partial f}{\partial x}$ where $z = f(x,y)$ is defined by $4x^2 + 9y^2 + 16z^2 = 100$.

We take the partial derivative with respect to x:

$$8x + 0 + 32z\frac{\partial z}{\partial x} = 0.$$

Solving for $\dfrac{\partial z}{\partial x}$, we have $\dfrac{\partial z}{\partial x} = -\dfrac{8x}{32z} = -\dfrac{x}{4z}$.

1) For $w = f(x,y,z)$, define $\dfrac{\partial f}{\partial z}$.

$$\frac{\partial f}{\partial z} = \lim_{h \to 0} \frac{f(x,y,z+h) - f(x,y,z)}{h}.$$

2) Find f_x and f_y for:

a) $f(x,y) = 10x^3 + 2x^2y^3 - 9y^4$.

$f_x = 30x^2 + 4xy^3$.
$f_y = 6x^2y^2 - 36y^3$.

b) $f(x,y) = e^{x^2 - y^2}$.

$f_x = e^{x^2-y^2}(2x) = 2xe^{x^2-y^2}$.
$f_y = e^{x^2-y^2}(-2y) = -2ye^{x^2-y^2}$.

c) $f(x,y) = xy \sec x$.

$f_x = (xy)(\sec x \tan x) + (\sec x)y$
$\quad = y \sec x(x \tan x + 1)$.
$f_y = x \sec x$.

3) Find f_y for $f(x, y, z) = \dfrac{x}{\sqrt{y+z}}$.

$f(x, y, z) = x(y+z)^{-1/2}$.

$f_y = x\left(-\dfrac{1}{2}\right)(y+z)^{-3/2}(1) = -\dfrac{x}{2(y+z)^{3/2}}$.

4) Find the slope of the tangent line in the y-direction to $f(x,y) = x^2 + 4xy + y^2$ at the point $(1, 2, 13)$.

$f_y = 4x + 2y$.

$f_y(1, 2) = 4(1) + 2(2) = 8$.

5) The level curves for $f(x, y)$ are as follows:

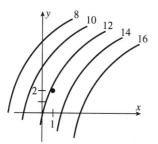

Is $f_x(1, 2)$ positive or negative?
Is $f_y(1, 2)$ positive or negative?

As we proceed from $(1, 2)$ to the right (in the x-direction), the function values increase. Therefore, $f_x(1, 2)$ is positive.

As we proceed from $(1, 2)$ upward (in the y-direction), the function values decrease. Therefore, $f_y(1, 2)$ is negative.

6) Find $\dfrac{\partial z}{\partial x}$, where $z = f(x, y)$ is given by

$x^2 + y^2 + z^2 = \sin(xyz)$.

Treat y as a constant:

$2x + 0 + 2z\dfrac{\partial z}{\partial x} = \cos(xyz)\left[xy\dfrac{\partial z}{\partial x} + yz\right]$.

$2x - yz\cos(xyz) = \dfrac{\partial z}{\partial x}[xy\cos(xyz) - 2z]$.

$\dfrac{\partial z}{\partial x} = \dfrac{2x - yz\cos(xyz)}{xy\cos(xyz) - 2z}$.

Page 930 (ET Page 906)

B. For a function $z = f(x, y)$ there are four **second partial derivatives**:

$$f_{xx}(x, y) = \dfrac{\partial^2 f}{\partial x^2} \qquad f_{yy}(x, y) = \dfrac{\partial^2 f}{\partial y^2}$$

$$f_{xy}(x, y) = \dfrac{\partial^2 f}{\partial y \partial x} \qquad f_{yx}(x, y) = \dfrac{\partial^2 f}{\partial x \partial y}.$$

For example, $f_{xy}(x, y)$ is determined by finding the partial derivative of f with respect to x and taking the derivative of that result with respect to y.

Example: Find all second partial derivatives for $f(x,y) = 7x + 2x^2y^3 - 10y^2$.
First, $f_x = 7 + 4xy^3$ and $f_y = 6x^2y^2 - 20y$. Then:

$$f_{xx} = 4y^3, \quad f_{xy} = 12xy^2, \quad f_{yx} = 12xy^2, \quad \text{and} \quad f_{yy} = 12x^2y - 20.$$

In the previous example, $f_{xy} = f_{yx}$. Clairaut's Theorem, stated next, says that in some cases you may interchange the order of the differentiations when finding second partial derivatives.

Clairaut's Theorem If f is defined on a disk containing (a, b) and both f_{xy} and f_{yx} are continuous on that disk, then $f_{xy} = f_{yx}$ at (a, b).

---

7) For $f(x,y) = \sin(x^2 + y^2)$, is $f_{xy}(x,y) = f_{yx}(x,y)$ for all (x,y)?

Yes, because both f_{xy} and f_{yx} are continuous for all (x, y).

8) Find all second partial derivatives for $f(x,y) = 3xy^3 + 8x^2y^4$.

$f_x = 3y^3 + 16xy^4$
$f_{xx} = 16y^4$
$f_{xy} = 9y^2 + 64xy^3$
$f_y = 9xy^2 + 32x^2y^3$
$f_{yy} = 18xy + 96x^2y^2$
$f_{yx} = 9y^2 + 64xy^3$ (same as f_{xy}).

9) True or False:

   a) If all third partial derivatives are continuous, then $f_{xyz} = f_{zxy}$.

True.

   b) If all fourth partial derivatives are continuous, then $f_{xxyz} = f_{xyzy}$.

False.

10) Let $f(x,y) = 4x^2y^5 + 3x^3y^2$.

   a) Compute f_{xyy}.

$f_x = 8xy^5 + 9x^2y^2$
$f_{xy} = 40xy^4 + 18x^2y$
$f_{xyy} = 160xy^3 + 18x^2$.

   b) Compute f_{xxy}.

$f_x = 8xy^5 + 9x^2y^2$
$f_{xx} = 8y^5 + 18xy^2$
$f_{xxy} = 40y^4 + 36xy$.

C. Technology Plus. Use a computer algebra system or a graphing calculator to solve.

T-1) Let $z = \sqrt{9 - x^2 - y^2}$.

    a) Find $f_x(1, 2)$ and the parametric equation for the tangent line to f at $(1, 2)$ in the x-direction.

$f(1, 2) = 2$.

$$f_x(x, y) = \frac{1}{2}(9 - x^2 - y^2)^{-1/2}(-2x)$$

$$= \frac{-x}{\sqrt{9 - x^2 - y^2}}.$$

$$f_x(1, 2) = -\frac{1}{2}.$$

Therefore, the line is

$$L_1: \quad x = 1 + t$$
$$y = 2$$
$$z = 2 - \frac{1}{2}t.$$

    b) Find $f_y(1, 2)$ and the parametric equations for the tangent line to f at $(1, 2)$ in the y-direction.

$$f_y(x, y) = \frac{1}{2}(9 - x^2 - y^2)^{-1/2}(-2y)$$

$$= \frac{-y}{\sqrt{9 - x^2 - y^2}}.$$

$$f_y(1, 2) = \frac{-2}{2} = -1.$$

Therefore, the line is

$$L_2: \quad x = 1$$
$$y = 2 + t$$
$$z = 2 - t.$$

    c) On the same screen draw the graphs of $f(x, y)$ and the lines in parts a) and b).

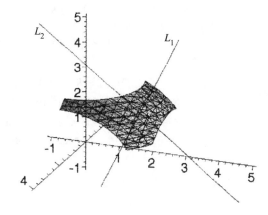

Section 14.4   Tangent Planes and Linear Approximations

This section generalizes the notion of a tangent line for a curve to that of a tangent plane for a surface. The tangent plane is determined by the tangent lines in the x- and y-directions. Tangents planes will be used to define a differential that, in turn, will be used to find a linear approximation of a functional value. The concepts of differentials and linear approximations may be extended to functions with more than two variables.

Concepts to Master

A. Tangent plane to $z = f(x, y)$

B. Differential; Differentiability

Summary and Focus Questions

Page 939 (ET Page 915)

A. For a surface given by $z = f(x, y)$, where f has continuous first partial derivatives, all the tangent lines at a given point on the surface form a plane called the **tangent plane.** If $P(x_0, y_0, z_0)$ is a point on $z = f(x, y)$, the tangent plane at P has equation

$$z - z_0 = f_x(x_0, y_0)(x - x_0) + f_y(x_0, y_0)(y - y_0).$$

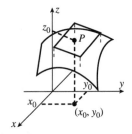

This equation looks familiar. For example, if y is the constant y_0, then the equation becomes $z - z_0 = f_x(x_0, y_0)(x - x_0)$, which is the equation of the tangent line to the curve $z = f(x, y_0)$ in the plane $y = y_0$.

Example: Let $f(x, y) = 8x^2 + 3xy$. Find the equation of the tangent line to the surface $z = f(x, y)$ at the point corresponding to $(2, 1)$.

$f_x = 16x + 3y$ and $f_y = 3x$. At the point $(2, 1)$, $f(2, 1) = 38$, $f_x(2, 1) = 35$, and $f_y(2, 1) = 6$. The tangent plane corresponding to $(2, 1)$ has equation

$$z - 38 = 35(x - 2) + 6(y - 1)$$
$$35x + 6y - z = 38.$$

1) Find the equation of the tangent plane to $f(x,y) = 4x^2 y$ at $(1,3)$.

$f(1, 3) = 4(1)^2 3 = 12$. Therefore, $(1, 3, 12)$ is a point on the plane.
$f_x = 8xy, f_x(1, 3) = 8(1)(3) = 24$.
$f_y = 4x^2, f_y(1, 3) = 4(1)^2 = 4$.

The plane is
$z - 12 = 24(x - 1) + 4(y - 3)$
$24x + 4y - z = 24$.

2) Where is the tangent plane horizontal for $f(x,y) = 4x^3 + 3x^2 y - 48y$?

In order for the tangent plane to be horizontal, both f_x and f_y must be zero.
$f_x = 12x^2 + 6xy = 6x(2x + y) = 0$
at $x = 0$ or $y = -2x$.
$f_y = 3x^2 - 48 = 0$ at $x^2 = 16$.
Thus $x = 4$ and $x = -4$.
At $x = 4, y = -8$ and $x = -4, y = 8$.
The tangent plane is horizontal at $(4, -8)$ and $(-4, 8)$.

Page
943
(ET Page
919)

B. If $z = f(x, y)$ then similar to functions of one variable, the **increment**, or **change**, in z is defined as

$$\Delta z = f(x + \Delta x, y + \Delta y) - f(x, y).$$

Another way to write this is

$$f(x + \Delta x, y + \Delta y) = f(x, y) + \Delta z.$$

If we let $dx = \Delta x$ and $dy = \Delta y$ then the **(total) differential** is

$$dz = f_x(x, y) dx + f_y(x, y) dy.$$

Example: Let $f(x, y) = 2xy^2$. The exact value of $f(3.02, 2.05)$ is $2(3.02)(2.05)^2 = 25.3831$. We will use the total differential to approximate $f(3.02, 2.05)$.

a) Let $\Delta x = 0.02$ and $\Delta y = 0.05$. $f(3, 2) = 2(3)2^2 = 24$. Thus $\Delta z = 25.3831 - 24 = 1.3831$.

b) $f_x = 2y^2$ and $f_y = 4xy$. Therefore, $f_x(3, 2) = 2(2)^2 = 8$ and $f_y(3, 2) = 4(3)(2) = 24$.

c) In general, $dz = 2y^2 dx + 4xy\ dy$. For $(x, y) = (3, 02, 2.05)$, $dx = \Delta x = 0.02$, $dy = \Delta y = 0.05$ and $dz = 8(0.02) + 24(0.05) = 1.3600$.

Since (x, y) is near $(3, 2)$, dz is a good approximation of Δz. In fact,
$\Delta z - dz = 1.3831 - 1.3600 = 0.0231$.

In other words, the number $f(3, 2) + dz = 24 + 1.3600 = 25.3600$ (which is easy to calculate because it is linear and uses only the values 3 and 2 for x and y) is a good approximation of $f(3.02, 2.05)$—within 0.0231 of the exact value.

The function $z = f(x, y)$ is **differentiable at** (a, b) if
$$\Delta z = f_x(a, b)\Delta x + f_y(a, b)\Delta y + \varepsilon_1 \Delta x + \varepsilon_2 \Delta y,$$
where $\varepsilon_1, \varepsilon_2$ are each functions of Δx and Δy and both $\varepsilon_1 \to 0$ and $\varepsilon_2 \to 0$ as $(\Delta x, \Delta y) \to (0, 0)$.

The functions ε_1 and ε_2 measure the difference between Δz and dz. Since their limits are 0 when f is differentiable, dz is a good linear approximation to Δz.

The total differential may be defined for functions of three or more variables. For example, if $s = f(w, x, y, z)$, then $ds = f_w dw + f_x dx + f_y dy + f_z dz$.

3) Let $z = f(x, y) = 2x^2 + y^3$.

   a) Find Δz and dz as (x, y) changes from $(2,1)$ to $(2.01, 1.03)$.

Let $\Delta x = 0.01$ and $\Delta y = 0.03$.

At $(2, 1)$, $z = 2(2)^2 + 1^3 = 9$.
At $(2.01, 1.03)$,
$z = 2(2.01)^2 + (1.03)^3 \approx 9.1729$.
$\Delta z = 9.1729 - 9 = 0.1729$.

$\dfrac{\partial z}{\partial x} = 4x$. At $(2, 1)$, $\dfrac{\partial z}{\partial x} = 4(2) = 8$.

$\dfrac{\partial z}{\partial y} = 3y^2$. At $(2, 1)$, $\dfrac{\partial z}{\partial y} = 3(1)^2 = 3$.

$dx = \Delta x = 0.01$.
$dy = \Delta y = 0.03$.
$dz = f_x dx + f_y dy$
$\quad = 8(0.01) + 3(0.03) = 0.17$.

   b) Use part a) to estimate $f(2.01, 1.03)$.

From part a), $f(2, 1) = 9$ and $dz = 0.17$.
Therefore, $f(2.01, 1.03) \approx 9 + 0.17 = 9.17$.
We saw in part a) that $f(2.01, 1.03) = 9.1729$ to four decimals. The estimate 9.17 is within 0.0029 of the exact value.

4) Find dz for $z = xe^{xy}$.

$dz = (xye^{xy} + e^{xy})dx + (x^2 e^{xy})dy$.

5) For $w = f(x, y, z) = xy^3 + yz^3$, find dw.

$dw = f_x dx + f_y dy + f_z dz$
$\quad = y^3 dx + (3xy^2 + z^3)dy + 3yz^2 dz$.

6) Estimate $\dfrac{(3.02)^2}{(0.99)^3}$ using a differential.

Let $z = f(x,y) = x^2 y^{-3}$. We estimate $f(3.02, 0.99)$ by calculating $f(3, 1)$ and dz.

$f(3, 1) = 3^2(1)^{-3} = 9$.

$f_x = 2xy^{-3} = 6$ at $(3, 1)$.

$f_y = -3x^2 y^{-4} = -27$ at $(3, 1)$. Thus

$dz = f_x dx + f_y dy$

$= 6(0.02) + (-27)(-0.01) = 0.39$.

Finally, $f(3.02, 0.99) = f(3, 1) + \Delta z$

$\approx f(3, 1) + dz = 9 + 0.39 = 9.39$.

Note: To 4 decimals, $\dfrac{(3.02)^2}{(0.99)^3} = 9.3996$.

7) True or False:

   a) If $f_x(x_0, y_0)$ and $f_y(x_0, y_0)$ exist, then f is differentiable at (x_0, y_0).

False. (f_x and f_y must also be continuous.) A counterexample at $(x_0, y_0) = (0, 0)$ is

$$f(x,y) = \begin{cases} \dfrac{xy}{x^2 + y^2} & (x,y) \neq (0,0) \\ 0 & (x,y) = (0,0). \end{cases}$$

Both f_x and f_y are zero at $(0, 0)$ but f is not continuous at $(0, 0)$.

   b) If f is differentiable at (x_0, y_0), then $f_x(x_0, y_0)$ and $f_y(x_0, y_0)$ exist.

True.

8) A circular pool is 7 m in diameter and 3 m deep. If the measurements are accurate to within 0.05 m, then use differentials to estimate the maximum error in calculating the volume of water.

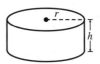

$V = \pi r^2 h$.

$dV = \dfrac{\partial V}{\partial r} dr + \dfrac{\partial V}{\partial h} dh$

$= 2\pi rh\, dr + \pi r^2 dh$

$\approx 2\pi rh\Delta r + \pi r^2 \Delta h$.

We are given $|\Delta r| \leq 0.05$ and $|\Delta h| \leq 0.05$.

Thus, at $r = 3.5$ and $h = 3$ we have

$dV \approx 2\pi(3.5)(3)(0.05) + \pi(3.5)^2(0.05)$

$= 1.6625\pi \approx 5.22 \text{ m}^3$.

Section 14.5   The Chain Rule

The Chain Rule in multidimensions has different versions, depending upon the number of variables involved. The Chain Rule is also useful for finding partial derivatives by implicit differentiation.

Concepts to Master

A. Forms of the Chain Rule

B. Implicit differentiation using the Chain Rule

Summary and Focus Questions

Page
948
(ET Page
924)

A. The Chain Rule (Case 1) Suppose x and y are differentiable functions of t and $z = f(x, y)$ is differentiable. Then z is differentiable with respect to t and

$$\frac{dz}{dt} = \frac{\partial z}{\partial x}\frac{dx}{dt} + \frac{\partial z}{\partial y}\frac{dy}{dt}.$$

Note that $\dfrac{dz}{dt}, \dfrac{dx}{dt}$, and $\dfrac{dy}{dt}$ are ordinary derivatives of functions of the variable t,

whereas $\dfrac{\partial z}{\partial x}$ and $\dfrac{\partial z}{\partial y}$ are partial derivatives.

Example: Let $z = 4x + y^2$, $x = t^2 + 2t$, and $y = 2t - 1$. Then z is a function

of t. Since $\dfrac{dx}{dt} = 2t + 2, \dfrac{dy}{dt} = 2, \dfrac{\partial z}{\partial x} = 4$, and $\dfrac{\partial z}{\partial y} = 2y$, by the Chain Rule:

$$\frac{dz}{dt} = 4(2t + 2) + 2y(2) = 4(2t + 2) + 2(2t - 1)2 = 8t + 8 + 8t - 4 = 16t + 4.$$

We can check this result by first finding z in terms of t,

$$z = 4x + y^2 = 4(t^2 + 2t) + (2t - 1)^2 = 4t^2 + 8t + 4t^2 - 4t + 1 = 8t^2 + 4t + 1$$

and finding directly that $\dfrac{dz}{dt} = 16t + 4.$

The Chain Rule (Case 2) Suppose $x = g(s, t)$, $y = h(s, t)$, and $z = f(s, t)$. In this case the version of the Chain rule for computing the partial derivatives of z with respect to s and t is:

$$\frac{\partial z}{\partial s} = \frac{\partial z}{\partial x}\frac{\partial x}{\partial s} + \frac{\partial z}{\partial y}\frac{\partial y}{\partial s} \quad \text{and} \quad \frac{\partial z}{\partial t} = \frac{\partial z}{\partial x}\frac{\partial x}{\partial t} + \frac{\partial z}{\partial y}\frac{\partial y}{\partial t}.$$

Example: Let $z = x^2 + 3xy + y^3, x = 2s - t^2$, and $y = 4st$.

$$\frac{\partial z}{\partial s} = \frac{\partial z}{\partial x}\frac{\partial x}{\partial s} + \frac{\partial z}{\partial y}\frac{\partial y}{\partial s} = (2x + 3y)(2) + (3x + 3y^2)(4t)$$
$$= [2(2s - t^2) + 3(4st)](2) + [3(2s - t^2) + 3(4st)^2](4t)$$
$$= 8s - 4t^2 + 48st - 12t^3 + 192s^2t^3.$$

$$\frac{\partial z}{\partial t} = \frac{\partial z}{\partial x}\frac{\partial x}{\partial t} + \frac{\partial z}{\partial y}\frac{\partial y}{\partial t} = (2x + 3y)(-2t) + (3x + 3y^2)(4s)$$
$$= (2(2s - t^2) + 3(4st))(-2t) + (3(2s - t^2) + 3(4st)^2)(4s)$$
$$= -8st + 4t^3 - 36st^2 + 24s^2 + 192s^3t^2.$$

Both versions of the Chain Rule may be generalized. For example, for $w = f(x, y, z), x = h(u, v, s, t), y = k(u, v, s, t), z = m(u, v, s, t)$, the function w has four partial derivatives that may be found by the Chain Rule. For instance:

$$\frac{\partial w}{\partial v} = \frac{\partial w}{\partial x}\frac{\partial x}{\partial v} + \frac{\partial w}{\partial y}\frac{\partial y}{\partial v} + \frac{\partial w}{\partial z}\frac{\partial z}{\partial v}.$$

There will be as many terms in the sum as there are intermediate variables (x, y, and z in this case).

1) Find $\dfrac{dz}{dt}$ where $z = x^2y, x = e^t, y = t^2$.

$$\frac{dz}{dt} = \frac{\partial z}{\partial x}\frac{dx}{dt} + \frac{\partial z}{\partial y}\frac{dy}{dt} = (2xy)e^t + x^2(2t)$$
$$= (2e^tt^2)e^t + (e^t)^2 2t = 2t^2e^{2t} + 2te^{2t}.$$

2) Find $\dfrac{\partial w}{\partial u}$ at $(u, v) = \left(\dfrac{\pi}{2}, \dfrac{\pi}{2}\right)$, where

$w = x^2yz, x = uv, y = u\sin v, z = v\sin u$.

$$\frac{\partial w}{\partial u} = \frac{\partial w}{\partial x}\frac{\partial x}{\partial u} + \frac{\partial w}{\partial y}\frac{\partial y}{\partial u} + \frac{\partial w}{\partial z}\frac{\partial z}{\partial u}$$
$$= (2xyz)v + x^2z(\sin v) + x^2y(v\cos u).$$

At $(u, v) = \left(\dfrac{\pi}{2}, \dfrac{\pi}{2}\right), x = \dfrac{\pi^2}{4}, y = \dfrac{\pi}{2}, z = \dfrac{\pi}{2}$.

Therefore, $\dfrac{\partial w}{\partial u}$

$$= \left(2\frac{\pi^2}{4}\frac{\pi}{2}\frac{\pi}{2}\right)\frac{\pi}{2} + \left(\frac{\pi^2}{4}\right)^2\frac{\pi}{2}(1) + \left(\frac{\pi^2}{4}\right)^2\frac{\pi}{2}\frac{\pi}{2}(0)$$
$$= \frac{3}{32}\pi^5.$$

Page
952
(ET Page
928)

B. Here is another way to perform implicit differentiation for a function of one variable: if $y = f(x)$ is defined implicitly by the equation $F(x, y) = 0$, then

$$\frac{dy}{dx} = -\frac{F_x}{F_y}.$$

Example: Let y be a function of x defined by $x^2 + y^2 = 1$. Let $F(x, y) = x^2 + y^2 - 1$. $F_x = 2x$ and $F_y = 2y$. The derivative of y with respect to x is:

$$y' = -\frac{F_x}{F_y} = -\frac{2x}{2y} = -\frac{x}{y}.$$

If $z = f(x, y)$ is defined implicitly by $F(x, y, z) = 0$, then:

$$\frac{\partial z}{\partial x} = -\frac{F_x}{F_z} \quad \text{and} \quad \frac{\partial z}{\partial y} = -\frac{F_y}{F_z}.$$

3) Find y' where y is defined by:

a) $2x^3 + 3xy + 4y^2 = 7$.

Let $F(x, y) = 2x^3 + 3xy + 4y^2 - 7$.
$F_x = 6x^2 + 3y$.
$F_y = 3x + 8y$.
$$y' = -\frac{F_x}{F_y} = -\frac{6x^2 + 3y}{3x + 8y}.$$

b) $x^2 y^3 + \cos(xy) = 0$.

Let $F(x, y) = x^2 y^3 + \cos(xy)$.
$F_x = 2xy^3 - y\sin(xy)$.
$F_y = 3x^2 y^2 - x\sin(xy)$.
$$y' = -\frac{F_x}{F_y} = \frac{y\sin(xy) - 2xy^3}{3x^2 y^2 - x\sin(xy)}.$$

4) Find the equation of the tangent plane to $x^2 + 2y^2 + z^2 = 6$ at the point $(2, -1, 1)$.

Instead of solving for z and differentiating, find $\dfrac{\partial z}{\partial x}$ and $\dfrac{\partial z}{\partial y}$ implicitly.
Let $F(x, y, z) = x^2 + 2y^2 + z^2 - 6$.
$$\frac{\partial z}{\partial x} = -\frac{F_x}{F_z} = -\frac{2x}{2z} = -\frac{x}{z}.$$
At $(2, -1, 1)$, $\dfrac{\partial z}{\partial x} = -2$.
$$\frac{\partial z}{\partial y} = -\frac{F_y}{F_z} = -\frac{4y}{2z} = -\frac{2y}{z}.$$
At $(2, -1, 1)$, $\dfrac{\partial z}{\partial y} = 2$.

The tangent plane is
$z - 1 = -2(x - 2) + 2(y + 1)$
$2x - 2y + z = 7$.

Section 14.6   Directional Derivatives and the Gradient Vector

The partial derivatives $f_x(x, y)$ and $f_y = (x, y)$ may be thought of as the derivatives of f in the x-direction and y-direction, respectively. This section defines the "directional" derivative for an arbitrary direction $\mathbf{u} = \langle a, b \rangle$ in the xy-plane. $f_x(x, y)$ and $f_y(x, y)$ will be special cases where the x-direction is specified by $\mathbf{i} = \langle 1, 0 \rangle$ and the y-direction, by $\mathbf{j} = \langle 0, 1 \rangle$. The vector that points in the direction for which the directional derivative is maximum is called the gradient.

Concepts to Master

A. Directional derivative

B. Gradient; Maximum value of directional derivative

C. Tangent plane to a level surface for $F(x, y, z)$

Summary and Focus Questions

A. Let $\mathbf{u}$ be a unit vector. The **directional derivative** of $f(x, y)$ at (x_0, y_0) in the direction of $\mathbf{u} = \langle a, b \rangle$ is:

$$D_{\mathbf{u}}f(x_0, y_0) = \lim_{h \to 0} \frac{f(x_0 + ah, y_0 + bh) - f(x_0, y_0)}{h}.$$

$D_{\mathbf{u}}f(x_0, y_0)$ may be interpreted as the slope of the tangent line to the graph of $z = f(x, y)$ in the $\mathbf{u}$ direction. It is the instantaneous rate of change of z at (x_0, y_0) in the $\mathbf{u}$ direction.

If f is a differentiable function of x and y, we may write the directional derivative in the direction $\mathbf{u} = \langle a, b \rangle$ in terms of partial derivatives:

$$D_{\mathbf{u}}f(x_0, y_0) = f_x(x_0, y_0)a + f_y(x_0, y_0)b.$$

Example: Find $D_{\mathbf{u}}f(1, 1)$, where $\mathbf{u} = \left\langle \dfrac{1}{2}, \dfrac{\sqrt{3}}{2} \right\rangle$ and $f(x, y) = 4x^2 y^3 - 7x^3 y^2$.

$f_x(x, y) = 8xy^3 - 21x^2 y^2$

$f_y(x, y) = 12x^2 y^2 - 14x^3 y$

$f_x(1, -1) = 8(1)(-1)^3 - 21(1)^2(-1)^2 = -29$

$f_y(1, -1) = 12(1)^2(-1)^2 - 14(1)^3(-1) = 26$

$D_{\mathbf{u}}f(1, -1) = f_x(1, -1)\dfrac{1}{2} + f_y(1, -1)\dfrac{\sqrt{3}}{2} = (-29)\dfrac{1}{2} + (26)\dfrac{\sqrt{3}}{2} = \dfrac{26\sqrt{3} - 29}{2}.$

Suppose $H(x, y)$ is the elevation above sea level of a point (x, y) in a hilly terrain and you are standing on a hillside at (x_0, y_0) holding a compass. Your compass is positioned so that north corresponds to the positive y-axis (the unit vector $\mathbf{j}$) and east corresponds to the positive x-axis (the unit vector $\mathbf{i}$). Let the unit vector $\mathbf{u} = \langle a, b \rangle = a\mathbf{i} + b\mathbf{j}$. The directional derivative $D_{\mathbf{u}}f(x_0, y_0)$ measures the slope of the hill in the direction that $\mathbf{u}$ is pointing.

1) If $\mathbf{u} = \mathbf{i}$, then what is $D_{\mathbf{u}}f(x_0 y_0)$?

The partial derivative $f_x(x_0, y_0)$.

2) Find the directional derivative of $f(x, y) = x^2 y^2 - xy^3$ in the direction $\mathbf{u} = \dfrac{1}{2}\mathbf{i} - \dfrac{\sqrt{3}}{2}\mathbf{j}$.

$f_x = 2xy^2 - y^3$,
$f_y = 2x^2 y - 3xy^2$.

$D_{\mathbf{u}}f = \dfrac{1}{2}(2xy^2 - y^3) - \dfrac{\sqrt{3}}{2}(2x^2 y - 3xy^2)$

$\quad\; = \dfrac{2 + 3\sqrt{3}}{2}xy^2 - \dfrac{1}{2}y^3 - \sqrt{3}x^2 y.$

B. The **gradient of** $f(x, y)$ at (x_0, y_0) is a vector in the xy-plane:

$$\nabla f(x_0, y_0) = \langle f_x(x_0, y_0), f_y(x_0, y_0) \rangle.$$

Example: Let $f(x, y) = x^2 + 2y^2 + 6xy$. Then $f_x(x, y) = 2x + 6y$ and $f_y = (x, y) = 4y + 6x$ and the gradient is $\nabla f(x, y) = \langle 2x + 6y, 4y + 6x \rangle$.

At the particular point $(2, 1)$, $\nabla f(2, 1) = \langle 2(2) + 6(1), 4(1) + 6(2) \rangle = \langle 10, 16 \rangle$.

A directional derivative in the direction $\mathbf{u}$ may be written as the dot product of the gradient with $\mathbf{u}$:

$$D_{\mathbf{u}}f(x_0, y_0) = \nabla f(x_0, y_0) \cdot \mathbf{u}.$$

For a fixed point (x_0, y_0), the value of $D_{\mathbf{u}}f(x_0, y_0)$, varies depending on the direction $\mathbf{u}$. The maximum value of $D_{\mathbf{u}}f(x_0, y_0)$ occurs in the direction of the gradient, given by the unit vector:

$$\mathbf{u} = \frac{\nabla f(x_0, y_0)}{|\nabla f(x_0, y_0)|}.$$

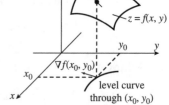

The maximum value of $D_{\mathbf{u}}f(x_0, y_0)$ in this direction is the length of the gradient: $|\nabla f(x_0, y_0)|$.

Since the gradient points in the direction of maximum increase for f, the gradient is normal to the level curve for f through the point (x_0, y_0).

The minimum value of $D_{\mathbf{u}}f(x_0, y_0)$ is $|-\nabla f(x_0, y_0)|$ and occurs when

$$\mathbf{u} = -\frac{\nabla f(x_0, y_0)}{|\nabla f(x_0, y_0)|}.$$

In terms of the hillside example, this means that if you are standing on the side of a hill at the point corresponding to (x_0, y_0), the gradient $\nabla f(x_0, y_0)$, will be the direction for which the hill rises the steepest. The opposite direction, $-\nabla f(x_0, y_0)$, will be the direction of steepest descent.

Example: Let $f(x, y) = x^2 + 2y^2 + 6xy$. In what direction **u** is $D_\mathbf{u}f(2, 1)$ maximum?

We saw previously that at $(2, 1)$, $\nabla f(2, 1) = \langle 10, 16 \rangle$. The maximum value of $D_\mathbf{u}f(2, 1)$ is $|\nabla f(2, 1)| = \sqrt{10^2 + 16^2} = \sqrt{356}$ and occurs in the direction:

$$\mathbf{u} = \left\langle \frac{10}{\sqrt{356}}, \frac{16}{\sqrt{356}} \right\rangle \approx \langle 0.53, 0.85 \rangle.$$

Directional derivatives and gradients for functions of three or more variables are defined similarly.

3) Find the gradient of $f(x, y) = e^{xy}$.

$f_x = ye^{xy}$ and $f_y = xe^{xy}$, so

$\nabla f(x, y) = ye^{xy}\mathbf{i} + xe^{xy}\mathbf{j}$.

4) Find the directional derivative of $f(x, y, z) = xy \sin z$ in the direction $\mathbf{u} = \left\langle \frac{1}{2}, -\frac{1}{\sqrt{2}}, \frac{1}{2} \right\rangle$ at the point $\left\langle 2, 1, \frac{\pi}{6} \right\rangle$.

$f_x = y \sin z$, $f_y = x \sin z$ and $f_z = xy \cos z$.

At $\left(2, 1, \frac{\pi}{6}\right)$, $f_x\left(2, 1, \frac{\pi}{6}\right) = 1\left(\frac{1}{2}\right) = \frac{1}{2}$,

$f_y\left(2, 1, \frac{\pi}{6}\right) = 2\left(\frac{1}{2}\right) = 1$, and

$f_z\left(2, 1, \frac{\pi}{6}\right) = 2(1)\frac{\sqrt{3}}{2} = \sqrt{3}$.

$\nabla f\left(2, 1, \frac{\pi}{6}\right) = \left\langle \frac{1}{2}, 1, \sqrt{3} \right\rangle$ and

$D_\mathbf{u}f\left(2, 1, \frac{\pi}{6}\right) = \left\langle \frac{1}{2}, 1, \sqrt{3} \right\rangle \cdot \left\langle \frac{1}{2}, -\frac{1}{\sqrt{2}}, \frac{1}{2} \right\rangle$

$= \frac{1}{2}\left(\frac{1}{2}\right) + 1\left(-\frac{1}{\sqrt{2}}\right) + \sqrt{3}\left(\frac{1}{2}\right)$

$= \frac{1 - 2\sqrt{2} + 2\sqrt{3}}{4}$.

5) Let $f(x, y, z) = xy^2 + yz^2$.

    a) What is the maximum value of $D_\mathbf{u}f(1, 2, 1)$ as **u** varies?

The maximum value is $|\nabla f(1, 2, 1)|$.

$\nabla f = \langle y^2, 2xy + z^2, 2yz \rangle$.

Thus, $\nabla f(1, 2, 1) = \langle 4, 5, 4 \rangle$ and

$|\nabla f(1, 2, 1)| = \sqrt{4^2 + 5^2 + 4^2} = \sqrt{57}$.

b) What direction **u** gives the maximum value of $D_{\mathbf{u}}f(1, 2, 1)$?

u is the direction $\nabla f(1, 2, 1)$. But **u** must be a unit vector, so $\mathbf{u} = \left\langle \dfrac{4}{\sqrt{57}}, \dfrac{5}{\sqrt{57}}, \dfrac{4}{\sqrt{57}} \right\rangle$.

c) Find $D_{\mathbf{v}}f(1, 2, 1)$ where

$$\mathbf{v} = \left\langle \frac{5}{6}, \frac{1}{2}, \frac{\sqrt{2}}{6} \right\rangle.$$

$$D_{\mathbf{v}}f(1, 2, 1) = \nabla f(1, 2, 1) \cdot \mathbf{v}$$
$$= 4\left\langle \frac{5}{6} \right\rangle + 5\left\langle \frac{1}{2} \right\rangle + 4\left\langle \frac{\sqrt{2}}{6} \right\rangle = \frac{35 + 4\sqrt{2}}{6}.$$

Page 964 (ET Page 940)

C. In three dimensions, we can relate the gradient to the tangent plane of a surface. Let S be a level surface $F(x, y, z) = k$ and $P(x_0, y_0, z_0)$ be a point on S. The gradient $\nabla F(x_0, y_0, z_0)$ is normal to the surface S at point P. The tangent plane to S at P has equation:

$$\nabla F(x_0, y_0, z_0) \cdot \left\langle x - x_0, y - y_0, z - z_0 \right\rangle = 0,$$

or

$$F_x(x_0, y_0, z_0)(x - x_0) + F_y(x_0, y_0, z_0)(y - y_0) + F_z(x_0, y_0, z_0)(z - z_0) = 0.$$

Likewise, the line normal to the surface at P_0 has the equation

$$\frac{x - x_0}{F_x(x_0, y_0, z_0)} = \frac{y - y_0}{F_y(x_0, y_0, z_0)} = \frac{z - z_0}{F_z(x_0, y_0, z_0)}.$$

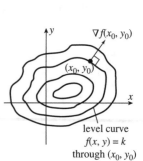

Dropping back to the two-dimensional case where $z = f(x, y)$, let $k = f(x_0, y_0)$. Then $\nabla f(x_0, y_0)$ is the vector in the xy-plane perpendicular to $f(x, y) = k$, which is the level curve for f at (x_0, y_0). This agrees with the discussion in Part B, where we saw that $\nabla f(x_0, y_0)$ is the direction in which f has the greatest increase.

Example: Let $f(x, y) = x^2 + y^2$. Then $\nabla f(x_0, y_0) = \left\langle 2x, 2y \right\rangle$. For $(x_0, y_0) = (3, 4)$, $f(3, 4) = 25$, so the level curve through $(3, 4)$ is the circle $x^2 + y^2 = 25$. The gradient $\nabla f(3, 4) = \left\langle 6, 8 \right\rangle$ is normal to this circle.

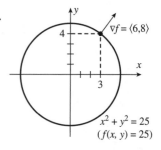

6) Find the equations of the tangent plane and normal line to $x^2 + 2y^2 - z^2 - 4xyz = 16$ at $(1, 4, 1)$.

Let $F(x, y, z) = x^2 + 2y^2 - z^2 - 4xyz - 16$.
$\nabla F = \langle 2x - 4yz, 4y - 4xz, -2z - 4xy \rangle$.
$\nabla F(1, 4, 1) = \langle -14, 12, -18 \rangle$.

The tangent plane is:
$-14(x - 1) + 12(y - 4) - 18(z - 1) = 0$
$7x - 6y + 9z = -8$.

The normal line is:
$$\frac{x - 1}{-14} = \frac{y - 4}{12} = \frac{z - 1}{-18}$$
$$\frac{x - 1}{7} = \frac{4 - y}{6} = \frac{z - 1}{9}.$$

7) Given the following level curve to $z = f(x, y)$, which vectors could be gradients?

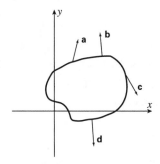

b and **d** appear to be normal to the level curve and are either gradients or the negatives of gradients.

Section 14.7   Maximum and Minimum Values

In this section partial derivatives are used to find the maxima and minima of functions of two variables. The terms, concepts, and procedures are very similar to those used with functions of one variable.

Concepts to Master

A. Local maximum; Local minimum; Critical point; Second Derivative Test

B. Absolute maximum and minimum; Absolute extrema on a closed, bounded set; Extreme value problems

Summary and Focus Questions

Page 970 (ET Page 946)

A. A function $z = f(x, y)$ has a **local maximum** at (a, b) if $f(a, b) \geq f(x, y)$ for all (x, y) in some open disk about (a, b). A **local minimum** at (a, b) is defined similarly but with $f(a, b) \leq f(x, y)$. These are **local extrema** of f.

A **critical point** (a, b) of f is a point in the domain of f for which both $f_x(a, b) = 0$ and $f_y(a, b) = 0$ or at least one of the partial derivatives does not exist.

If (a, b) is a local extremum for a continuous $z = f(x, y)$, then (a, b) is a critical point. The converse is false. If f has continuous second partials, a critical point that is not a local extremum is a **saddle point**.

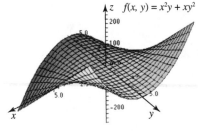

Example: For $f(x, y) = x^2 y + xy^2$ both $f_x(x, y) = 2xy + y^2$ and $f_y(x, y) = x^2 + 2xy$ are zero at $(0, 0)$. Thus, $(0, 0)$ is a critical point. However, along the line $y = 2x$, $f(x, y)$ becomes $f(x, y) = x^2(2x) + x(2x)^2 = 6x^3$. Thus, $(0, 0)$ is not a local maximum or local minimum. It is a saddle point. The graph of f is at the right.

The method of finding some critical points for $z = f(x, y)$ involves setting $f_x = 0$ and $f_y = 0$ and solving these two equations in two variables simultaneously. Often the system $f_x = 0$ and $f_y = 0$ is not linear and may be very difficult to solve. There are no general methods. One approach is to solve for one variable in terms of the other variable in one equation and substitute that value into the other equation.

Example: Find the critical points for $f(x, y) = x^3 - 24xy + 8y^3$.

Since f is a polynomial, the critical points are where both partial derivatives are zero. $f_x = 3x^2 - 24y = 0$ when $x^2 - 8y = 0$, and $f_y = -24x + 24y^2 = 0$ when $-x + y^2 = 0$. Substituting $x = y^2$ from the second equation into the first, we have $y^4 - 8y = y(y^3 - 8) = 0$. The solutions are $y = 0$ and $y = 2$. At $y = 0$, $x = 0$ and at $y = 2$, $x = 4$. The two critical points are $(0, 0)$ and $(4, 2)$.

Second Derivative Test Suppose $z = f(x, y)$ has continuous second partial derivatives in a disk with center (a, b). Suppose $f_x(a, b) = 0$ and $f_y(a, b) = 0$ and let $D(a, b) = f_{xx}(a, b)f_{yy}(a, b) - [f_{xy}(a, b)]^2$.

Then the following holds:

Conditions	Conclusion
$D > 0$, $f_{xx} > 0$	f has a local minimum at (a, b).
$D > 0$, $f_{xx} < 0$	f has a local maximum at (a, b).
$D < 0$	f has a saddle point at (a, b).
$D = 0$	No conclusion can be made.

A way to remember how to calculate D is as a determinant:

$$D = \begin{vmatrix} f_{xx} & f_{xy} \\ f_{yx} & f_{yy} \end{vmatrix} = f_{xx}f_{yy} - (f_{xy})^2.$$

Example: Classify the critical points for $f(x, y) = x^3 - 24xy + 8y^3$.

From the previous example, the critical points for f are $(0, 0)$ and $(4, 2)$. The second partials are $f_{xx} = 6x$, $f_{yy} = 48y$, and $f_{xy} = -24$.

At $(0, 0)$, $f_{xx} = 0$, $f_{yy} = 0$, and $f_{xy} = -24$, so $D = 0(0) - (-24)^2 = -576$. The point $(0, 0)$ is a saddle point.

At $(4, 2)$, $f_{xx} = 24$, $f_{yy} = 96$, and $f_{xy} = -24$, so $D = 24(96) - (-24)^2 = 1728$. The point $(4, 2)$ is a relative minimum.

---

1) Let $f(x, y) = y^3 - 24x - 3x^2y$.

    a) Find the critical points of f.

Since f is a polynomial, critical points occur only where $f_x = 0$ and $f_y = 0$:

$$f_x = -24 - 6xy = 0$$
$$f_y = 3y^2 - 3x^2 = 0.$$

To solve this system, we observe from the second equation that $3x^2 = 3y^2$, so $x = y$ or $x = -y$.

Substituting $x = y$ in the first:

$$-24 - 6x(x) = 0$$
$$-24 - 6x^2 = 0$$
$$6x^2 = -24.$$

This has no solution.

Substituting $x = -y$ in the first:

$$-24 - 6x(-x) = 0$$
$$-24 + 6x^2 = 0$$
$$6x^2 = 24$$
$$x^2 = 4; \ x = 2, -2.$$

When $x = 2$, $y = -2$, and when $x = -2$, $y = 2$. The critical points are $(2, -2)$ and $(-2, 2)$.

b) Find the local extrema of $f(x, y)$.

The extrema are among the critical points.
$f_{xx} = -6y$, $f_{xy} = -6x$, $f_{yy} = 6y$.
$$D = (-6y)(6y) - (-6x^2) = -36y^2 - 36x^2$$
At both $(2, -2)$ and $(-2, 2)$,
$$D = -288 < 0.$$
Both $(2, -2)$ and $(-2, 2)$ are saddle points. There are no local extrema.

2) Suppose f has continuous second derivatives and has four critical points with the following information about the second derivatives.

Point	f_{xx}	f_{yy}	f_{xy}
P: (7, 1)	8	2	−4
Q: (1, 2)	1	9	2
R: (0, 4)	−6	3	1
S: (−1, 3)	−2	−5	3

Classify each critical point.

We calculate $D = f_{xx} f_{yy} - (f_{xy})^2$ for each:

Point	D
P	$8(2) - (-4)^2 = 0$
Q	$1(9) - 2^2 = 5$
R	$-6(3) - 1^2 = -19$
S	$-2(-5) - 3^2 = 1$

By the Second Derivative Test:
Q is a local minimum, R is a saddle point, S is a local maximum.
We cannot conclude anything about P from the information given.

3) a) Find all critical points of:

$$f(x, y) = x^2 - 4xy + 2x - \frac{4}{15}(5y + 41)^{3/2}.$$

$f_x = 2x - 4y + 2 = 2(x - 2y + 1).$
$f_y = -4x - \frac{4}{15}\left(\frac{3}{2}\right)(5y + 41)^{1/2}(5)$
$= -2\left(2x + \sqrt{5y + 41}\right).$

Set both f_x and f_y to 0.
$$x - 2y + 1 = 0$$
$$2x + \sqrt{5y + 41} = 0.$$

Solve for x in the first:
$$x = 2y - 1$$

Substitute in the second equation:

$$2(2y-1) + \sqrt{5y+41} = 0$$
$$2 - 4y = \sqrt{5y+41}$$
$$4 - 16y + 16y^2 = 5y + 41$$
$$16y^2 - 21y - 37 = 0$$
$$(16y - 37)(y + 1) = 0$$
$$y = \frac{37}{16}, \; y = -1.$$

If $y = \dfrac{37}{16}$, $x = 2\left(\dfrac{37}{16}\right) - 1 = \dfrac{29}{8}$.

If $y = -1$, $x = 2(-1) - 1 = -3$.

The possible critical points are $\left(\dfrac{29}{8}, \dfrac{37}{16}\right)$ and $(-3, -1)$.

$f_x(-3, -1) = f_y(-3, -1) = 0$, but $f_y\left(\dfrac{29}{8}, \dfrac{37}{16}\right) \neq 0$. Thus, $(-3, -1)$ is the only critical point.

b) Classify the critical point in part a).

$f_{xx} = 2$, $f_{xy} = -4$ and

$$f_{yy} = -2\left(\frac{1}{2}\right)(5y+41)^{-1/2}(5) = \frac{-5}{\sqrt{5y+41}}.$$

At $(-3, -1)$, $f_{yy} = \dfrac{-5}{6}$.

$f_{xx}f_{yy} - (f_{xy})^2 = 2\left(\dfrac{-5}{6}\right) - (-4)^2$, which is negative. Thus, $(-3, -1)$ is a saddle point.

Page 975
(ET Page 951)

B. The **absolute maximum** of f is the value $f(a, b)$ for some (a, b) such that $f(a, b) \geq f(x, y)$ for all (x, y) in the domain of f. **Absolute minimum** is defined in a similar way but with $f(a, b) \leq f(x, y)$. These are the **extreme values** of f.

If f is a continuous function whose domain is closed (contains all its boundary points) and bounded, then f has an absolute maximum and an absolute minimum.

To find the absolute extrema of continuous $f(x, y)$ on a closed and bounded set D, compute $f(x, y)$ at critical points in D and along the boundary of D.

Example: Find the point on the plane $2x + y + 2z = 14$ that is closest to the point $(5, 4, 6)$.

The distance from $(5, 4, 6)$ to a point (x, y, z) on the plane is $\sqrt{(x-5)^2 + (y-4)^2 + (z-6)^2}$. To make the problem a little easier, we note that this distance is minimum when its square is minimum.

Using the equation of the plane to solve for y, we have $y = 14 - 2x - 2z$. The squared distance to minimize is:

$$d = (x-5)^2 + (14 - 2x - 2z - 4)^2 + (z-6)^2$$
$$= (x-5)^2 + (10 - 2x - 2z)^2 + (z-6)^2.$$

$$\frac{\partial d}{\partial x} = 2(x-5) + 2(10 - 2x - 2z)(-2) = 10x + 8z - 50$$

$$\frac{\partial d}{\partial z} = 2(10 - 2x - 2z)(-2) + 2(z-6) = 8x + 10z - 52.$$

The resulting system of two equations in two variables is:
$$10x + 8z - 50 = 0$$
$$8x + 10z - 52 = 0.$$

The solution to this system is $x = \frac{7}{3}$, $z = \frac{10}{3}$. Substituting these values into $y = 14 - 2x - 2z$ gives $y = \frac{8}{3}$. The point on the plane $2x + y + 2z = 14$ that is closest to the point $(5, 4, 6)$ is $\left(\frac{7}{3}, \frac{8}{3}, \frac{10}{3}\right)$.

4) Does the absolute extrema exist for $f(x, y) = x^2 + y^2$ with domain $D = \{(x, y) \mid 0 \le x \le 2, 0 \le y \le 3\}$?

Yes, f is continuous and D is both closed and bounded. (D is a rectangle.)

5) Find the extreme values of $f(x, y) = 10 - x^2 - 2y^2 + 2x + 8y$ with domain D as graphed.

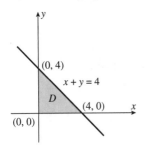

First find the critical points:

$f_x = -2x + 2 = 0$ at $x = 1$.
$f_y = -4y + 8 = 0$ at $y = 2$.

$(1, 2)$ is the only critical point, and it is within D. D has three boundary lines:
L_1 (x-axis): $y = 0, 0 \le x \le 4$.

Here $f(x, y) = f(x, 0) = 10 - x^2 + 2x$ has a maximum at $(1, 0)$ and a minimum at $(4, 0)$.

L_2 (y-axis): $x = 0, 0 \le y \le 4$.

Here $f(x, y) = f(0, y) = 10 - 2y^2 + 8y$ has a maximum at $(0, 2)$ and a minimum at $(0, 0)$ and $(0, 4)$.

L_3 ($x + y = 4$): $y = 4 - x, 0 \le x \le 4$.
Here $f(x, y) = f(x, 4 - x)$
$$= 10 - x^2 - 2(4 - x)^2 + 2x + 8(4 - x)$$
$$= 10 + 10x - 3x^2$$

5) *(continued)*

has a maximum at $\left(\frac{5}{3}, \frac{7}{3}\right)$ and a minimum at $(4, 0)$.

We compute the corresponding $f(x, y)$ values and summarize in a table:

(x, y)	How found?	f(x, y)
(1, 0)	max on L_1	11
(4, 0)	min on both L_1 and L_3	2
(0, 2)	max on L_2	18
(0, 4)	min on L_2	10
$\left(\frac{5}{3}, \frac{7}{3}\right)$	max on L_3	18.33
(1, 2)	critical point	19

The absolute maximum is 19 and occurs at $(1, 2)$. The absolute minimum is 2 and occurs at $(4, 0)$.

6) Show that among all rectangular parallelepipeds with volume 1 cubic meter, the one with smallest surface area is a cube.

We are given $xyz = 1$ and must minimize $S = 2xy + 2xz + 2yz$.

From $xyz = 1$, $z = \dfrac{1}{xy}$.

Thus, $S = 2xy + 2x\left(\dfrac{1}{xy}\right) + 2y\left(\dfrac{1}{xy}\right)$

$$= 2xy + \frac{2}{y} + \frac{2}{x}.$$

$$S_x = 2y - \frac{2}{x^2} = 0$$

$$2y = \frac{2}{x^2}, \, y = \frac{1}{x^2}.$$

$$S_y = 2x - \frac{2}{y^2} = 0$$

$$2x = \frac{2}{y^2}, \, x = \frac{1}{y^2}.$$

From $y = \dfrac{1}{x^2}$ and $x = \dfrac{1}{y^2}$, $y = y^4$.

Thus, $y = 0$ or $y = 1$.

$y = 0$ is not possible so $y = 1$.

Thus, $x = \dfrac{1}{y^2} = 1$ and $z = \dfrac{1}{xy} = \dfrac{1}{1 \cdot 1} = 1$.

Therefore, the object is a cube with edges 1 m long.

Section 14.8 Lagrange Multipliers

Extreme value problems are frequently stated in terms of finding the maximum (or minimum) of a function given a constraint on the variables in the form of an equation. For example, in problem 6 of the previous section we found the minimum of $S = 2xy + 2xy + 2yz$ under the condition that $xyz = 1$. The method of solution there involved solving for one variable in terms of the others, substituting that into the function to reduce the number of variables. This section gives a different method for solving extreme value problems.

Concepts to Master

A. Solution to extreme value problems using Lagrange Multipliers

B. Technology Plus

Summary and Focus Questions

Page
981
(ET Page
957)

A. If $f(x, y)$ and $g(x, y)$ have continuous partial derivatives and (a, b) is a local extremum for f when restricted to $g(x, y) = k$ (a constraint), then there is a number λ, called a **Lagrange multiplier,** such that:

$$\nabla f(a, b) = \lambda \nabla g(a, b).$$

Thus, solving the preceding constrained extremum problem is the same as solving the equations $\nabla f = \lambda \nabla g$ and $g(x, y) = k$.

Lagrange multipliers are very handy for solving the constrained extrema problems of the form "find the maximum (or minimum) of $z = f(x, y)$ such that $g(x, y) = k$."

Example: The function $f(x, y) = xy$ has no maximum value (just let x and y get larger and larger.) However, if we add the constraint that $x + 3y = 24$, then $f(x, y) = xy$ does have a maximum value:

Let $g(x, y) = x + 3y$. Then $\nabla f = \langle y, x \rangle$ and $\nabla g = \langle 1, 3 \rangle$. The equations $\nabla f = \lambda \nabla g$ and $g(x, y) = k$ become $\langle y, x \rangle = \lambda \langle 1, 3 \rangle$ and $x + 3y = 24$.

Therefore:

$$y = \lambda$$
$$x = 3\lambda$$
$$x + 3y = 24$$

Substituting the first and second equations into the third equation:

$$3\lambda + 3(\lambda) = 24$$
$$\lambda = 4.$$

Therefore, $x = 3(4) = 12$ and $y = 4$.

The solution to the constrained optimum problem involving three variables is determined by solving $\nabla f(x, y, z) = \lambda \nabla g(x, y, z)$ and $g(x, y, z) = k$.

Example: Find the minimum of $f(x, y, z) = 2xy + 2xz + 2yz$ subject to $xyz = 1$.

$$\nabla f(x, y, z) = \langle 2y + 2z, 2x + 2z, 2x + 2y \rangle$$
$$\nabla g(x, y, z) = \langle yz, xz, xy \rangle.$$

From $\nabla f(x, y, z) = \lambda \nabla g(x, y, z)$:

$$
\begin{aligned}
2y + 2z &= \lambda yz \\
2x + 2z &= \lambda xz \qquad \rightarrow \\
2x + 2y &= \lambda xy
\end{aligned}
\qquad
\begin{aligned}
2xy + 2xz &= \lambda xyz \\
2xy + 2yz &= \lambda xyz \\
2xz + 2yz &= \lambda xyz.
\end{aligned}
$$

From the first and second equations $2xz = 2yz$ and, therefore, $x = y$. From the second and third equations, $2xy = 2xz$ and, therefore, $y = z$. Thus, $x = y = z$. Since $xyz = 1$, we have $x = 1$, $y = 1$, and $z = 1$.

For problems with two constraints, $g_1(x, y, z) = k_1$, $g_2(x, y, z) = k_2$, use

$$\nabla f(x, y, z) = \lambda_1 \nabla g_1(x, y, z) + \lambda_2 \nabla g_2(x, y, z).$$

1) Find the extrema of
$f(x, y) = x^2 - 4xy + 2y^2$ subject to
$2x - 3y = 14$.

$\nabla f = \langle 2x - 4y, \ -4x + 4y \rangle.$

Let $g(x, y) = 2x - 3y.$

Then $\nabla g = \langle 2, -3 \rangle.$

From $\nabla f = \lambda \nabla g$, $g(x, y) = 14$, we have three equations with variables x, y, λ.

(1) $2x - 4y = 2\lambda$

(2) $-4x + 4y = -3\lambda$

(3) $2x - 3y = 14.$

Add (1) and (2) to get $-2x = -\lambda$, $x = \dfrac{1}{2}\lambda$

Multiply (1) by 2 and add (2):

$-4y = \lambda$, $y = -\dfrac{1}{4}\lambda.$

From (3), $2\left(\dfrac{1}{2}\lambda\right) - 3\left(-\dfrac{1}{4}\lambda\right) = 14$

$\dfrac{7}{4}\lambda = 14$, so $\lambda = 8.$

Thus, $x = \frac{1}{2}\lambda = 4$, $y = -\frac{1}{4}\lambda = -2$.

$f(4, -2) = 56$.

If $2x - 3y = 14$, then $y = \frac{2}{3}x - \frac{14}{3}$. Thus, for large values of x, y is also large. For large values of x (and, therefore, y) $f(x, y)$ is negative. For large negative x, $f(x, y)$ is negative. There is no absolute minimum; 56 is the absolute maximum.

2) Find the system of equations for solving:
Maximize
$f(x, y, z) = x^2 + 6xy^2 + 3xz^2 + yz^3$ subject to $x + y + z^2 = 2$ and $x^2 + y + z = 3$.

$\nabla f = \langle 2x + 6y^2 + 3z^2, 12xy + z^3, 6xz + 3yz^2 \rangle$.
Let $g_1(x, y, z) = x + y + z^2$, $\nabla g_1 = \langle 1, 1, 2z \rangle$.
Let $g_2(x, y, z) = x^2 + y + z$, $\nabla g_2 = \langle 2x, 1, 1 \rangle$.

The system of five equations with variables x, y, z, λ_1 and λ_2 from $\nabla f = \lambda_1 \nabla g_1 + \lambda_2 \nabla g_2$, $g_1(x, y, z) = 2$, $g_2(x, y, z) = 3$ is:

$$2x + 6y^2 + 3z^2 = \lambda_1 + 2x\lambda_2$$
$$12xy + z^3 = \lambda_1 + \lambda_2$$
$$6xz + 3yz^2 = 2z\lambda_1 + \lambda_2$$
$$x + y + z^2 = 2$$
$$x^2 + y + z = 3.$$

3) Find the volume of the largest rectangular box in the first octant with three faces in the coordinate planes and one vertex in the plane $3x + y + 2z = 12$.

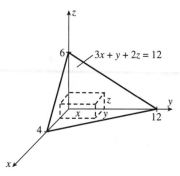

We need to find the maximum volume of the box, $V = xyz$, subject to $3x + y + 2z = 12$.

Let $g(x, y, z) = 3x + y + 2z$.
$\nabla V(x, y, z) = \langle yz, xz, xy \rangle$
$\nabla g(x, y, z) = \langle 3, 1, 2 \rangle$.

Thus, $\nabla f = \lambda \nabla g$ becomes:
$$yz = 3\lambda$$
$$xz = \lambda$$
$$xy = 2\lambda.$$

Multiply by x, y, z respectively:
$$xyz = 3x\lambda$$
$$xyz = y\lambda$$
$$xyz = 2z\lambda.$$

Therefore, $3x\lambda = y\lambda = 2z\lambda$ and for $\lambda \neq 0$:
$$3x = y = 2z.$$
Then $3x + y + 2z = 12$ becomes
$y + y + y = 12$, so $y = 4$.

Thus, $3x = 4$, so $x = \frac{4}{3}$. From $2z = 4$, $z = 2$.

The maximum volume occurs when the box is $\frac{4}{3}$ by 4 by 2. The maximum volume is

$$\left(\frac{4}{3}\right)(4)(2) = \frac{32}{3}.$$

B. Technology Plus. Use a computer algebra system or a graphing calculator to solve.

T-1) a) Use Lagrange multipliers to solve "maximize xy subject to $\dfrac{x^2}{4} + y^2 = 1$." Use $x \geq 0$, $y \geq 0$.

For $f(x, y) = xy$, $\nabla f = \langle y, x \rangle$.

For $g(x, y) = \dfrac{x^2}{4} + y^2$, $\nabla g = \left\langle \dfrac{x}{2}, 2y \right\rangle$.

From $\nabla f = \lambda \nabla g$,

(1) $y = \lambda\left(\dfrac{x}{2}\right)$

(2) $x = \lambda(2y)$.

Multiply (1) by x and (2) by y:
$$xy = \lambda\frac{x^2}{2}$$
$$xy = \lambda 2y^2$$

Thus, $\lambda\dfrac{x^2}{2} = \lambda 2y^2$. If $\lambda = 0$ then we have a minimum.

Hence, $\dfrac{x^2}{2} = 2y^2$, $x = 4y^2$, and $x = 2y$.

(Note $x = -2y$ would result in a minimum.)
$$\frac{(2y)^2}{4} + y^2 = 1$$
$$2y^2 = 1, \text{ so } y = \frac{1}{\sqrt{2}}.$$

Therefore, $x = 2\left(\dfrac{1}{\sqrt{2}}\right) = \sqrt{2}$ and the maximum of xy is $(\sqrt{2})\left(\dfrac{1}{\sqrt{2}}\right) = 1$.

b) Confirm your results in part a) by graphing on the same screen:

$$\frac{x^2}{4} + y^2 = 1 \text{ and}$$

$xy = k$ for $k = 0.33, 0.67, 1, 1.33,$ and 1.67.

For what value of k does $xy = k$ just touch the ellipse?

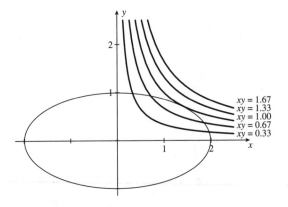

$k = 1$. The curve $xy = 1$ is the only curve tangent to the ellipse. Also note that 1 is the maximum value of xy subject to

$$\frac{x^2}{4} + y^2 = 1.$$

Chapter 15 — Multiple Integrals

Section 15.1   Double Integrals of Rectangles

Recall that definite integrals are the limits of Riemann sums and may be used to determine the area under a curve. This section defines double integrals as the limits of "double" Riemann sums and uses them to find the volume of a solid under a surface.

Concepts to Master

A. Double Riemann sum; Double integral; Integrable

B. Midpoint Rule; Average value of a function over a rectangle

C. Interpretation of the double integral as a volume

D. Technology Plus

Summary and Focus Questions

Page 998 (ET Page 974)*

A. For a function $z = f(x, y)$ defined on a rectangular region $R = [a, b] \times [c, d] = \{(x, y) \mid a \leq x \leq b, c \leq y \leq d\}$, divide the interval $[a, b]$ into m subintervals of width $\Delta x = \dfrac{b - a}{m}$ and the interval $[c, d]$ into n subintervals of width $\Delta y = \dfrac{d - c}{n}$. This divides the region R into mn subrectangles R_{ij}, each of which has area $\Delta A = \Delta x \Delta y$. For each i, j where $1 \leq i \leq m$ and $1 \leq j \leq n$, choose a sample point (x_{ij}^*, y_{ij}^*) from the subrectangle R_{ij}. The resulting **double Riemann sum** is

$$\sum_{i=1}^{m} \sum_{j=1}^{n} f(x_{ij}^*, y_{ij}^*) \Delta A.$$

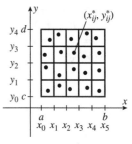

R, divided into 20 subrectangles

The **double integral of f over R** is the limit of Riemann sums:

$$\iint_R f(x, y)\, dA = \lim_{m,n \to \infty} \sum_{i=1}^{m} \sum_{j=1}^{n} f(x_{ij}^*, y_{ij}^*) \Delta A.$$

If this limit exists, then f is **integrable over R.**

---

*When using the Early Transcendentals text, use the page number in parentheses.

Example: Estimate $\displaystyle\iint\limits_{R} 2xy^2\, dA$, where R is the rectangle $[1, 4] \times [0, 2]$.

Let $f(x, y) = 2xy^2$. If we partition $[1, 4]$ into three subintervals $[1, 2]$, $[2, 3]$, and $[3, 4]$, and partition the interval $[0, 2]$ into two subintervals $[0, 1]$ and $[1, 2]$, then we have divided the rectangle $[1, 4] \times [0, 2]$ into six subrectangles. Choose the upper-right corner of each rectangle for sample points (x_{ij}^*, y_{ij}^*).

Since $\Delta x = 1$ and $\Delta y = 1$, $\Delta A = 1$.

R_{ij}	(x_{ij}^*, y_{ij}^*)	$f(x_{ij}^*, y_{ij}^*)$	$f(x_{ij}^*, y_{ij}^*)\Delta A$
$R_{11} = [1, 2] \times [0, 1]$	$(2, 1)$	4	4
$R_{12} = [1, 2] \times [1, 2]$	$(2, 2)$	16	16
$R_{21} = [2, 3] \times [0, 1]$	$(3, 1)$	6	6
$R_2 = [2, 3] \times [1, 2]$	$(3, 2)$	24	24
$R_{31} = [3, 4] \times [0, 1]$	$(4, 1)$	8	8
$R_{32} = [3, 4] \times [1, 2]$	$(4, 2)$	32	32
			90

The Riemann sum for $\displaystyle\iint\limits_{R} 2xy^2\, dA$ is 90.

1) Find the Riemann sum for $f(x, y) = x^2 + y^2$ over $R = [1, 9] \times [-1, 3]$ with $m = 4$ and $n = 2$. Use the left side midpoint of each subrectangle.

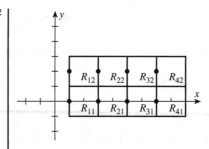

$\Delta A = \Delta x \Delta y = 2(2) = 4.$

R_{ij}	(x_{ij}^*, y_{ij}^*)	$f(x_{ij}^*, y_{ij}^*)$	$f(x_{ij}^*, y_{ij}^*)\Delta A$
R_{11}	$(1, 0)$	1	4
R_{12}	$(1, 2)$	5	20
R_{21}	$(3, 0)$	9	36
R_{22}	$(3, 2)$	13	52
R_{31}	$(5, 0)$	25	100
R_{32}	$(5, 2)$	29	116
R_{41}	$(7, 0)$	49	196
R_{42}	$(7, 2)$	53	212
			736

The Riemann sum is 736.

2) For a given set of subrectangles, what choice for (x_{ij}^*, y_{ij}^*) yields the largest Riemann sum

for $\displaystyle\iint_R e^{x-y}\,dA$?

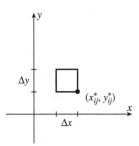

e^{x-y} is largest for large x and small y. Choose the lower-right corner of R_{ij} for (x_{ij}^*, y_{ij}^*).

Page 1002 (ET Page 978)

B. If we choose one x_i from each subinterval of $[a, b]$, one y_j from each subinterval of $[c, d]$ and let $(x_{ij}^*, y_{ij}^*) = (x_i, y_j)$ for all i and j, then the Riemann sum has a simpler form:

$$\sum_{i=1}^{m}\sum_{j=1}^{n} f(x_i, y_j)\,\Delta A.$$

This applies to a version of the Midpoint Rule for double integrals.

Midpoint Rule If $\bar{x}_i$ is the midpoint of $[x_{i-1}, x_i]$ for $i = 1, 2, \ldots, m$ and $\bar{y}_j$ is the midpoint of $[y_{j-1}, y_j]$ for $j = 1, 2, \ldots, n$, then:

$$\iint_R f(x,y)\,dA \approx \sum_{i=1}^{m}\sum_{j=1}^{n} f(\bar{x}_i, \bar{y}_j)\,\Delta A.$$

The **average value** of $f(x,y)$ over a rectangular region R, $a \le x \le b$, $c \le y \le d$, is:

$$f_{ave} = \frac{1}{(b-a)(d-c)}\iint_R f(x,y)\,dA.$$

3) Use the Midpoint Rule with $m = 2$ and $n = 3$ to approximate:

$\displaystyle\iint_R xy\,dA$, where R is $[1, 2] \times [2, 5]$.

$\Delta x = 0.5,\ \Delta y = 1,\ \Delta A = 0.5.$

$(\bar{x}_i, \bar{y}_j)$	$f(\bar{x}_i, \bar{y}_j)$	$f(\bar{x}_i, \bar{y}_j)\Delta A$
(1.25,2.5)	3.125	1.5625
(1.25,3.5)	4.375	2.1875
(1.25,4.5)	5.625	2.8125
(1.75,2.5)	4.375	2.1875
(1.75,3.5)	6.125	3.0625
(1.75,4.5)	7.875	3.9375
		15.7500

The Riemann sum is 15.75.

4) If $R = [1, 2] \times [3, 5]$ and

$\displaystyle\iint_R (x+y)\,dA = 11$, what is the average value of $f(x,y) = x + y$ over R?

$$f_{ave} = \frac{1}{(2-1)(5-3)}(11) = \frac{11}{2}.$$

Page 1000 (ET Page 976)

C. Let $R = [a, b] \times [c, d]$. For $f(x, y) \geq 0$, the double integral $\iint\limits_{R} f(x, y)\, dA$ may be interpreted as the **volume** of the solid above R and under the surface $z = f(x, y)$.

Think of the solid as a rectangular box with an irregular top formed by the surface.

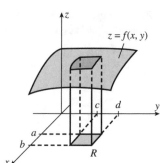

Some properties of double integrals include:

$$\iint\limits_{R} [f(x, y) + g(x, y)]\, dA = \iint\limits_{R} f(x, y)\, dA + \iint\limits_{R} g(x, y)\, dA.$$

$$\iint\limits_{R} [f(x, y) - g(x, y)]\, dA = \iint\limits_{R} f(x, y)\, dA - \iint\limits_{R} g(x, y)\, dA.$$

$$\iint\limits_{R} cf(x, y)\, dA = c \iint\limits_{R} f(x, y)\, dA, \text{ for } c \text{ a constant.}$$

If $f(x, y) \geq g(x, y)$ for all $(x, y) \in R$, then $\iint\limits_{R} f(x, y)\, dA \geq \iint\limits_{R} g(x, y)\, dA.$

5) Write a double integral expression for the volume of the pictured solid.

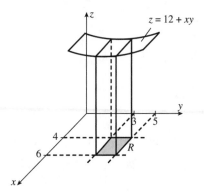

Let R be the rectangular region $4 \leq x \leq 6$, $3 \leq y \leq 5$. The volume pictured is

$$\iint\limits_{R} (12 + xy)\, dA.$$

6) Let $R = [1, 4] \times [5, 9]$. Evaluate $\iint\limits_{R} 2\, dA.$

$\iint\limits_{R} 2\, dA$ is the volume of the solid under $f(x, y) = 2$. The solid is a rectangular block with a 3 by 4 base and height 2.

$$\iint\limits_{R} 2\, dA = 3(4)2 = 24.$$

7) True or False:

a) $\iint\limits_R xy\,dA = \iint\limits_R x\,dA \iint\limits_R y\,dA.$

False.

b) $\iint\limits_R 6xy\,dA = 6\iint\limits_R xy\,dA.$

True.

c) If $R_1 = [1,2] \times [6,7]$,
$R_2 = [2,3] \times [7,8]$, and
$R_3 = [1,3] \times [6,8]$, then

$$\iint\limits_{R_3} xy\,dA = \iint\limits_{R_1} xy\,dA + \iint\limits_{R_2} xy\,dA.$$

False. The region R_3 is composed of more than just R_1 and R_2.

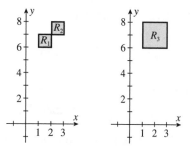

D. Technology Plus. Use a computer algebra system or a graphing calculator to solve.

T-1) Estimate $\displaystyle\int_2^4 \int_1^5 (x + xy + y)\,dA$ using the Midpoint Rule with $\Delta x = 0.5$ and $\Delta y = 1$.

Let $f(x,y) = x + xy + y$.
$\Delta x = 0.5$
$\Delta y = 1$
$\Delta A = 0.5$

Compute $f(\overline{x}_i, \overline{y}_j)$ using a spreadsheet or a table:

f	1.5	2.5	3.5	4.5
2.25	7.125	10.375	13.625	16.875
2.75	8.375	12.125	15.875	19.625
3.25	9.625	13.875	18.125	22.375
3.75	10.875	15.625	20.375	25.125

with $\overline{y}_j$ labeling the column headers and $\overline{x}_i$ labeling the rows.

$$\sum_{i=1}^4 \sum_{j=1}^4 f(\overline{x}_i, \overline{y}_j) = 240.$$

$$\sum_{i=1}^4 \sum_{j=1}^4 f(\overline{x}_i, \overline{y}_j)\Delta A = 120.$$

Section 15.2 Iterated Integrals

This section includes Fubini's Theorem—a method for calculating a double integral as two successive single "partial" integrals using the Fundamental Theorem of Calculus. Just as we needed to hold a variable constant when performing partial differentiation, we will need to hold a variable constant for the "partial integration" in each step.

Concepts to Master

Iterated integrals; Fubini's Theorem (double integrals written as iterated integrals)

Summary and Focus Questions

Page 1006 (ET Page 982)

An **iterated integral** for the function f over the region $R = [a, b] \times [c, d]$ is defined as

$$\int_a^b \int_c^d f(x, y)\, dy\, dx = \int_a^b \left[\int_c^d f(x, y)\, dy \right] dx.$$

This is evaluated by first performing the partial integration of $f(x, y)$ with respect to y (holding x constant) and then integrating that result with respect to x.

Example: $\displaystyle \int_1^2 \int_0^1 (2xy + x)\, dy\, dx$

$$= \int_1^2 \left[\int_0^1 (2xy + x)\, dy \right] dx = \int_1^2 \left[(xy^2 + xy) \right]_0^1 dx$$

$$= \int_1^2 (x + x - 0)\, dx = \int_1^2 2x\, dx = x^2 \Big]_1^2 = 4 - 1 = 3.$$

The iterated integral with the order of the differentials reversed is

$$\int_c^d \int_a^b f(x, y)\, dA = \int_c^d \left[\int_a^b f(x, y)\, dx \right] dy.$$

This one is evaluated by first integrating with respect to x and then integrating the result with respect to y.

The next theorem says the two iterated integrals are equal and either one may be used to calculate the double integral of $f(x, y)$ over R.

Fubini's Theorem If f is continuous on $R = [a, b] \times [c, d]$, then

$$\iint\limits_R f(x, y)\, dA = \int_a^b \int_c^d f(x, y)\, dy\, dx = \int_c^d \int_a^b f(x, y)\, dx\, dy.$$

Example: Evaluate $\iint\limits_{R}(2xy+3y^2)\,dA$, where R is the rectangle $[1,2]\times[0,1]$.

The two iterated integrals for the double integral are:

$$\int_0^1\int_1^2(2xy+3y^2)\,dx\,dy = \int_0^1(x^2y+3xy^2)\Big]_1^2\,dy = \int_0^1((4y+6y^2)-(y+3y^2))\,dy$$

$$= \int_0^1(3y+3y^2)\,dy = \left(\frac{3}{2}y^2+y^3\right)\Big]_0^1 = \left(\frac{3}{2}+1\right)-(0) = \frac{5}{2}.$$

$$\int_1^2\int_0^1(2xy+3y^2)\,dy\,dx = \int_1^2(xy^2+y^3)\Big]_0^1\,dx = \int_1^2((x+1)-0)\,dx$$

$$= \int_1^2(x+1)\,dx = \left(\frac{1}{2}x^2+x\right)\Big]_1^2 = \left(\frac{4}{2}+2\right)-\left(\frac{1}{2}+1\right) = \frac{5}{2}.$$

Therefore, $\iint\limits_{R}(2xy+3y^2)\,dA = \frac{5}{2}.$

1) Evaluate:

a) $\int_1^3\int_0^2(x^2+4y)\,dy\,dx.$

$$\int_1^3\left[\int_0^2(x^2+4y)\,dy\right]dx$$

$$= \int_1^3\left[(x^2y+2y^2)\Big]_0^2\right]dx$$

$$= \int_1^3(2x^2+8)\,dx$$

$$= \left(\frac{2}{3}x^3+8x\right)\Big]_1^3 = \frac{100}{3}.$$

b) $\int_0^2\int_0^3 8xy\,dx\,dy.$

$$\int_0^2\int_0^3 8xy\,dx\,dy = \int_0^2\left[\int_0^3 8xy\,dx\right]dy$$

$$= \int_0^2\left(4x^2y\Big]_0^3\right)dy = \int_0^2 36y\,dy$$

$$= 18y^2\Big]_0^2 = 72-0 = 72.$$

c) $\int_0^3\int_0^2 8xy\,dy\,dx.$

72. This is the iterated integral in part b), written in reverse order. By Fubini's Theorem, these iterated integrals are the same.

2) Write an iterated integral for:

a) $\iint\limits_{R}(x^2+y^2)\,dA$ where R is the rectangle $0\le x\le 4, 3\le y\le 5.$

$$\int_0^4\int_3^5(x^2+y^2)\,dy\,dx \text{ or } \int_3^5\int_0^4(x^2+y^2)\,dx\,dy.$$

b) the volume of the following solid:

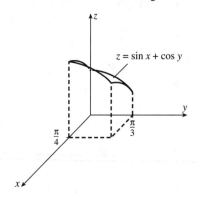

$z = \sin x + \cos y$

$\dfrac{\pi}{4}$ $\dfrac{\pi}{3}$

$$\int_0^{\pi/4} \int_0^{\pi/3} (\sin x + \cos y)\, dy\, dx \text{ or}$$

$$\int_0^{\pi/3} \int_0^{\pi/4} (\sin x + \cos y)\, dx\, dy.$$

3) Find $\displaystyle\iint\limits_R (xy + e^x)\, dA$, where
$R = [0, 1] \times [0, 2]$.

$$\iint\limits_R (xy + e^x)\, dA = \int_0^2 \int_0^1 (xy + e^x)\, dx\, dy$$

$$= \int_0^2 \left(\frac{x^2 y}{2} + e^x \right) \Big]_0^1 dy$$

$$= \int_0^2 \left(\frac{y}{2} + e - 1 \right) dy$$

$$= \left(\frac{y^2}{4} + (e - 1)y \right) \Big]_0^2$$

$$= 2e - 1.$$

4) True or False:
$$\int_0^3 \int_0^4 f(x, y)\, dy\, dx = \int_0^3 \int_0^4 f(x, y)\, dx\, dy.$$

False. There are a few functions for which this is true. For example,
$$\int_0^3 \int_0^4 4xy\, dy\, dx = \int_0^3 \int_0^4 4xy\, dx\, dy = 144.$$
However, in general this is false.

Section 15.3   Double Integrals over General Regions

This section extends the idea of a double integral to bounded regions other than rectangles. Fubini's Theorem may be applied to evaluate them using iterated integrals.

Concepts to Master

A. Evaluating a double integral over a general region as an iterated integral

B. Changing the order of integration of an iterated integral

C. Properties of double integrals

D. Technology Plus

Summary and Focus Questions

Page 1012 (ET Page 988)

A. Let D be a region in the plane and $f(x, y)$ be continuous on D. If D may be described in either of the following ways, then $\iint_D f(x, y) \, dA$ may be evaluated using an iterated integral:

Type I

$$y = g_2(x)$$
$$D$$
$$y = g_1(x)$$
$$a \qquad b$$

$$a \le x \le b$$
$$g_1(x) \le y \le g_2(x)$$
$$\iint_D f(x, y) \, dA = \int_a^b \int_{g_1(x)}^{g_2(x)} f(x, y) \, dy \, dx.$$

Type II

$$x = h_2(y)$$
$$d$$
$$D$$
$$c$$
$$x = h_1(y)$$

$$c \le y \le d$$
$$h_1(y) \le x \le h_2(y)$$
$$\iint_D f(x, y) \, dA = \int_c^d \int_{h_1(y)}^{h_2(y)} f(x, y) \, dx \, dy.$$

The choice of type depends on how you decide to describe the boundaries of D. In some cases, both types are appropriate.

Example: Evaluate $\iint\limits_{D} 6x^2 y\, dA$, where D is the region in the figure at the right.

The region D is bounded by $y = 3$, $y = 3x$, and $x = 2$ and may be described both as Type I and Type II.

Type I: The upper boundary of D is the "top" function $y = 3x$ and the lower boundary or "bottom" function for D is $y = 3$. The region D is $1 \leq x \leq 2$, $3 \leq y \leq 3x$.

Therefore, $\iint\limits_{D} 6x^2 y\, dA = \int_{1}^{2}\int_{3}^{3x} 6x^2 y\, dy\, dx$

$$= \int_{1}^{2} 3x^2 y^2 \Big]_{3}^{3x} dx = \int_{1}^{2}(3x^2(3x)^2 - 3x^2(3)^2)\, dx$$

$$= \int_{1}^{2}(27x^4 - 27x^2)\, dx = \left(\frac{27}{5}x^5 - 9x^3\right)\Big]_{1}^{2}$$

$$= \left(\frac{27}{5}2^5 - 9(2^3)\right) - \left(\frac{27}{5}1^5 - 9(1^3)\right) = \frac{522}{5}.$$

Type II: From $y = 3x$, $x = \frac{1}{3}y$. Thus, the "left" (lower) boundary of D is $x = \frac{1}{3}y$ and the upper boundary or "right" function is $x = 2$. The region D is $3 \leq y \leq 6, \frac{1}{3}y \leq x \leq 2$.

Therefore, $\iint\limits_{D} 6x^2 y\, dA = \int_{3}^{6}\int_{\frac{1}{3}y}^{2} 6x^2 y\, dx\, dy = \int_{3}^{6} 2x^3 y \Big]_{\frac{1}{3}y}^{2} dy$

$$= \int_{3}^{6}\left(2(2^3)y - 2\left(\frac{1}{3}y\right)^3 y\right) dy = \int_{3}^{6}\left(16y - \frac{2}{27}y^4\right) dy = \left(8y^2 - \frac{2}{27(5)}y^5\right)\Big]_{3}^{6}$$

$$= \left(8(6)^2 - \frac{2}{27(5)}(6^5)\right) - \left(8(3)^2 - \frac{2}{27(5)}(3^5)\right) = \frac{522}{5}.$$

1) Write each as an iterated integral:

a) $\iint\limits_{D} xy^2\, dA$, where D is the following region.

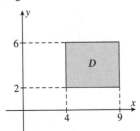

$\int_{2}^{6}\int_{4}^{9} xy^2\, dx\, dy$ or $\int_{4}^{9}\int_{2}^{6} xy^2\, dy\, dx$.

b) $\iint_D e^{xy} \, dA$, where D is the following region.

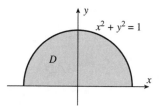

The top of the region is bounded by the semicircle. From $x^2 + y^2 = 1$,

$$y = \sqrt{1 - x^2}.$$

The region may be described as Type I:

$$-1 \le x \le 1.$$
$$0 \le y \le \sqrt{1 - x^2}.$$

This integral is $\int_{-1}^{1} \int_{0}^{\sqrt{1-x^2}} e^{xy} \, dy \, dx$.

c) $\iint_D (x + y) \, dA$, where D is the following region.

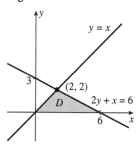

Describing D as Type I would be difficult (the top function would be defined piecewise). It is easier to describe D as Type II. The left function is $h_1(y) = y$ and from $2y + x = 6$, the right function is $h_2(y) = 6 - 2y$. The double integral is

$$\int_0^2 \left[\int_y^{6-2y} (x + y) \, dx \right] dy.$$

2) Evaluate

a) $\int_1^2 \int_y^{y^2} 4xy \, dx \, dy$.

$$\int_1^2 \int_y^{y^2} 4xy \, dx \, dy = \int_1^2 2x^2 y \Big]_y^{y^2} dy$$

$$= \int_1^2 \left[2(y^2)^2 y - 2y^2 y \right] dy$$

$$= \int_1^2 (2y^5 - 2y^3) \, dy$$

$$= \left(\frac{y^6}{3} - \frac{y^4}{2} \right) \Big]_1^2$$

$$= \left(\frac{64}{3} - \frac{16}{2} \right) - \left(\frac{1}{3} - \frac{1}{2} \right) = \frac{27}{2}.$$

b) $\int_0^{\pi/6} \int_0^x \cos(x + y) \, dy \, dx$

$$\int_0^{\pi/6} \int_0^x \cos(x + y) \, dy \, dx = \int_0^{\pi/6} \sin(x + y) \Big]_0^x \, dx$$

$$= \int_0^{\pi/6} (\sin 2x - \sin x) \, dx = \left(-\frac{1}{2} \cos 2x + \cos x \right) \Big]_0^{\pi/6}$$

$$= \left(-\frac{1}{2} \cos \frac{\pi}{3} + \cos \frac{\pi}{6} \right) - \left(-\frac{1}{2} \cos 0 + \cos 0 \right)$$

$$= \left(-\frac{1}{4} + \frac{\sqrt{3}}{2} \right) - \left(-\frac{1}{2} + 1 \right) = \frac{2\sqrt{3} - 3}{4}.$$

Page
1015
(ET Page
991)

B. An iterated integral corresponds to a double integral over a region D. If D may be described as both Type I and Type II it is possible to write the given form of the iterated integral in reverse order.

Example: Rewrite $\int_0^3 \int_0^{6-2x} (x+y)\, dy\, dx$ in the form $\iint \cdots dx\, dy$.

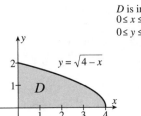

$\int_0^3 \int_0^{6-2x} (x+y)\, dy\, dx$ corresponds to the Type I region

$D: 0 \le x \le 3,\ 0 \le y \le 6 - 2x$.

From $y = 6 - 2x$, $x = 3 - \dfrac{y}{2}$.

D may be described as a Type II region: $0 \le y \le 6,\ 0 \le x \le 3 - \dfrac{y}{2}$.

Therefore, $\int_0^3 \int_0^{6-2x} (x+y)\, dy\, dx = \int_0^6 \int_0^{3-y/2} (x+y)\, dx\, dy$.

3) Change the order of integration of

$\int_0^4 \int_0^{\sqrt{4-x}} xy\, dy\, dx$.

D is initially:
$0 \le x \le 4$
$0 \le y \le \sqrt{4-x}$

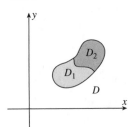

The given iterated integral corresponds to the double integral $\iint_D xy\, dA$, where D is shown in the figure. From $y = \sqrt{4-x}$, $y^2 = 4 - x$, so $x = 4 - y^2$.
D may be written as a Type II region:
$0 \le y \le 2$
$0 \le x \le 4 - y^2$.
The double integral is $\int_0^2 \int_0^{4-y^2} xy\, dx\, dy$.

Page
1017
(ET Page
993)

C. If a region D may be described as the union of two regions D_1 and D_2 that do not overlap, except perhaps at their boundaries, then:

$$\iint_D f(x,y)\, dA = \iint_{D_1} f(x,y)\, dA + \iint_{D_2} f(x,y)\, dA.$$

In particular, if D is neither Type I or II, it may be possible to partition D into subregions of these types. Then the double integral of D is the sum of the double integrals over the subregions.

Example: Find $\displaystyle\iint_D (2x + 4y)\,dA$, where D is the shaded "bow tie" region at the right.

The region D is neither Type I nor Type II. However, $D = D_1 \cup D_2$, with D_1 being the part of D left of the y-axis and D_2 the part to the right of the y-axis. Both D_1 and D_2 are Type I regions:

$$D_1:\ -1 \le x \le 0 \qquad\qquad D_2:\ 0 \le x \le 2$$
$$x + 1 \le y \le 1 \qquad\qquad\qquad 1 \le y \le x + 1.$$

$$\iint_D (2x + 4y)\,dA = \iint_{D_1} (2x + 4y)\,dA + \iint_{D_2} (2x + 4y)\,dA$$

$$= \int_{-1}^{0}\int_{x+1}^{1} (2x + 4y)\,dy\,dx + \int_{0}^{2}\int_{1}^{x+1} (2x + 4y)\,dy\,dx$$

$$= \int_{-1}^{0} (2xy + 2y^2)\Big]_{x+1}^{1}\,dx + \int_{0}^{2} (2xy + 2y^2)\Big]_{1}^{x+1}\,dx$$

$$= \int_{-1}^{0} (-4x^2 - 4x)\,dx + \int_{0}^{2} (4x^2 + 4x)\,dx$$

$$= \left(-\frac{4}{3}x^3 - 2x^2\right)\Big]_{-1}^{0} + \left(\frac{4}{3}x^3 + 2x^2\right)\Big]_{0}^{2} = \frac{2}{3} + \frac{56}{3} = \frac{58}{3}.$$

The area of a region D in the plane may be written as $\displaystyle\iint_D 1\,dA$, which may be evaluated using iterated integrals.

If f is integrable over D and $m \le f(x, y) \le M$ for all $(x, y) \in D$, then $mA \le \displaystyle\iint_D f(x, y)\,dA \le MA$, where A is the area of D.

4) Write an iterated integral for $\displaystyle\iint_D x^2 y^2\,dA$ where:

   a) the region D is bounded by
   $y = 5x$, $y = x$, $x + y = 6$.

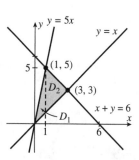

D may be divided into two Type I regions by the line $x = 1$.

D_1 is $0 \le x \le 1$, $x \le y \le 5x$.
D_2 is $1 \le x \le 3$, $x \le y \le 6 - x$.
The integral is

$$\int_0^1 \int_x^{5x} x^2 y^2 dy\, dx + \int_1^3 \int_x^{6-x} x^2 y^2\, dy\, dx.$$

b) the region D is the first three quadrants of the unit circle.

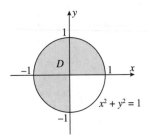

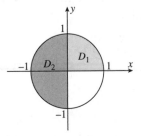

Let D_1 be the first quadrant (Type I):
$0 \le x \le 1$, $0 \le y \le \sqrt{1 - x^2}$.

Let D_2 be the second and third quadrants (Type II):
$-1 \le y \le 1, -\sqrt{1 - y^2} \le x \le \sqrt{1 - y^2}$.

The integral is

$$\int_0^1 \int_0^{\sqrt{1-x^2}} x^2 y^2 dy\, dx + \int_{-1}^1 \int_{-\sqrt{1-y^2}}^{\sqrt{1-y^2}} x^2 y^2\, dx\, dy.$$

5) Use an iterated integral to find the shaded area.

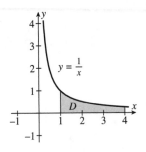

D may be described as $1 \le x \le 4$,

$0 \le y \le \dfrac{1}{x}$. The area is

$$\iint_D 1\, dA = \int_1^4 \int_0^{1/x} 1\, dy\, dx = \int_1^4 y\Big]_0^{1/x}\, dx$$

$$= \int_1^4 \frac{1}{x}\, dx = \ln x\big]_1^4 = \ln 4.$$

6) True or False:
$$\int_0^1 \int_3^5 x^2 y\, dy\, dx$$
$$= \int_0^1 \int_3^4 x^2 y\, dy\, dx + \int_0^1 \int_4^5 x^2 y\, dy\, dx.$$

True.

7) Let $D = \{(x,y)| 1 \le x \le y^2, 1 \le y \le 2\}$.

Using $1 \le x^2 y \le 32$, find bounds on

$\iint\limits_{D} x^2 y \, dA.$

$m = 1$ and $M = 32$.

The area of D is

$$\iint\limits_{D} dx \, dy = \int_1^2 \int_1^{y^2} dx \, dy$$

$$= \int_1^2 x \Big]_1^{y^2} dy$$

$$= \int_1^2 (y^2 - 1) dy = \left(\frac{y^3}{3} - y\right)\Big]_1^2$$

$$= \frac{4}{3}.$$

Thus,

$$1\left(\frac{4}{3}\right) \le \iint\limits_{D} x^2 y \, dA \le 32\left(\frac{4}{3}\right)$$

$$\frac{4}{3} \le \iint\limits_{D} x^2 y \, dA \le \frac{128}{3}.$$

D. Technology Plus. Use a computer algebra system or a graphing calculator to solve.

T-1) Draw the solid bounded by

$0 \le x \le \dfrac{\pi}{2},\ 0 \le y \le x,\ 0 \le z \le \sin x \cos y.$

Then find the volume of the solid.

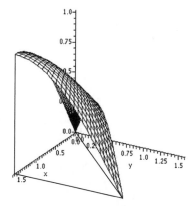

The volume is $\displaystyle\int_0^{\pi/2} \int_0^x \sin x \cos y \, dy \, dx$

$$= \int_0^{\pi/2} (\sin x \sin y) \Big]_0^x dx$$

$$= \int_0^{\pi/2} \sin^2 x \, dx$$

$$= \left(\frac{x}{2} - \frac{\sin x \cos x}{2}\right)\Big]_0^{\pi/2} = \frac{\pi}{4}.$$

T-2) Use a CAS to help write $\iint\limits_{R} xy\,dA$ as an iterated integral, where R is the larger region bounded by $x = 0$, $y = x^2$ and $x = 9 - y^2$.

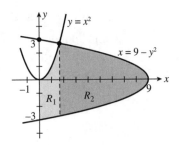

The region R is neither Type I nor Type II, but dividing it as indicated gives two regions, R_1 of Type I and R_2 of Type II.

The curves intersect when

$$x = 9 - (x^2)^2$$
$$x^4 + x - 9 = 0.$$

Using CAS, $x \approx 1.65$. Then $y \approx 2.71$.
R_1 is $0 \le x \le 1.65$,
$$-\sqrt{9 - x} \le y \le x^2$$

R_2 is $-2.71 \le y \le 2.71$,
$$1.65 \le x \le 9 - y^2.$$

The double integral is

$$\iint\limits_{R_1} xy\,dA + \iint\limits_{R_2} xy\,dA$$

$$= \int_{0}^{1.65} \int_{-\sqrt{9-x}}^{x^2} xy\,dy\,dx + \int_{-2.71}^{2.71} \int_{1.65}^{9-y^2} xy\,dx\,dy.$$

Section 15.4 Double Integrals in Polar Coordinates

This section shows you how to change a double integral into an iterated integral when the region of integration is described with polar coordinates.

Concept to Master

Write a double integral as an iterated integral in polar coordinates

---

Summary and Focus Questions

Page 1021 (ET Page 997)

Let D be a region in the plane and $f(x,y)$ be continuous on D. If D may be described using polar coordinates ($x = r\cos\theta$, $y = r\sin\theta$) in either of the following ways, then $\iint\limits_{D} f(x,y)\,dA$ may be converted to an iterated integral in polar coordinates:

Type I

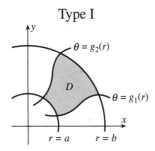

$$a \leq r \leq b$$
$$g_1(r) \leq \theta \leq g_2(r)$$
$$\iint\limits_{D} f(x,y)\,dA$$
$$= \int_a^b \int_{g_1(r)}^{g_2(r)} f(r\cos\theta, r\sin\theta)\, r\, d\theta\, dr.$$

Type II

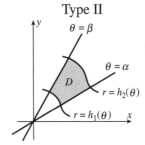

$$\alpha \leq \theta \leq \beta$$
$$h_1(\theta) \leq r \leq h_2(\theta)$$
$$\iint\limits_{D} f(x,y)\,dA$$
$$= \int_\alpha^\beta \int_{h_1(\theta)}^{h_2(\theta)} f(r\cos\theta, r\sin\theta)\, r\, dr\, d\theta.$$

Don't forget to include the additional factor of r in the integrand.

Example: Find $\iint\limits_{D} 4(x^2 + y^2)\,dA$, where D is shaded the shaded region at the right.

The region D may be described with polar coordinates:

$$2\theta \leq r \leq 4, \frac{\pi}{4} \leq \theta \leq \frac{\pi}{3}.$$

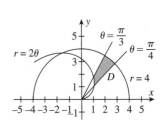

In polar coordinates, the double integral is

$$\int_{\pi/4}^{\pi/3} \int_{2\theta}^{4} 4[(r\cos\theta)^2 + (r\sin\theta)^2] r\,dr\,d\theta = \int_{\pi/4}^{\pi/3} \int_{2\theta}^{4} 4r^3(\cos^2\theta + \sin^2\theta)dr d\theta = \int_{\pi/4}^{\pi/3} \int_{2\theta}^{4} 4r^3\,dr\,d\theta$$

$$= \int_{\pi/4}^{\pi/3} r^4 \Big]_{2\theta}^{4} d\theta = \int_{\pi/4}^{\pi/3} (256 - 16\theta^4)\,d\theta = \left(256\theta - \frac{16}{5}\theta^5\right)\Big]_{\pi/4}^{\pi/3}$$

$$= \left(256\frac{\pi}{3} - \frac{16}{5}\left(\frac{\pi}{3}\right)^5\right) - \left(256\frac{\pi}{4} - \frac{16}{5}\left(\frac{\pi}{4}\right)^5\right) = \frac{64\pi}{3} - \frac{16\pi^5}{1215} + \frac{\pi^5}{320} \approx 63.9.$$

1) Rewrite using polar coordinates.

a) $\iint\limits_{D} x\,dA$, where D is the following region:

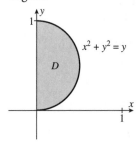

$x^2 + y^2 = y$

D

Converting $x^2 + y^2 = y$ to a polar equation gives $r^2 = r\sin\theta$, so $r = \sin\theta$.

The region D is Type II:

$$0 \le \theta \le \frac{\pi}{2}, 0 \le r \le \sin\theta.$$

The double integral is

$$\int_{0}^{\pi/2} \int_{0}^{\sin\theta} (r\cos\theta)r\,dr\,d\theta$$

(remember the "extra" r factor)

$$= \int_{0}^{\pi/2} \int_{0}^{\sin\theta} r^2 \cos\theta\,dr\,d\theta.$$

b) $\int_{0}^{4} \int_{0}^{x} xy\,dy\,dx.$

The corresponding region D is given by

$$0 \le x \le 4, 0 \le y \le x.$$

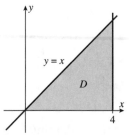

$y = x$

D

D may be described as a Type II region in polar coordinates:

$$0 \le \theta \le \frac{\pi}{4}, 0 \le r \le 4\sec\theta.$$

(The line $x = 4$ is $r\cos\theta = 4$, or $r = 4\sec\theta$.)

The double integral is

$$\int_{0}^{\pi/4} \int_{0}^{4\sec\theta} (r\cos\theta)(r\sin\theta)r\,dr\,d\theta$$

$$= \int_{0}^{\pi/4} \int_{0}^{4\sec\theta} \frac{r^3 \sin 2\theta}{2}\,dr\,d\theta.$$

2) Write a polar iterated integral for the
volume under $z = x^2 + y^2$ and above the
region D given here.

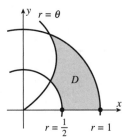

The region is Type I:
$$\frac{1}{2} \le r \le 1, \ 0 \le \theta \le r.$$

The volume under $z = x^2 + y^2$ is

$$\iint\limits_D (x^2 + y^2)\, dA = \int_{1/2}^1 \int_0^r r^2 r\, d\theta\, dr$$

$$= \int_{1/2}^1 \int_0^r r^3\, d\theta\, dr.$$

Section 15.5 Applications of Double Integrals

This section includes some applications of double integrals to physical phenomena and to probability distributions involving two variables.

Concepts to Master

A. Mass of a lamina; Center of mass; Moments of inertia

B. Probabilities involving joint density functions; Expected values

Summary and Focus Questions

Page 1027
(ET Page 1003)

A. Recall that a **lamina** is a flat planar area D representing a distribution of matter. In Section 8.3, with the density assumed constant throughout D, a moment about an axis was defined to be the tendency of the lamina to rotate about that axis.

Suppose the lamina has varying density $\rho(x, y)$ at each point (x, y) in D. If ρ is continuous on D, then:

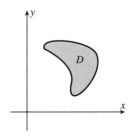

Mass of the lamina is:	Moment with respect to the *x*-axis is:	Moment with respect to the *y*-axis is:
$m = \displaystyle\iint_D \rho(x, y)\, dA$	$M_x = \displaystyle\iint_D y\rho(x, y)\, dA$	$M_y = \displaystyle\iint_D x\rho(x, y)\, dA$

Mass is measured in units of kilograms or pounds. The units for moments of inertia are kg · m or lb · ft.

The **center of mass** is $(\bar{x}, \bar{y})$, where $\bar{x} = \dfrac{M_y}{m}$ and $\bar{y} = \dfrac{M_x}{m}$.

Example: Find the center of mass for the triangular lamina at the right whose density at (x, y) is $\rho(x, y) = x + y$ kg/m². The region may be described as:

$$0 \le x \le 1$$
$$0 \le y \le x.$$

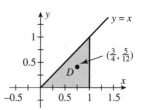

$$m = \iint_D (x + y)\, dA = \int_0^1 \int_0^x (x + y)\, dy\, dx$$

$$= \int_0^1 \left(xy + \frac{y^2}{2} \right)\Bigg]_0^x dx = \int_0^1 \left(x^2 + \frac{x^2}{2} \right) dx = \int_0^1 \frac{3x^2}{2}\, dx = \frac{x^3}{2}\Bigg]_0^1 = \frac{1}{2} \text{ kg.}$$

$$M_x = \iint_D y(x+y)\, dA = \int_0^1 \int_0^x y(x+y)\, dy\, dx = \int_0^1 \left(\frac{xy^2}{2} + \frac{y^3}{3} \right) \Big]_0^x dx$$

$$= \int_0^1 \frac{5x^3}{6}\, dx = \frac{5x^4}{24} \Big]_0^1 = \frac{5}{24}\ \text{kg} \cdot \text{m}.$$

$$M_y = \iint_D x(x+y)\, dA = \int_0^1 \int_0^x x(x+y)\, dy\, dx = \int_0^1 \left(x^2 y + \frac{xy^2}{2} \right) \Big]_0^x dx$$

$$= \int_0^1 \frac{3x^3}{2}\, dx = \frac{3x^4}{8} \Big]_0^1 = \frac{3}{8}\ \text{kg} \cdot \text{m}.$$

Therefore, $\bar{x} = \dfrac{\frac{3}{8}}{\frac{1}{2}} = \dfrac{3}{4}$ and $\bar{y} = \dfrac{\frac{5}{24}}{\frac{1}{2}} = \dfrac{5}{12}$. This is the balance point of the lamina D. It is closer to $(1, 1)$ than the center of mass found in Chapter 8 because the density is not uniform.

Moments of inertia (or **second moments**) measure the effort necessary to overcome rotational inertia. Think of a spinning skater: the moment of inertia about the axis of rotation lessens (he or she spins faster with less effort) as the arms are drawn inward.

Moment of inertia about the x-axis is:	**Moment of inertia about the y-axis is:**	**Moment of inertia about the origin is:**
$I_x = \iint_D y^2 \rho(x, y)\, dA$	$I_y = \iint_D x^2 \rho(x, y)\, dA$	$I_0 = I_x + I_y$

The units for moments of inertia about axes are $\text{kg} \cdot \text{m}^2$ or $\text{lb} \cdot \text{ft}^2$.

1) Find the mass of the following lamina, whose density at each point (x, y) is $72xy$.

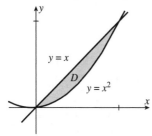

D may be described by $0 \le x \le 1$, $x^2 \le y \le x$.

$$m = \iint_D 72xy\, dA = \int_0^1 \int_{x^2}^x 72xy\, dy\, dx$$

$$= \int_0^1 36xy^2 \Big]_{x^2}^x dx$$

$$= \int_0^1 (36x^3 - 36x^5)\, dx$$

$$= (9x^4 - 6x^6) \Big]_0^1 = 3.$$

2) Find the y-coordinate of the center of mass of the lamina in question 1.

$$M_x = \iint_D y(72xy)dA = \iint_D 72xy^2 \, dA$$

$$= \int_0^1 \int_{x^2}^x 72xy^2 \, dy \, dx$$

$$= \int_0^1 (24x^4 - 24x^7) \, dx = \frac{9}{5}.$$

$$\bar{y} = \frac{M_x}{m} = \frac{9/5}{3} = \frac{3}{5}.$$

3) Find an iterated integral for the moment of inertia about the y-axis of the lamina in question 1.

$$I_y = \iint_D x^2(72xy) \, dA = \int_0^1 \int_{x^2}^x 72x^3 y \, dy \, dx = \left(\frac{3}{2}\right).$$

B. A **joint density function** f is a function of two variables x and y such that:

1) $f(x,y) \geq 0$.

Page 1032 (ET Page 1008)

2) $\iint_{\mathbb{R}^2} f(x,y)dA = 1$.

3) the probability that (x, y) lies in a region D is:

$$P((x,y) \in D) = \iint_D f(x,y)dA.$$

If x and y are **independent variables,** then there are functions f_1 and f_2 so that:

$$f(x,y) = (f_1(x))(f_2(y)).$$

The **X-mean** for a joint density function f is $\mu_1 = \iint_{\mathbb{R}^2} x f(x,y) \, dA$.

The **Y-mean** for a joint density function f is $\mu_2 = \iint_{\mathbb{R}^2} y f(x,y) \, dA$.

The X-mean and Y-mean are also called the **expected values** of X and Y, respectively.

4) Let $f(x,y) = \begin{cases} \dfrac{e^{x+y}}{(e-1)^2} & 0 \leq x \leq 1 \\ & 0 \leq y \leq 1 \\ 0 & \text{otherwise.} \end{cases}$

    a) Show that f is a joint probability function.

First, note that $f(x,y) \geq 0$ for all x and y.

$$\iint_{\mathbb{R}^2} f(x,y)dA = \int_0^1 \int_0^1 \frac{e^{x+y}}{(e-1)^2} \, dy \, dx$$

$$= \int_0^1 \left(\frac{e^{x+y}}{(e-1)^2} \right]_0^1 \, dx = \int_0^1 \frac{e^{x+1} - e^x}{(e-1)^2} \, dx$$

$$= \frac{e^{x+1} - e^x}{(e-1)^2} \right]_0^1 \, dx = \frac{e^2 - e - (e-1)}{(e-1)^2}$$

$$= 1.$$

b) Are x and y independent?

Yes, because we may write $f(x, y)$ as

$$\frac{e^{x+y}}{(e-1)^2} = \left(\frac{e^x}{e-1}\right)\left(\frac{e^y}{e-1}\right).$$

c) Find $P\left(0 \le x \le \frac{1}{2}, 0 \le y \le \frac{1}{3}\right)$.

$$\int_0^{1/2}\int_0^{1/3} \frac{e^{x+y}}{(e-1)^2}\, dy\, dx = \int_0^1 \left(\frac{e^{x+y}}{(e-1)^2}\right)\Bigg]_0^{1/3} dx$$

$$= \int_0^1 \frac{e^{x+1/3} - e^x}{(e-1)^2}\, dx = \frac{e^{x+1/3} - e^x}{(e-1)^2}\Bigg]_0^{1/2} dx$$

$$= \frac{e^{5/6} - e^{1/2} - (e^{1/3} - e^0)}{(e-1)^2}$$

$$= \frac{e^{5/6} - e^{1/2} - e^{1/3} + 1}{(e-1)^2} \approx 0.087.$$

d) Find the X-mean.

$$\mu_1 = \iint_{\mathbb{R}^2} x\, f(x, y)\, dA = \int_0^1\int_0^1 \frac{xe^{x+y}}{(e-1)^2}\, dy\, dx$$

$$= \int_0^1 \left(\frac{xe^{x+y}}{(e-1)^2}\right)\Bigg]_0^1 dx = \int_0^1 \frac{x(e^{x+1} - e^x)}{(e-1)^2}\, dx$$

$$= \int_0^1 \frac{xe^x(e-1)}{(e-1)^2}\, dx$$

$$= \int_0^1 \frac{xe^x}{e-1}\, dx = \text{(integration by parts)}$$

$$= \frac{xe^x - e^x}{e-1}\Bigg]_0^1 = \frac{e^1 - e^1}{e-1} - \frac{0-1}{e-1}$$

$$= \frac{1}{e-1} \approx 0.582.$$

Section 15.6 Surface Area

In Chapter 8 we found the areas of surfaces of revolution. In this section we find the areas of more general surfaces.

Concept to Master

Area of a surface in three dimensions

Summary and Focus Questions

Page
1307
(ET Page
1013)

The **area** of a surface S given by a function $z = f(x, y)$, for $(x, y) \in D$, with continuous partial derivatives is:

$$A(S) = \iint_D \sqrt{1 + \left[f_x(x, y) \right]^2 + \left[f_y(x, y) \right]^2} \, dA = \iint_D \sqrt{1 + \left(\frac{\partial z}{\partial x} \right)^2 + \left(\frac{\partial z}{\partial y} \right)^2} \, dA.$$

It helps to remember this formula by noting the similarity of this equation with the equation for the length of an arc (one lower dimension).

Example: Find the area of the surface given by $f(x, y) = x^{\frac{3}{2}} + \frac{\sqrt{5}}{2} y$ that is above the region D shown at the right.

The region D may be described as:

$$0 \le x \le 1$$
$$0 \le y \le 1 + x.$$

The partials of f are:

$$f_x = \frac{3}{2} x^{\frac{1}{2}} \text{ and } f_y = \frac{\sqrt{5}}{2}.$$

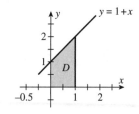

$$A(S) = \int_0^1 \int_0^{1+x} \sqrt{1 + \left(\frac{3}{2} x^{\frac{1}{2}} \right)^2 + \left(\frac{\sqrt{5}}{2} \right)^2} \, dy \, dx = \int_0^1 \int_0^{1+x} \sqrt{1 + \frac{9}{4} x + \frac{5}{4}} \, dy \, dx$$

$$= \int_0^1 \int_0^{1+x} \frac{3}{2} \sqrt{1 + x} \, dy \, dx = \int_0^1 \frac{3}{2} y \sqrt{1 + x} \Big]_0^{1+x} \, dx$$

$$= \int_0^1 \frac{3}{2} (1 + x)^{\frac{3}{2}} \, dx = \frac{3}{2} \frac{2}{5} (1 + x)^{\frac{5}{2}} \Big]_0^1 = \frac{3}{5} \left(2^{\frac{5}{2}} - 1^{\frac{5}{2}} \right) = \frac{12\sqrt{2} - 3}{5}.$$

1) Find the area of that part of the surface $z = 4 - x^2 - y^2$ that lies in the first octant.

The surface S is shown in the figure and lies above the region D given by
$$0 \le x \le 2$$
$$0 \le y \le \sqrt{4 - x^2}.$$

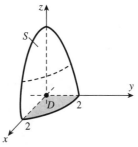

The surface area of S is

$$\int_0^2 \int_0^{\sqrt{4-x^2}} \sqrt{1 + (-2x)^2 + (-2y)^2} \, dy \, dx$$

$$= \int_0^2 \int_0^{\sqrt{4-x^2}} \sqrt{1 + 4x^2 + 4y^2} \, dy \, dx.$$

To evaluate this integral, switch to polar coordinates. Then D is
$$0 \le r \le 2,$$
$$0 \le \theta \le \pi/2.$$
$$\sqrt{1 + 4(r\cos\theta)^2 + 4(r\sin\theta)^2} = \sqrt{1 + 4r^2}.$$
When we rewrite the integral, the area is

$$\int_0^2 \int_0^{\frac{\pi}{2}} r\sqrt{1 + 4r^2} \, d\theta \, dr$$

$$= \int_0^2 r\sqrt{1 + 4r^2}\, \theta \Big|_0^{\frac{\pi}{2}} \, dr = \frac{\pi}{2} \int_0^2 r\sqrt{1 + 4r^2}\, dr$$

$$= \frac{\pi}{24}(1 + 4r^2)^{\frac{3}{2}} \Big|_0^2 = \frac{\pi}{24}\left(17^{\frac{3}{2}} - 1\right).$$

$$\approx 9.044.$$

Section 15.7 Triple Integrals

This section extends all the concepts of double integrals to functions of three variables.

Concepts to Master

A. Triple Integrals; Fubini's Theorem for triple integrals

B. Applications of the triple integral as volume, mass, center of mass, and moment of inertia

---

Summary and Focus Questions

A. A **triple integral** of a function $f(x, y, z)$ over a region E that is a solid rectangular block given by $a \leq x \leq b, c \leq y \leq d, r \leq z \leq s$ in a three-dimensional space is defined as a limit of Riemann sums:

Page 1041
(ET Page 1017)

$$\iiint_E f(x, y, z)\,dV = \lim_{l,m,n\to\infty} \sum_{i=1}^{l}\sum_{j=1}^{m}\sum_{k=1}^{n} f(x_{ijk}^*, y_{ijk}^*, z_{ijk}^*)\Delta V,$$

where we partition E into subboxes having equal volume $\Delta V = \Delta x \Delta y \Delta z$, and $(x_{ijk}^*, y_{ijk}^*, z_{ijk}^*)$ is chosen from the ijk subbox.

If we choose one x_i from each subinterval of $[a, b]$, one y_j from each subinterval of $[c, d]$, one z_k from each subinterval of $[r, s]$ and let $(x_{ijk}^*, y_{ijk}^*, z_{ijk}^*) = (x_i, y_j, z_k)$ for all i, j, and k, then we can write the simpler Riemann sum form:

$$\sum_{i=1}^{l}\sum_{j=1}^{m}\sum_{k=1}^{n} f(x_i, y_j, z_k)\Delta V.$$

The triple integral may be evaluated by **Fubini's Theorem**:

$$\iiint_E f(x, y, z)\,dV = \int_r^s \int_c^d \int_a^b f(x, y, z)\,dx\,dy\,dz.$$

This is evaluated by holding y and z constant and integrating with respect to x, and then continuing the calculation as a double iterated integral. Moreover, the triple integral may be evaluated by any of five other iterated

integrals, such as $\int_r^s \int_a^b \int_c^d f(x, y, z)\,dy\,dx\,dz$, $\int_c^d \int_r^s \int_a^b f(x, y, z)\,dx\,dz\,dy$,

$\int_a^b \int_c^d \int_r^s f(x, y, z)\,dz\,dy\,dx$, and so on. Just be sure that the limits of integration and the differential symbols match up.

Example: If $E = \{(x, y, z) \mid 0 \le x \le 2, 1 \le y \le 3, 0 \le z \le 1\}$, then:

$$\iiint_E xyz \, dV = \int_0^1 \int_1^3 \int_0^2 xyz \, dx \, dy \, dz = \int_0^1 \int_1^3 \frac{1}{2} x^2 yz \Big]_0^2 dy \, dz$$

$$= \int_0^1 \int_1^3 2yz \, dy \, dz = \int_0^1 y^2 z \Big]_1^3 dz$$

$$= \int_0^1 8z \, dz = 4z^2 \Big]_0^1 = 4 - 0 = 4.$$

More generally, if E can be described by $a \le x \le b$, $g_1(x) \le y \le g_2(x)$, $h_1(x, y) \le z \le h_2(x, y)$, then:

$$\iiint_E f(x, y, z) \, dV = \int_a^b \int_{g_1(x)}^{g_2(x)} \int_{h_1(x,y)}^{h_2(x,y)} f(x, y, z) \, dz \, dy \, dx.$$

Just like Fubini's Theorem, this is evaluated "from the inside out"; that is, integrate with respect to z and substitute the z-limits of integration, then integrate with respect to y and substitute the y-limits, and, finally, integrate with respect to x and substitute the x-limits.

Example: Evaluate $\int_0^1 \int_0^{y+2} \int_y^{x+y} (2z - x) \, dz \, dx \, dy.$

$$\int_0^1 \int_0^{y+2} \int_y^{x+y} (2z - x) \, dz \, dx \, dy = \int_0^1 \int_0^{y+2} (z^2 - zx) \Big]_y^{x+y} dx \, dy$$

$$= \int_0^1 \int_0^{y+2} \left[\left((x+y)^2 - (x+y)x \right) - \left(y^2 - yx \right) \right] dx \, dy$$

$$= \int_0^1 \int_0^{y+2} 2xy \, dx \, dy = \int_0^1 x^2 y \Big]_0^{y+2} dy = \int_0^1 \left[(y+2)^2 y - 0^2 y \right] dy$$

$$= \int_0^1 (y^3 + 4y^2 + 4y) dy = \left(\frac{1}{4} y^4 + \frac{4}{3} y^3 + 2y^2 \right) \Big]_0^1 = \left(\frac{1}{4} + \frac{4}{3} + 2 \right) - 0 = \frac{43}{12}.$$

1) Find the missing limits of integration on the second triple integral:

$$\int_1^4 \int_2^5 \int_3^7 xyz \, dx \, dz \, dy = \iiint xyz \, dz \, dy \, dx.$$

$$\int_1^4 \int_2^5 \int_3^7 xyz \, dx \, dz \, dy = \int_3^7 \int_1^4 \int_2^5 xyz \, dz \, dy \, dx.$$

2) Evaluate $\int_1^2 \int_1^y \int_0^{x+y} 12x\, dz\, dx\, dy.$

$\int_1^2 \int_1^y \int_0^{x+y} 12x\, dz\, dx\, dy$

$$= \int_1^2 \int_1^y \left(12xz \big]_0^{x+y} \right) dx\, dy$$

$$= \int_1^2 \int_1^y (12x^2 + 12xy)\, dx\, dy$$

$$= \int_1^2 (4x^3 + 6x^2 y)\big]_1^y\, dy$$

$$= \int_1^2 (4y^3 + 6y^3) - (4 + 6y)\, dy$$

$$= \int_1^2 (10y^3 - 6y - 4)\, dy$$

$$= \left(\frac{10}{4} y^4 - 3y^2 - 4y \right)\Big]_1^2 = \frac{49}{2}.$$

3) Write two iterated integrals for $\iiint\limits_E x\, dV$, where E is the region in the first octant bounded by the plane $x + 2y + 4z = 8$.

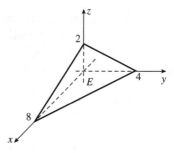

There are six possible iterated integrals. The limits of integration depend on how the solid E is described.

a) First write E as
$0 \le z \le 2$
$0 \le y \le 4 - 2z$
$0 \le x \le 8 - 2y - 4z.$

$$\iiint\limits_E x\, dV = \int_0^2 \int_0^{4-2z} \int_0^{8-2y-4z} x\, dx\, dy\, dz.$$

b) Write E as
$0 \le x \le 8$
$0 \le y \le 4 - \dfrac{x}{2}$
$0 \le z \le 2 - \dfrac{x}{4} - \dfrac{y}{2}.$

$$\iiint\limits_E x\, dV = \int_0^8 \int_0^{4-\frac{x}{2}} \int_0^{2-\frac{x}{4}-\frac{y}{2}} x\, dz\, dy\, dx.$$

4) Write $\displaystyle\iiint\limits_{E} (x+y)\,dV$ as an iterated integral, where E is the solid ball of radius 2 about the origin.

The region E is the solid sphere $x^2 + y^2 + z^2 = 4$. E may be described by:

$$-2 \le x \le 2$$
$$-\sqrt{4-x^2} \le y \le \sqrt{4-x^2}$$
$$-\sqrt{4-x^2-y^2} \le z \le \sqrt{4-x^2-y^2}.$$

$$\iiint\limits_{E} (x+y)\,dV$$
$$= \int_{-2}^{2} \int_{-\sqrt{4-x^2}}^{\sqrt{4-x^2}} \int_{-\sqrt{4-x^2-y^2}}^{\sqrt{4-x^2-y^2}} (x+y)\,dz\,dy\,dx.$$

5) Write $\displaystyle\int_0^1 \int_0^y \int_0^{1-y} y^2 \, dx\,dz\,dy$ as an iterated integral in the form $\displaystyle\int\int\int y^2 \, dz\,dy\,dx.$

The given iterated integral corresponds to the space
$$0 \le y \le 1,\ 0 \le z \le y,\ 0 \le x \le 1 - y.$$
The space for the given iterated integral is the following shaded solid.

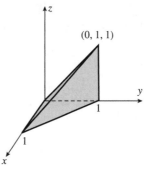

For the integral's differentials to have the order $dz\,dy\,dx$, the space needs to be described with constant limits bounding x, limits on y involving just x, and then limits on z involving x and y. Therefore, we use:

$$0 \le x \le 1$$
$$0 \le y \le 1 - x$$
$$0 \le z \le y.$$

The desired iterated integral is
$$\int_0^1 \int_0^{1-x} \int_0^y y^2 \, dz\,dy\,dx.$$

Page
1046
(ET Page
1022)

B. The volume of a solid E may be found by evaluating $\iiint\limits_{E} 1\,dV$.

If a solid E has a continuous density $\rho(x, y, z)$, the **mass of solid** is:

$$m = \iiint\limits_{E} \rho(x, y, z)\,dV.$$

The three **moments** about the coordinate planes are:

Moment about xy-plane: **Moment about xz-plane:**   **Moment about yz-plane:**

$$M_{xy} = \iiint\limits_{E} z\rho(x,y,z)\,dV \qquad M_{xz} = \iiint\limits_{E} y\rho(x,y,z)\,dV \qquad M_{yz} = \iiint\limits_{E} x\rho(x,y,z)\,dV$$

The **moments of inertia** about the axes are:

Moment about x-axis: **Moment about y-axis:**   **Moment about z-axis:**

$$I_x = \iiint\limits_{E} (y^2 + z^2)\rho(x,y,z)\,dV \qquad I_y = \iiint\limits_{E} (x^2 + z^2)\rho(x,y,z)\,dV \qquad I_z = \iiint\limits_{E} (x^2 + y^2)\rho(x,y,z)\,dV$$

6) Find a triple iterated integral for the volume inside the cone $z^2 = x^2 + y^2$ bounded by $z = 0$, $z = 8$.

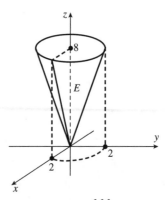

The volume is $\iiint\limits_{E} 1\,dV$, where the solid E may be described as

$$0 \le z \le 8$$
$$-\frac{z}{4} \le y \le \frac{z}{4}$$
$$-\sqrt{z^2 - y^2} \le x \le \sqrt{z^2 - y^2}.$$

Then $\iiint\limits_{E} 1\,dV = \int_0^8 \int_{-z/4}^{z/4} \int_{-\sqrt{z^2 - y^2}}^{\sqrt{z^2 - y^2}} 1\,dx\,dy\,dz.$

7) A solid E has density xyz at each point (x, y, z) in E and is described as bounded by the parabolic cylinder $x = z^2$ and the planes $y = 0$, $x = 0$, $y = 5$, $z = 2$.

   a) Find the z-coordinate of the center mass of E.

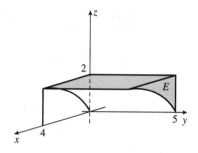

The solid E may be described as
$$0 \le y \le 5$$
$$0 \le z \le 2$$
$$0 \le x \le z^2.$$

The mass $m = \iiint\limits_E xyz \, dV$

$$= \int_0^5 \int_0^2 \int_0^{z^2} xyz \, dx \, dz \, dy = \frac{200}{3}.$$

$$M_{xy} = \iiint\limits_E z(xyz) \, dV$$

$$= \int_0^5 \int_0^2 \int_0^{z^2} xyz^2 \, dx \, dz \, dy = \frac{800}{7}.$$

$$\bar{z} = \frac{M_{xy}}{m} = \frac{\frac{800}{7}}{\frac{200}{3}} = \frac{12}{7} \approx 1.71.$$

   b) Find the moment of inertia about the x-axis.

$$I_x = \iiint\limits_E (y^2 + z^2) xyz \, dV$$

$$= \int_0^5 \int_0^2 \int_0^{z^2} (xy^3 z + xyz^3) \, dx \, dz \, dy$$

$$= \frac{3100}{3} \approx 1033.3.$$

Section 15.8 Triple Integrals in Cylindrical Coordinates

The next two sections describe two other coordinate systems for three-dimensional space. In this section cylindrical coordinates are simply polar coordinates in the xy-plane with a z-axis added. The concept of double integrals in polar coordinates is extended to triple integrals in cylindrical coordinates.

Concepts to Master

A. Cylindrical coordinates; Conversion from xyz-coordinates to cylindrical coordinates and vice versa

B. Triple integrals as iterated integrals in cylindrical coordinates

Summary and Focus Questions

Page
1051
(ET Page
1027)

A. The **cylindrical coordinates** (r, θ, z) of a point P in space are polar coordinates (r, θ) in the xy-plane and the usual z-coordinate. They are called cylindrical because the equation $r = a$ is a circular cylinder about the z-axis.

If the standard three-dimensional coordinate system is placed upon the cylindrical coordinate system as in the figure, then to change from cylindrical coordinates to rectangular:

$$x = r\cos\theta$$
$$y = r\sin\theta$$
$$z = z.$$

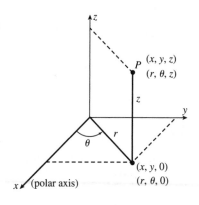

To change from rectangular coordinates to cylindrical:

$$r^2 = x^2 + y^2$$
$$\tan\theta = \frac{y}{x}$$
$$z = z.$$

Examples: a) Plot the point $(2, \frac{\pi}{3}, 4)$ in cylindrical coordinates and convert to rectangular coordinates. The point is plotted by locating 2 on the polar axis, turning an angle $\frac{\pi}{3}$, then going upward 4 units.

$$x = 2 \, \cos\frac{\pi}{3} = 2\left(\frac{1}{2}\right) = 1.$$

$$y = 2 \, \sin\frac{\pi}{3} = 2\left(\frac{\sqrt{3}}{2}\right) = \sqrt{3}.$$

The point has rectangular coordinates $(1, \sqrt{3}, 4)$.

b) Plot the point $(1, 1, 3)$ in rectangular coordinates and convert to cylindrical coordinates.

$r^2 = 1^2 + 1^2 = 2$, so $r = \sqrt{2}$.

$\tan\theta = \dfrac{1}{1} = 1$, so $\theta = \dfrac{\pi}{4}$.

$z = 3$.

The point has cylindrical coordinates $\left(\sqrt{2}, \dfrac{\pi}{4}, 3\right)$.

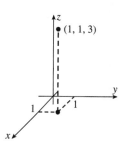

1) Plot each of these points in cylindrical coordinates:

$A\,(2, 0, 2)$

$B\left(3, -\dfrac{\pi}{2}, -1\right)$

$C\left(2, \dfrac{\pi}{4}, 3\right)$

$D\left(4, \dfrac{\pi}{3}, 5\right)$

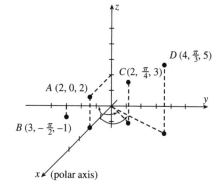

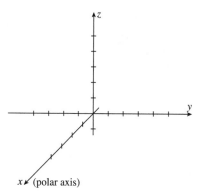

2) Find the indicated coordinates of each point:

a) Rectangular: $(2, 2\sqrt{3}, 6)$.

Cylindrical: _____.

$\left(4, \dfrac{\pi}{3}, 6\right)$.

$r^2 = 2^2 + (2\sqrt{3})^2 = 16$, $r = 4$.

$\tan\theta = \dfrac{2\sqrt{3}}{2} = \sqrt{3}$, $\theta = \dfrac{\pi}{3}$.

$z = 6$.

b) Cylindrical: $\left(4, \dfrac{\pi}{4}, 3 \right)$.

  Rectangular: _____.

$(2\sqrt{2},\, 2\sqrt{2},\, 3)$.

$$x = 4\cos\dfrac{\pi}{4} = 4\dfrac{\sqrt{2}}{2} = 2\sqrt{2}.$$

$$y = 4\sin\dfrac{\pi}{4} = 4\dfrac{\sqrt{2}}{2} = 2\sqrt{2}.$$

$$z = 3.$$

3) Identify the surface with equation:

  a) $z^2 = r^2$ in cylindrical coordinates.

Switch to rectangular: $z^2 = x^2 + y^2$.
This is an elliptic cone.

  b) $\theta = \dfrac{\pi}{3}$.

This is a plane through the z-axis along the
line $\theta = \dfrac{\pi}{3}$ in the xy-plane (the $r\theta$-plane).

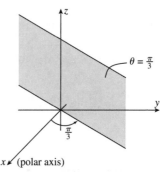

B. If f is continuous on a solid region E and E may be described in cylindrical
coordinates by $\alpha \le \theta \le \beta,\ h_1(\theta) \le r \le h_2(\theta),\ u_1(x,y) \le z \le u_2(x,y)$, then:

Page
1053
(ET Page
1029)

$$\iiint\limits_{E} f(x,y,z)\,dV = \int_{\alpha}^{\beta}\int_{h_1(\theta)}^{h_2(\theta)}\int_{u_1(r\cos\theta,\, r\sin\theta)}^{u_2(r\cos\theta,\, r\sin\theta)} f(r\cos\theta,\, r\sin\theta,\, z)\,r\,dz\,dr\,d\theta.$$

Consistent with polar coordinates, the integrand has an additional factor of r.

4) Evaluate $\displaystyle\iiint\limits_{E} 2\,dV$ where E is the solid

  between the paraboloids:

  $z = 4x^2 + 4y^2$

  $z = 80 - x^2 - y^2$.

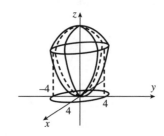

The paraboloids intersect in the circle
$4x^2 + 4y^2 = 80 - x^2 - y^2$ in the xy-plane.
This simplifies to $x^2 + y^2 = 16$.

E may be described in cylindrical
coordinates as
$0 \le \theta \le 2\pi$
$0 \le r \le 4$.
Since $4x^2 + 4y^2 \le z \le 80 - x^2 - y^2$,
$4r^2 \le z \le 80 - r^2$.

$$\iiint\limits_{E} 2\, dV = \int_0^{2\pi}\int_0^4\int_{4r^2}^{80-r^2} 2r\, dz\, dr\, d\theta$$

$$= \int_0^{2\pi}\int_0^4 2zr \Big]_{4r^2}^{80-r^2} dr\, d\theta$$

$$= \int_0^{2\pi}\int_0^4 (160r - 10r^3)\, dr\, d\theta$$

$$= \int_0^{2\pi} \left(80r^2 - \frac{5}{2}r^4\right)\Big]_0^4 d\theta$$

$$= \int_0^{2\pi} 640\, d\theta = 640\theta \Big]_0^{2\pi} = 1280\pi.$$

Section 15.9 Triple Integrals in Spherical Coordinates

Spherical coordinates are a generalization of polar coordinates to three dimensions, where a distance and two angles are necessary to specify a point. They provide another way to describe solid regions and evaluate triple integrals.

Concepts to Master

A. Spherical coordinates; Conversions between rectangular, cylindrical and spherical coordinates

B. Triple integrals as iterated integrals in spherical coordinates

Summary and Focus Questions

Page 1057 (ET Page 1033)

A. The **spherical coordinates** (ρ, θ, ϕ) of a point P in space are defined as:

ρ = distance from P to the origin.

θ = angle between the positive x-axis and point in xy-plane (same as cylindrical coordinates).

ϕ = angle between the positive z-axis and the line from P to the origin.

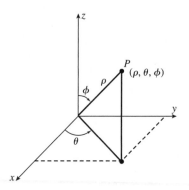

These are called spherical coordinates because the equation $\rho = a$ is a sphere about the origin with radius a.

We now have three different coordinate systems for three-dimensional space. The following table shows how to convert coordinates from one system to the others.

From	To Rectangular (x, y, z)	To Cylindrical (r, θ, z)	To Spherical (ρ, θ, ϕ)
Rectangular (x, y, z)		$r^2 = x^2 + y^2$ $\tan\theta = \dfrac{y}{x}$ $z = z$	$\rho^2 = x^2 + y^2 + z^2$ $\tan\theta = \dfrac{y}{x}$ $\cos\phi = \dfrac{z}{\rho}$
Cylindrical (r, θ, z)	$x = r\cos\theta$ $y = r\sin\theta$ $z = z$		$\rho^2 = r^2 + z^2$ $\theta = \theta$ $\cos\phi = \dfrac{z}{\rho}$
Spherical (ρ, θ, ϕ)	$x = \rho\sin\phi\cos\theta$ $y = \rho\sin\phi\sin\theta$ $z = \rho\cos\phi$	$r = \rho\sin\phi$ $\theta = \theta$ $z = \rho\cos\phi$	

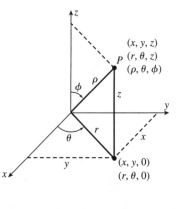

Examples: Convert the rectangular coordinates $(1, 2, 3)$ to cylindrical coordinates and to spherical coordinates.

Cylindrical:

$r^2 = 1^2 + 2^2 = 5$, so $r = \sqrt{5}$.

$\tan\theta = \dfrac{2}{1} = 2$, so $\theta = \tan^{-1}2 \approx 63.4°$.

$z = 3$.

The point has cylindrical coordinates $(\sqrt{5}, 63.4°, 3)$.

Spherical:

$\rho^2 = 1^2 + 2^2 + 3^2 = 14$, so $\rho = \sqrt{14}$.

$\theta = \tan^{-1}2 \approx 63.4°$.

$\cos\phi = \dfrac{3}{\sqrt{14}}$, $\phi = \cos^{-1}\dfrac{3}{\sqrt{14}} \approx 36.7°$.

The point has spherical coordinates $(\sqrt{14}, 63.4°, 36.7°)$.

1) Plot each of these points in spherical coordinates:

$A(4, 0, 0)$

$B\left(3, -\dfrac{\pi}{2}, \dfrac{\pi}{3}\right)$

$C\left(2, \dfrac{\pi}{4}, \dfrac{3\pi}{4}\right)$

$D\left(2, \dfrac{2\pi}{3}, \dfrac{\pi}{6}\right)$

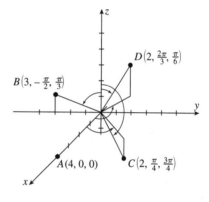

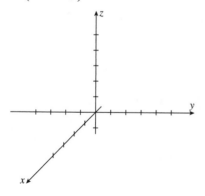

2) Find the indicated coordinates of each point:

a) Spherical: $\left(8, \dfrac{\pi}{3}, \dfrac{\pi}{6}\right)$

Rectangular: _____.

$(2, 2\sqrt{3}, 4\sqrt{3})$.

$x = 8\sin\dfrac{\pi}{6}\cos\dfrac{\pi}{3} = 8\cdot\dfrac{1}{2}\cdot\dfrac{1}{2} = 2$.

$y = 8\sin\dfrac{\pi}{6}\sin\dfrac{\pi}{3} = 2\sqrt{3}$.

$z = 8\cos\dfrac{\pi}{6} = 8\dfrac{\sqrt{3}}{2} = 4\sqrt{3}$.

b) Cylindrical: $\left(\sqrt{3}, \dfrac{\pi}{6}, 1 \right)$.

Spherical: _____.

$\left(2, \dfrac{\pi}{6}, \dfrac{\pi}{3} \right)$.

$\rho^2 = \left(\sqrt{3} \right)^2 + 1^2 = 4$, so $\rho = 2$.

$\theta = \dfrac{\pi}{6}$.

$\cos \phi = \dfrac{1}{2}$, so $\phi = \dfrac{\pi}{3}$.

c) Spherical: $\left(10, \dfrac{\pi}{12}, \dfrac{\pi}{3} \right)$.

Cylindrical: _____.

$\left(5\sqrt{3}, \dfrac{\pi}{6}, 5 \right)$.

$r = 10 \sin \dfrac{\pi}{3} = 10 \left(\dfrac{\sqrt{3}}{2} \right) = 5\sqrt{3}$.

$\theta = \dfrac{\pi}{12}$

$z = 10 \cos \dfrac{\pi}{3} = 10 \left(\dfrac{1}{2} \right) = 5$.

d) Rectangular: $(1, 2, 2)$.

Spherical: _____.

$(3, 1.107, 0.841)$.

$\rho^2 = 1^2 + 2^2 + 2^2 = 9$, $\rho = 3$.

$\tan \theta = \dfrac{2}{1}$, $\theta = \tan^{-1} 2 \approx 1.107$.

$\cos \phi = \dfrac{2}{3}$, $\phi = \cos^{-1} \dfrac{2}{3} \approx 0.841$.

3) Identify the surface with equation:

a) $\rho = \cos \phi$ in the spherical coordinates.

Switch to rectangular coordinates.

$\rho = \cos \phi$

$\rho^2 = \rho \cos \phi$

$x^2 + y^2 + z^2 = z$

$x^2 + y^2 + \left(z - \dfrac{1}{2} \right)^2 = \dfrac{1}{4}$.

This is a sphere with radius $\dfrac{1}{2}$ and center $\left(0, 0, \dfrac{1}{2} \right)$.

b) $\phi = \dfrac{\pi}{4}$

This is a cone with vertex at the origin.

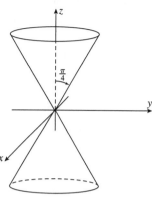

B. If f is continuous on a solid region E and E may be described in spherical coordinates by $\alpha \le \theta \le \beta$, $c \le \phi \le d$, $g_1(\theta, \phi) \le \rho \le g_2(\theta, \phi)$, then

Page
1058
(ET Page
1034)

$$\iiint\limits_{E} f(x, y, z)\, dV$$

$$= \int_c^d \int_\alpha^\beta \int_{g_1(\theta, \phi)}^{g_2(\theta, \phi)} f(\rho \sin \phi \cos \theta, \rho \sin \phi \sin \theta, \rho \cos \phi)\rho^2 \sin \phi \, d\rho \, d\theta \, d\phi.$$

Note that the iterated integral has a factor of $\rho^2 \sin \phi$.

4) Write as an iterated integral in spherical coordinates:

$\iiint\limits_{E} (10 - x^2 - y^2 - z^2)\, dV$, where E is the top half of the sphere $x^2 + y^2 + z^2 = 9$.

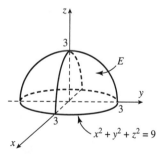

E may be described in spherical coordinates as

$0 \le \rho \le 3$

$0 \le \theta \le 2\pi$

$0 \le \phi \le \dfrac{\pi}{2}$.

$$\iiint\limits_{E} (10 - x^2 - y^2 - z^2)\, dV$$

$$= \int_0^{\pi/2} \int_0^{2\pi} \int_0^3 (10 - \rho^2)\rho^2 \sin \phi \, d\rho \, d\theta \, d\phi.$$

5) Which system, spherical or cylindrical, would be more appropriate to evaluate $\iiint\limits_{E} x\,dV$ where E is the solid between the cone $z^2 = x^2 + y^2$ and the planes $z = 0$ and $z = 4$?

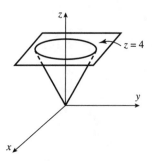

Cylindrical, since E may be written as:

$0 \leq \theta \leq 2\pi$

$0 \leq r \leq 2$

$\sqrt{x^2 + y^2} \leq z \leq 4.$

Section 15.10 Change of Variables in Multiple Integrals

For a single-variable integral we saw that substitution is an important technique for evaluating the integral when the integrand has the form $y = f(g(x))$. The key step involves determining the intermediate function $u = g(x)$ and the corresponding $du = g'(x)\,dx$. For example, to evaluate $\int (x^2 + 1)^3 2x\,dx$, let $u = x^2 + 1$ and $du = 2x\,dx$. The function $u = x^2 + 1$ may be thought of as a transformation of the reals—that is, x is transformed into $x^2 + 1$. In higher dimensions, several variables will be involved in a similar transformation. The concept corresponding to du is a determinant of partial derivatives called the Jacobian.

Concepts to Master

A. C^1 transformations; Jacobians in two and three dimensions

B. Change of variable for double and triple integrals

Summary and Focus Questions

Page 1064 (ET Page 1040)

A. A **transformation in the plane** is a function T from the uv-plane to the xy-plane
$$T(u,v) = (x, y), \text{ where } x = g(u,v) \text{ and } y = h(u,v).$$

Example: The transformation T given by $T(u,v) = (x,y)$, where $x = 2u - v$, $y = u + 3v$, maps the uv-plane to the xy-plane. We write $T(u, v) = (2u - v, u + 3v)$ and calculate some corresponding points:

	(u, v)	(x, y) = T(u, v)
A	(0, 0)	(0, 0)
B	(0, 1)	(−1, 3)
C	(2, 1)	(3, 5)
D	(2, 0)	(4, 2)

The rectangular region $S = \{(u,v) \mid 0 \le u \le 2, 0 \le v \le 1\}$ has corresponding region R, called the **image of S**. The graphs of S and R follow.

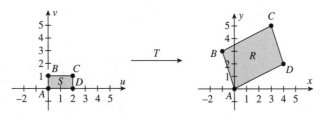

In a similar manner we may define a **transformation in space** to be a function from uvw-space to xyz-space with three component functions.

A transformation T is C^1 means that all the component functions have continuous first partial derivatives. The transformation $T(u,v) = (2u - v, u + 3v)$ previously given is C^1 because each of its four partial derivatives is constant (and therefore continuous).

For a C^1 transformation T from uv-space to xy-space, the **Jacobian of T** is the determinant

$$\frac{\partial(x,y)}{\partial(u,v)} = \begin{vmatrix} \dfrac{\partial x}{\partial u} & \dfrac{\partial x}{\partial v} \\[2mm] \dfrac{\partial y}{\partial u} & \dfrac{\partial y}{\partial v} \end{vmatrix} = \frac{\partial x}{\partial u}\frac{\partial y}{\partial v} - \frac{\partial x}{\partial v}\frac{\partial y}{\partial u}.$$

In three dimensions for a C^1 transformation T from uvw-space to xyz-space, the **Jacobian of T** is

$$\frac{\partial(x,y,z)}{\partial(u,v,w)} = \begin{vmatrix} \dfrac{\partial x}{\partial u} & \dfrac{\partial x}{\partial v} & \dfrac{\partial x}{\partial w} \\[2mm] \dfrac{\partial y}{\partial u} & \dfrac{\partial y}{\partial v} & \dfrac{\partial y}{\partial w} \\[2mm] \dfrac{\partial z}{\partial u} & \dfrac{\partial z}{\partial v} & \dfrac{\partial z}{\partial w} \end{vmatrix}.$$

Example: Find the Jacobian of the transformation T given by $T(u,v,w) = (x,y,z)$, where $x = 2u - vw$, $y = 3v + uw$, $z = w + 2uv$.

$$\frac{\partial(x,y,z)}{\partial(u,v,w)} = \begin{vmatrix} 2 & -w & -v \\ w & 3 & u \\ 2v & 2u & 1 \end{vmatrix} = (2)(3 - 2u^2) - (-w)(w - 2uv) + (-v)(2uw - 6v)$$

$$= 6 - 4u^2 + 6v^2 + w^2 - 4uvw.$$

1) Find the Jacobian of each transformation:

a) $x = 2u + v$
$y = u - 2v$

$$\frac{\partial(x,y)}{\partial(u,v)} = \begin{vmatrix} 2 & 1 \\ 1 & -2 \end{vmatrix} = (2)(-2) - (1)(1) = -5.$$

b) $x = uv$
$y = u + v$

$$\frac{\partial(x,y)}{\partial(u,v)} = \begin{vmatrix} v & u \\ 1 & 1 \end{vmatrix} = v - u.$$

c) $x = 2u + v$
$y = uw$
$z = u^2 + v^2 + w^2$

$$\frac{\partial(x,y,z)}{\partial(u,v,w)} = \begin{vmatrix} 2 & 1 & 0 \\ w & 0 & u \\ 2u & 2v & 2w \end{vmatrix}$$

$$= 2\begin{vmatrix} 0 & u \\ 2v & 2w \end{vmatrix} - 1\begin{vmatrix} w & u \\ 2u & 2w \end{vmatrix} + 0\begin{vmatrix} w & 0 \\ 2u & 2v \end{vmatrix}$$

$$= 2(-2uv) - 1(2w^2 - 2u^2) + 0$$

$$= 2u^2 - 4uv - 2w^2.$$

2) Let T be the transformation given by $x = u + v$, $y = 2u - v$. Find the image of the region

$$S = \{(u,v) \mid -1 \le u \le 1, 0 \le v \le 2\}.$$

The graph of S is:

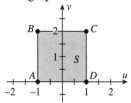

We calculate some points:

	(u, v)	(x, y)
A	(−1, 0)	(−1, −2)
B	(−1, 2)	(1, −4)
C	(1, 2)	(3, 0)
D	(1, 0)	(1, 2)

The graph of the image of S is

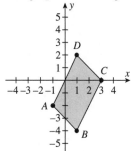

3) Find the region S that maps to the region R under the transformation T given by

$$x = \frac{2}{3}(v - u)$$

$$y = \frac{1}{3}(u + 2v).$$

R is the shaded region:

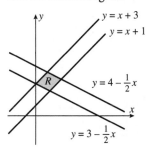

We find T^{-1} by solving for u and v.

Multiplying each of $x = \frac{2}{3}(u - v)$ and $y = \frac{1}{3}(u + 2v)$ by 3 gives:

$$3x = 2v - 2u$$
$$3y = u + 2v.$$

Subtracting the equations gives $3x - 3y = -3u$. Thus $u = y - x$.

Then $3x = 2v - 2(y - x)$, so $v = \frac{1}{2}x + y$.

The four boundary lines of R are transformed by T^{-1} to four lines in the uv-plane as follows:

xy	uv
$y = 4 - \frac{1}{2}x$	$v = 4$
$y = 3 - \frac{1}{2}x$	$v = 3$
$y = x + 3$	$u = 3$
$y = x + 1$	$u = 1$

$S = \{(u, v) \mid 1 \le u \le 3, 3 \le v \le 4\}$
$= [1, 3] \times [3, 4]$. S is mapped onto R by T.

Page
1068
(ET Page
1044)

B. The two-dimensional version of the substitution technique uses the Jacobian in the same manner as du is used in the single-variable case. Let T be a one-to-one C^1 transformation that maps a Type I or II region S in the uv-plane onto a Type I or II region R in the xy-plane. If T has a nonzero Jacobian and f is continuous, then:

$$\iint_R f(x,y)\, dA = \iint_S f(x(u,v),\, y(u,v)) \left| \frac{\partial(x,y)}{\partial(u,v)} \right| du\, dv.$$

A similar result holds for transformations in three dimensions.

4) Use question 3 to change the variable in

$$\iint_R (6x + 3y)\, dA \text{ and evaluate.}$$

Let T be the transformation:

$$x = \frac{2}{3}(v - u),\, y = \frac{1}{3}(u + 2v).$$

Let $S = \{(u, v) \mid 1 \le u \le 3,\, 3 \le v \le 4\}$
$= [1, 3] \times [3, 4]$. In question 3 we saw that $T(S) = R$—that is, S is transformed into the region R by T.

The Jacobian of T is:

$$\begin{vmatrix} \dfrac{\partial x}{\partial u} & \dfrac{\partial x}{\partial v} \\[2mm] \dfrac{\partial y}{\partial u} & \dfrac{\partial y}{\partial v} \end{vmatrix} = \begin{vmatrix} -\dfrac{2}{3} & \dfrac{2}{3} \\[2mm] \dfrac{1}{3} & \dfrac{2}{3} \end{vmatrix} = -\dfrac{4}{9} - \dfrac{2}{9} = -\dfrac{2}{3}.$$

$$\iint_R (6x + 3y)\, dA$$

$$= \iint_S \left(6\left[\frac{2}{3}(v - u)\right] + 3\left[\frac{1}{3}(u + 2v)\right]\right) \left|-\frac{2}{3}\right| du\, dv$$

$$= \iint_S (4v - 2u)\, du\, dv$$

$$= \int_3^4 \int_1^3 (4v - 2u)\, du\, dv$$

$$= \int_3^4 (4uv - u^2)\Big]_1^3\, dv$$

$$= \int_3^4 (8v - 8)\, dv$$

$$= (4v^2 - 8v)\Big]_3^4 = 20.$$

5) Use a change of variable to evaluate $\iint\limits_{R}(x^2 - y^2)\,dA$ where R is the shaded region.

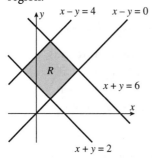

The region is bounded by lines whose equations include the expressions $x + y$ and $x - y$.

We let $u = x + y$, $v = x - y$.

Then $2 \le u \le 6$, $0 \le v \le 4$.

We must write x, y in terms of u, v:

$$u = x + y$$
$$v = x - y$$

Adding the equations, we get

$$u + v = 2x$$
$$x = \frac{u + v}{2}.$$

Substitute this result in the first equation:

$$u = \frac{u + v}{2} + y$$
$$y = \frac{u - v}{2}.$$

The integrand $f(x, y) = x^2 - y^2$

$$= \left(\frac{u + v}{2}\right)^2 - \left(\frac{u - v}{2}\right)^2 = uv.$$

The Jacobian is

$$\begin{vmatrix} \dfrac{\partial x}{\partial u} & \dfrac{\partial x}{\partial v} \\[2mm] \dfrac{\partial y}{\partial u} & \dfrac{\partial y}{\partial v} \end{vmatrix} = \begin{vmatrix} \dfrac{1}{2} & \dfrac{1}{2} \\[2mm] \dfrac{1}{2} & -\dfrac{1}{2} \end{vmatrix} = -\frac{1}{4} - \frac{1}{4} = -\frac{1}{2}.$$

Finally, make the change of variable in the double integral:

$$\iint\limits_{R}(x^2 - y^2)\,dA = \iint\limits_{S} uv\left|-\frac{1}{2}\right|\,dA$$

$$= \int_{2}^{6}\int_{0}^{4} uv\left|-\frac{1}{2}\right|\,dv\,du$$

$$= \int_{2}^{6}\int_{0}^{4}\frac{uv}{2}\,dv\,du = \int_{2}^{6}\frac{uv^2}{4}\bigg]_{0}^{4}\,du$$

$$= \int_{2}^{6} 4u\,du = 2u^2\bigg]_{2}^{6} = 64.$$

Chapter 16 — Vector Calculus

Section 16.1   Vector Fields

This section describes functions from $\mathbb{R}^2$ to $\mathbb{R}^2$ and from $\mathbb{R}^3$ to $\mathbb{R}^3$. Such functions are known as vector fields because we think of the domain elements as points in a set (a field of points in a plane or three dimensions) and range elements as two- or three-dimensional vectors associated with those points. A conservative vector field is a vector field that is the gradient of a function.

Concepts to Master

A. Vector fields in two and three dimension

B. Conservative vector fields

Summary and Focus Questions

Page 1080 (ET Page 1057)*

A. A two-dimensional **vector field** whose domain is a set D in $\mathbb{R}^2$ is a function **F** that assigns to each point (x, y) in D a vector $\mathbf{F}(x, y)$. You should think of domain elements as points and range elements as vectors. When points in D are thought of as vectors, $\mathbf{x} = \langle x, y \rangle$, we write $\mathbf{F}(\mathbf{x})$.

A two-dimensional vector field may be represented by drawing the xy-plane, choosing several points and sketching the vectors associated with those points. The vector $\mathbf{F}(x, y)$ may be written in terms of its **scalar field** component functions:

$$\mathbf{F}(x, y) = P(x, y)\mathbf{i} + Q(x, y)\mathbf{j}.$$

Example: Let $\mathbf{F}(x, y) = x\mathbf{i} + 2\mathbf{j}$. The component functions are $P(x, y) = x$ and $Q(x, y) = 2$. For the four domain points $(0, 0)$, $(1,-4)$, $(3, 1)$, and $(-1, -1)$, their associated vectors are

$$\mathbf{F}(0, 0) = 0\mathbf{i} + 2\mathbf{j}$$
$$\mathbf{F}(1, -4) = 1\mathbf{i} + 2\mathbf{j}$$
$$\mathbf{F}(3, 1) = 3\mathbf{i} + 2\mathbf{j}$$
$$\mathbf{F}(-1, -1) = -1\mathbf{i} + 2\mathbf{j}$$

*When using the Early Transcendentals text, use the page number in parentheses.

Suppose the *xy*-plane represents a map of a large flat field and imagine that **F**(*x*, *y*) = *x***i** + 2**j** describes the wind direction and speed at (*x*, *y*) in the field. We can draw a field of vectors graph representing wind velocities:

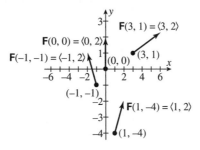

Vector field in three dimensions are described in a similar way using three component functions.

Example: **F**(*x*, *y*, *z*) = (*y* + *z*)**i** + (*x* + *z*)**j** + (*x* + *y*)**k** is a three-dimensional vector field with component functions *P*(*x*, *y*, *z*) = *y* + *z*, *Q*(*x*, *y*, *z*) = *x* + *z*, *R*(*x*, *y*, *z*) = *x* + *y*. **F**(1, 3, 4) = 7**i** + 5**j** + 4**k**.

1) Determine the scalar field components of:

   a) **F**(*x*, *y*) = cos*x***i** + sin*xy***j**.

       $P(x, y) = \cos x$
       $Q(x, y) = \sin xy$.

   b) **F**(*x*, *y*, *z*) = (*x*² + *y*)**i** + (*y*² + *z*)**k**.

       $P(x, y, z) = x^2 + y$
       $Q(x, y, z) = 0$
       $R(x, y, z) = y^2 + z$.

2) For the vector field **F**(*x*, *y*) = |*x*|**i** + |*y*|**j**, find the vector values and sketch the images of (0, 0), (−2, 0), (−3, 1), (1, 2), (2, −2).

$$\mathbf{F}(0, 0) = |0|\mathbf{i} + |0|\mathbf{j} = \mathbf{0}$$
$$\mathbf{F}(-2, 0) = |-2|\mathbf{i} + |0|\mathbf{j} = 2\mathbf{i}$$
$$\mathbf{F}(-3, 1) = 3\mathbf{i} + \mathbf{j}$$
$$\mathbf{F}(1, 2) = \mathbf{i} + 2\mathbf{j}$$
$$\mathbf{F}(2, -2) = 2\mathbf{i} + 2\mathbf{j}.$$

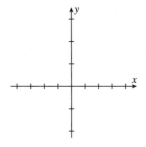

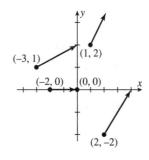

3) Describe and sketch the vector field
$F(x, y, z) = 2i + j + k$.

To every point in $\mathbb{R}^3$, we associate the (constant) vector $\langle 2, 1, 1 \rangle$. A sketch with the vector values of the points $(0, 0, 0), (3, 2, 5)$, and $(3, 3, 3)$ is given next.

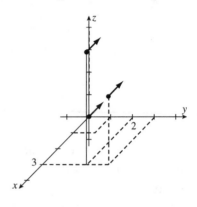

B. For a scalar function $z = f(x, y)$, the gradient ∇f (from Chapter 14) is actually a vector field called the **gradient vector field**. In two dimensions:

$$\nabla f(x, y) = f_x(x, y)i + f_y(x, y)j.$$

A vector field **F** is **conservative** means **F** is a gradient vector field; that is, $F = \nabla f$ for some scalar function f called the **potential function** of **F**.

Example: Is $F(x, y) = 2xyi + x^2j$ conservative?

We need to try to find a scalar function f such that $f_x = 2xy$ and $f_y = x^2$. We see that $f(x, y) = x^2y$ is one such potential function. Therefore, **F** is conservative.

Section 16.3 has a simple method for determining whether a vector field is conservative.

4) Find the gradient vector field of:

    a) $f(x, y) = x^2 + 3xy$.

    b) $f(x, y, z) = x^2y + xz^2 + 2yz$.

$\nabla f = (2x + 3y)i + 3xj$.

$\nabla f = (2xy + z^2)i + (x^2 + 2z)j + (2xz + 2y)k$.

5) If **F** is a conservative vector field, then _____ for some scalar function f.

$\nabla f = F$.

6) Is $F(x, y) = 2xi + 2yj$ conservative?

Yes. If $\nabla f = F$ for some f, then $f_x = 2x$ and $f_y = 2y$. The scalar function $f(x, y) = x^2 + y^2$ is one such function.

Section 16.2 Line Integrals

This section generalizes the idea of integrating a single-variable real function over an interval $[a, b]$ in two ways: (1) as the integral of a function of two or three variables over a curve in two- or three-dimensional space, and (2) as the integral of a vector field over a curve. Applications of these line integrals include calculating work done moving an object along a curve and finding the mass and center of mass of a wire whose density varies along its length.

Concepts to Master

A. Line integrals in two and three dimensions; Piecewise smooth curve; Orientation of a curve

B. Mass and center of mass of a wire

C. Line integral of a vector field; Work

D. Technology Plus

Summary and Focus Questions

Page
1087
(ET Page
1063)

A. Let f be a function of two variables and C be a smooth curve given by $\mathbf{r}(t) = x(t)\mathbf{i} + y(t)\mathbf{j}$, for $t \in [a, b]$. Partitioning $[a, b]$ into n subintervals divides the curve C into subarcs of length Δs_i. By choosing t_i^* in each subinterval, we obtain sample points (x_i^*, y_i^*) on each subarc. The corresponding **Riemann sum** is

$$\sum_{i=1}^{n} f(x_i^*, y_i^*)\Delta s_i.$$

The **line integral of f along C** is:

$$\int_C f(x, y)\, ds = \lim_{n \to \infty} \sum_{i=1}^{n} f(x_i^*, y_i^*)\Delta s_i.$$

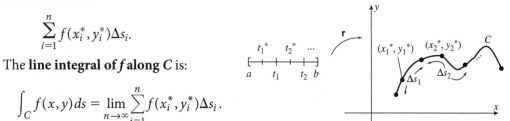

When f is positive, the line integral $\int_C f(x, y)\, ds$ may be thought of as the area under that part of the function f that lies above the curve C.

If f is continuous, calculate the line integral as follows:

$$\int_C f(x, y)\, ds = \int_a^b f(x(t),\, y(t)) \sqrt{\left(\frac{dx}{dt}\right)^2 + \left(\frac{dy}{dt}\right)^2}\; dt.$$

Line integrals in three-dimensional space are defined and computed similarly:

$$\int_C f(x, y, z)\, ds = \int_a^b f(x(t), t(t), z(t))\sqrt{\left(\frac{dx}{dt}\right)^2 + \left(\frac{dy}{dt}\right)^2 + \left(\frac{dz}{dt}\right)^2}\, dt.$$

Example: Let C be the line in the plane from $(0, 1)$ to $(1, 0)$ and $f(x, y) = x^2 + y^2$. Find the line integral $\int_C f(x, y)\, ds$.

The curve (line) C is given parametrically by the vector function

$\mathbf{r}(t) = \langle t, 1 - t \rangle$ with component functions $x = t$ and $y = 1 - t$, for

$0 \le t \le 1$. $\int_C (x^2 + y^2)\, ds$ is the area above the line segment C and below

$f(x, y) = x^2 + y^2$. See the shaded area in the figure.

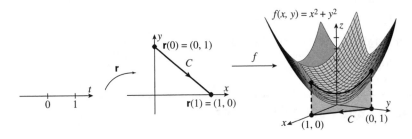

Since $\dfrac{dx}{dt} = 1$ and $\dfrac{dy}{dt} = -1$:

$$\int_C (x^2 + y^2)\, ds = \int_0^1 \left[t^2 + (1 - t)^2\right]\sqrt{1^2 + (-1)^2}\, dt$$

$$= \int_0^1 \sqrt{2}(2t^2 - 2t + 1)\, dt = \frac{2\sqrt{2}}{3}.$$

A curve C is **piecewise smooth** if C is a union of a finite number of smooth subarcs.

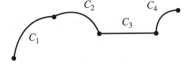

If C is piecewise smooth and composed of smooth curves C_1 and C_2, then

$$\int_C f\, ds = \int_{C_1} f\, ds + \int_{C_2} f\, ds.$$

Two other line integrals may be defined. The **line integral of f along C with respect to x**:

$$\int_C f(x, y)\, dx = \int_a^b f(x(t), y(t))x'(t)\, dt$$

and the **line integral of f along C with respect to y**:

$$\int_C f(x, y)\, dy = \int_a^b f(x(t), y(t))y'(t)\, dt.$$

A parameterization $x = x(t), y = y(t), t \in [a, b]$ of a curve C determines an **orientation**, or direction, along the curve as t increases from a to b. The curve denoted $-C$ is the same set of points as C but with the opposite orientation.

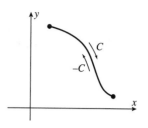

1) Evaluate $\int_C y \, ds$, where C is given by
$x(t) = t, \ y(t) = t^3, \ 0 \le t \le \sqrt[4]{7}$.

$x'(t) = 1$ and $y'(t) = 3t^2$, so

$$\int_C y \, ds = \int_0^{\sqrt[4]{7}} t^3 \sqrt{1^2 + (3t^2)^2} \, dt$$

$$= \int_0^{\sqrt[4]{7}} t^3 \sqrt{1 + 9t^4} \, dt$$

$$= \frac{1}{54}(1 + 9t^4)^{3/2} \Big]_0^{\sqrt[4]{7}} = \frac{511}{54}.$$

2) Evaluate $\int_C (x - y) \, dx$ and $\int_C (x - y) \, dy$, where C is the curve given by $x = 1 + e^t$, $y = 1 - e^{-t}, 0 \le t \le 1$.

$x'(t) = e^t, \ y'(t) = e^{-t}$.

$$\int_C (x - y) \, dx = \int_0^1 (1 + e^t - (1 - e^{-t}))e^t \, dt$$

$$= \int_0^1 (e^{2t} + 1) \, dt = \frac{e^2}{2} + \frac{1}{2}.$$

$$\int_C (x - y) \, dy = \int_0^1 (1 + e^t - (1 - e^{-t}))e^{-t} \, dt$$

$$= \int_0^1 (1 + e^{-2t}) \, dt = \frac{3}{2} - \frac{1}{2e^2}.$$

3) Write a definite integral for $\int_C xyz \, ds$, where C is the helix $x = t, y = \sin t, z = \cos t$, $0 \le t \le \pi$.

$$\int_0^\pi (t \sin t \cos t) \sqrt{1^2 + \cos^2 t + (-\sin t)^2} \, dt$$

$$= \int_0^\pi \sqrt{2} \, t \sin t \cos t \, dt = \int_0^\pi \frac{t \sin 2t}{\sqrt{2}} \, dt.$$

Page 1089 (ET Page 1065)

B. If $\rho(x, y)$ is the density of the material at the point (x, y) along a thin wire shaped like a curve C, then the **mass** of the wire is:

$$m = \int_C \rho(x, y) \, ds.$$

The **center of mass** of the wire is located at the point $(\overline{x}, \overline{y})$, where:

$$\overline{x} = \frac{\int_C x \rho(x, y) \, ds}{m} \quad \text{and} \quad \overline{y} = \frac{\int_C y \rho(x, y) \, ds}{m}.$$

The point where the center of mass is located need not be a point on the wire.

4) Find the center of mass of a piece of wire along the line $y = 2x \, (0 \leq x \leq 1)$ whose density at (x, y) is $1 + xy$.

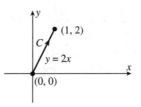

The curve (line) C may be represented by $x = t$, $y = 2t$, $0 \leq t \leq 1$. $\rho(x, y) = 1 + xy$

The mass is $m = \displaystyle\int_C (1 + xy)\, ds$

$$= \int_0^1 (1 + t(2t))\sqrt{1^2 + 2^2}\, dt$$

$$= \int_0^1 \sqrt{5}(1 + 2t^2)\, dt = \sqrt{5}\left(t + \frac{2}{3}t^3 \right)\Big]_0^1$$

$$= \sqrt{5}\left(\left(1 + \frac{2}{3}\right) - 0 \right) = \frac{5\sqrt{5}}{3}.$$

Now find $\bar{x}$ and $\bar{y}$.

$$\int_C x\rho(x, y)\, ds = \int_0^1 t(1 + 2t^2)\sqrt{5}\, dt$$

$$= \int_0^1 \sqrt{5}(t + 2t^3)\, dt = \sqrt{5}\left(\frac{1}{2}t^2 + \frac{1}{2}t^4 \right)\Big]_0^1$$

$$= \sqrt{5}\left(\left(\frac{1}{2} + \frac{1}{2}\right) - 0 \right) = \sqrt{5}.$$

$$\bar{x} = \frac{\sqrt{5}}{\dfrac{5\sqrt{5}}{3}} = \frac{3}{5}.$$

$$\int_C y\rho(x, y)\, ds = \int_0^1 (2t)(1 + 2t^2)\sqrt{5}\, dt$$

$$= \int_0^1 \sqrt{5}(2t + 4t^3)\, dt = \sqrt{5}(t^2 + t^4)\Big]_0^1$$

$$= \sqrt{5}[(1 + 1) - 0] = 2\sqrt{5}.$$

$$\bar{y} = \frac{2\sqrt{5}}{\dfrac{5\sqrt{5}}{3}} = \frac{6}{5}.$$

Because the curve in this example happens to be a straight line, the center of mass of the wire is a point on the curve.

Page
1094
(ET Page
1070)

C. If **F** is a continuous vector field on a smooth curve C given by $\mathbf{r}(t), t \in [a, b]$, the line **integral of F along C** is:

$$\int_C \mathbf{F} \cdot d\mathbf{r} = \int_a^b \mathbf{F}(\mathbf{r}(t)) \cdot \mathbf{r}'(t)\,dt.$$

This integral may also be written as $\int_C \mathbf{F} \cdot \mathbf{T}\,ds$ where **T** is the unit tangent vector function C and $\mathbf{F} \cdot \mathbf{T}$ is the dot product. The differential in this expression is ds, not dt.

Example: Let C be that part of the parabola $x = 1 + y^2$ between $(1, 0)$ and $(2, 1)$ and **F** be the vector field $\mathbf{F}(x, y) = xy\mathbf{i} + y^2\mathbf{j}$. Find the line integral of **F** along C.

The curve C is $x = 1 + y^2$ for $0 \le y \le 1$. Thus, a parameterization of C is

$$x = 1 + t^2, y = t, 0 \le t \le 1.$$

So, we let $\mathbf{r}(t) = (1 + t^2)\mathbf{i} + t\mathbf{j}$. Then:

$$\mathbf{r}'(t) = 2t\mathbf{i} + 1\mathbf{j}$$
$$\mathbf{F}(\mathbf{r}(t)) = (1 + t^2)t\mathbf{i} + t^2\mathbf{j} = (t + t^3)\mathbf{i} + t^2\mathbf{j}$$
$$\mathbf{F}(\mathbf{r}(t)) \cdot \mathbf{r}'(t) = (t + t^3)2t + (t^2)1 = 2t^2 + 2t^4 + t^2 = 3t^2 + 2t^4.$$

Then $\int_C \mathbf{F} \cdot d\mathbf{r} = \int_0^1 \mathbf{F}(\mathbf{r}(t)) \cdot \mathbf{r}'(t)\,dt = \int_0^1 (3t^2 + 2t^4)\,dt = \left(t^3 + \frac{2}{5}t^5\right)\Big|_0^1 = \frac{7}{5}.$

If we evaluate the line integral by calculating $\int_C \mathbf{F} \cdot \mathbf{T}\,ds$, then:

$$|\mathbf{r}'(t)| = \sqrt{4t^2 + 1}, \text{ so } \mathbf{T} = \frac{2t}{\sqrt{4t^2 + 1}}\mathbf{i} + \frac{1}{\sqrt{4t^2 + 1}}\mathbf{j} \text{ and}$$

$$\mathbf{F} \cdot \mathbf{T} = \frac{(t + t^3)2t}{\sqrt{4t^2 + 1}} + \frac{t^2(1)}{\sqrt{4t^2 + 1}} = \frac{3t^2 + 2t^4}{\sqrt{4t^2 + 1}}.$$

The line integral is (note the differential ds):

$$\int_C \mathbf{F} \cdot \mathbf{T}\,ds = \int_0^1 \frac{3t^2 + 2t^4}{\sqrt{4t^2 + 1}}\sqrt{4t^2 + 1}\,dt = \int_0^1 (3t^2 + 2t^4)\,dt = \frac{7}{5},$$

the same as previously calculated.

A **line integral of a vector field** $\mathbf{F} = P\mathbf{i} + Q\mathbf{j} + R\mathbf{k}$ may be written in terms of line integrals of its components:

$$\int_C \mathbf{F} \cdot d\mathbf{r} = \int_C P\,dx + Q\,dy + R\,dz,$$

where $\int_C P\,dx + Q\,dy + R\,dz$ is shorthand for $\int_C P\,dx + \int_C Q\,dy + \int_C R\,dz.$

If $\mathbf{F(x)}$ is interpreted as a force field ($\mathbf{F}$ is the vector force at the point $\mathbf{x}$), then $\int_C \mathbf{F} \cdot d\mathbf{r}$ is the **work** done by the force $\mathbf{F}$ moving a particle along the curve C.

5) Evaluate $\int_C \mathbf{F} \cdot d\mathbf{r}$, where C is the line segment from $(0, 0, 0)$ to $(1, 3, 2)$ and $\mathbf{F}(x, y, z) = (y + z)\mathbf{i} + (2x + z)\mathbf{j} + (x + 3y)\mathbf{k}$.

C may be parameterized by $x = t$, $y = 3t$, $z = 2t$ for $t \in [0, 1]$.

We have $dx = dt$, $dy = 3\,dt$, $dz = 2\,dt$.

$$\int_C \mathbf{F} \cdot d\mathbf{r} = \int_C P\,dx + Q\,dy + R\,dz$$

$$= \int_C (y + z)\,dx + (2x + z)\,dy + (x + 3y)\,dz$$

$$= \int_0^1 (3t + 2t)\,dt + (2t + 2t)3t\,dt + (t + 9t)\,dt$$

$$= \int_0^1 37t\,dt = \frac{37}{2}.$$

6) In the force field $\mathbf{F}(x, y) = x\mathbf{i} + (x + y)\mathbf{j}$, find the amount of work done moving a particle along the parabola $y = x^2$ from $(1, 1)$ to $(2, 4)$.

The parabolic curve C may be described by $x = t$, $y = t^2$, $t \in [1, 2]$.
$x'(t) = 1$, $y'(t) = 2t$.
$dx = dt$, $dy = 2t\,dt$.

The work done is

$$\int_C \mathbf{F} \cdot d\mathbf{r} = \int_C x\,dx + (x + y)\,dy$$

$$= \int_1^2 t(1)\,dt + (t + t^2)2t\,dt$$

$$= \int_1^2 (t + 2t^2 + 2t^3)\,dt = \frac{41}{3}.$$

D. Technology Plus. Use a computer algebra system or a graphing calculator to solve.

T-1) a) Evaluate the line integral $\int_C \mathbf{F} \cdot d\mathbf{r}$, where

$\mathbf{F}(x, y) = xy\mathbf{i} + \dfrac{e^{x^2}}{4}\mathbf{j}$ and C is the curve

$\mathbf{r}(t) = t\mathbf{i} + t^2\mathbf{j}, 0 \le t \le 2$.

$\mathbf{F}(\mathbf{r}(t)) = t(t^2)\mathbf{i} + \dfrac{e^{t^2}}{4}\mathbf{j} = t^3\mathbf{i} + \dfrac{e^{t^2}}{4}\mathbf{j}$.

$\mathbf{r}'(t) = 1\mathbf{i} + 2t\mathbf{j}$

$\mathbf{F}(\mathbf{r}(t)) \cdot \mathbf{r}'(t) = t^3 + \dfrac{te^{t^2}}{2}$.

$$\int_C \mathbf{F} \cdot d\mathbf{r} = \int_0^2 \left(t^3 + \frac{te^{t^2}}{2} \right) dt$$

$$= \left(\frac{t^4}{4} + \frac{e^{t^2}}{4} \right)\Bigg]_0^2 = \frac{e^4 + 15}{4}.$$

b) Illustrate part a) by graphing, on the same screen, C and $\mathbf{F}(\mathbf{r}(t))$ for $t = \dfrac{1}{2}, 1,$ and $\dfrac{3}{2}$.

$\mathbf{r}\left(\dfrac{1}{2}\right) = \left\langle \dfrac{1}{2}, \dfrac{1}{4} \right\rangle.$

$\mathbf{F}\left(\mathbf{r}\left(\dfrac{1}{2}\right)\right) = \left\langle \dfrac{1}{8}, \dfrac{e^{1/4}}{4} \right\rangle \approx \langle 0.125, 0.321 \rangle.$

$\mathbf{r}(1) = \langle 1, 1 \rangle.$

$\mathbf{F}(\mathbf{r}(1)) = \left\langle 1, \dfrac{e}{4} \right\rangle \approx \langle 1, 0.680 \rangle$

$\mathbf{r}\left(\dfrac{3}{2}\right) = \left\langle \dfrac{3}{2}, \dfrac{9}{4} \right\rangle.$

$\mathbf{F}\left(\mathbf{r}\left(\dfrac{3}{2}\right)\right) = \left\langle \dfrac{27}{8}, \dfrac{e^{9/4}}{4} \right\rangle \approx \langle 3.375, 2.372 \rangle.$

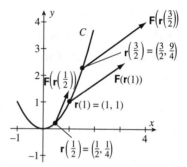

Section 16.3 The Fundamental Theorem for Line Integrals

This section contains a theorem to evaluate certain types of line integrals that generalizes the Fundamental Theorem of Calculus. It also describes "independence of path," whereby the value of a line integral depends only upon the two endpoints of the curve and not on any particular path in between the points.

Concepts to Master

A. Line integrals of gradients

B. Closed path; Simple path; Connected; Simply connected

C. Independence of path; Test for path independence

D. Technology Plus

Summary and Focus Questions

Page 1099 (ET Page 1075)

A. The Fundamental Theorem of Calculus for functions of one variable says $\int_a^b f'(x)\,dx = f(b) - f(a)$. Let $z = f(x, y)$ be a function of two variables. If you think of the gradient ∇f as a higher-dimensional version of $f'(x)$, then for line integrals of the form $\int_C \nabla f \cdot d\mathbf{r}$, we have a result similar to the Fundamental Theorem of Calculus:

The Fundamental Theorem for Line Integrals

If C is a piecewise smooth curve given by $\mathbf{r}(t)$, $a \le t \le b$, and f is a differentiable function with ∇f continuous on C, then:

$$\int_C \nabla f \cdot d\mathbf{r} = f(\mathbf{r}(b)) - f(\mathbf{r}(a)).$$

Example: Let C be that portion of the parabola $y = x^2$ for $0 \le x \le 1$ and let $f(x, y) = 2xy$. There are two ways to evaluate $\int_C \nabla f \cdot d\mathbf{r}$.

a) From the previous section, we first parameterize C with $x = t$, $y = t^2$, for $0 \le t \le 1$, and calculate:

$$\nabla f = \langle 2y, 2x \rangle, \mathbf{r}(t) = \langle t, t^2 \rangle, \mathbf{r}'(t) = \langle 1, 2t \rangle$$
$$\nabla f(\mathbf{r}(t)) \cdot \mathbf{r}'(t) = \langle 2t^2, 2t \rangle \cdot \langle 1, 2t \rangle = 2t^2 + 4t^2 = 6t^2.$$

Then:

$$\int_C \nabla f \cdot d\mathbf{r} = \int_C \nabla f(\mathbf{r}(t)) \cdot \mathbf{r}'(t)\, dt = \int_0^1 6t^2 dt = 2t^3 \Big]_0^1 = 2.$$

b) Use the Fundamental Theorem for Line Integrals:

$$\int_C \nabla f \cdot d\mathbf{r} = f(\mathbf{r}(1)) - f(\mathbf{r}(0)) = f(1,1) - f(0,0) = 2(1)(1) - 2(0)(0) = 2.$$

---

1) Let $f(x, y) = x^2 + xy + 2y^2$ and C be any smooth curve from $(0, 1)$ to $(3, 0)$. Find $\int_C \nabla f \cdot d\mathbf{r}$.

$$\int_C \nabla f \cdot d\mathbf{r} = f(3, 0) - f(0, 1)$$
$$= 9 - 2 = 7.$$

2) If C is a smooth curve given by $\mathbf{r}(t)$ with $\mathbf{r}(a) = \mathbf{r}(b)$ and f is differentiable and ∇f is continuous on C, then $\int_C \nabla f \cdot d\mathbf{r} = $ _____ .

0. Since $\mathbf{r}(a) = \mathbf{r}(b)$,
$f(\mathbf{r}(a)) = f(\mathbf{r}(b))$.
Therefore, $f(\mathbf{r}(b)) - f(\mathbf{r}(a)) = 0$.

B. A curve C given by $\mathbf{r}(t), a \le t \le b$ is **closed** means $\mathbf{r}(a) = \mathbf{r}(b)$. A curve C is **simple** means C does not intersect itself except, perhaps, at its endpoints.

Page 1102 (ET Page 1078)

Simple, not closed

Closed, not simple

Closed and simple

A domain D is **open** if every point is an interior point.
D is **connected** if any two points of D can be connected by a path entirely in D.
D is **simply-connected** means every simple closed curve in D encloses only points of D.

Open
(no boundary points in D).

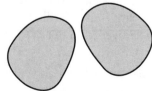

Not Connected

Connected, not
simply-connected

3)

True or False:

a) C is simple.

b) C is closed.

False.

False.

4) True or False:

The following region is simply-connected.

True.

C. Let C be a path from point A to point B and let $\mathbf{F}$ be a continuous vector field on an open connected region D containing A and B.

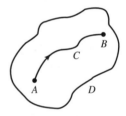

We say the line integral $\int_C \mathbf{F} \cdot d\mathbf{r}$ is **independent of path** if its value depends only upon the endpoints A and B (and not on the path C from A to B.)

We have already seen some line integrals that are independent of path: a line integral of the form $\int_C \nabla f \cdot d\mathbf{r}$ can be evaluated by the Fundamental Theorem for Line Integrals with just its endpoints and, therefore, is independent of path.

This also means that $\int_C \nabla f \cdot d\mathbf{r} = 0$ whenever C is a closed curve (begins and ends at the same point.)

Example: Let $\mathbf{F}$ be the vector field $\mathbf{F}(x, y) = (3x^2 + 6xy)\mathbf{i} + (3x^2 + 4y)\mathbf{j}$. In general, it may be difficult to use the definition to determine whether a given integral is independent of path (but we will soon have a simple test not involving the definition). However, in this case we note that for $f(x, y) = x^3 + 3x^2 y + 2y^2$, $f_x = 3x^2 + 6xy$ and $f_y = 3x^2 + 4y$. Therefore, $\mathbf{F} = \nabla f$. Thus, $\int_C \mathbf{F} \cdot d\mathbf{r}$ is independent of path for any curve C parameterized by $\mathbf{r}$.

So, we now know every line integral whose integrand is a gradient is independent of path. Under certain conditions, the converse of this statement is true:

> If **F** is a continuous vector field on an open simply connected region D, and $\int_C \mathbf{F} \cdot d\mathbf{r}$ is independent of path, then **F** must be conservative (that is, $\mathbf{F} = \nabla f$, for some differentiable f).

There is an easy test to determine whether a given $\mathbf{F} = P\mathbf{i} + Q\mathbf{j}$ is conservative:

$$\mathbf{F} = \nabla f \text{ for some differentiable } f \text{ if and only if } \frac{\partial P}{\partial y} = \frac{\partial Q}{\partial x}.$$

Example: $\mathbf{F}(x, y) = (3x^2 + 6xy)\mathbf{i} + (3x^2 + 4y)\mathbf{j}$ (see the previous example) is the integrand of a path independent line integral because $\frac{\partial P}{\partial y} = 6x$ and $\frac{\partial Q}{\partial x} = 6x$.

If $\mathbf{F} = P\mathbf{i} + Q\mathbf{j} = \nabla f$, to reconstruct f, integrate $\int P\,dx$, differentiate the result with respect to y and then compare that result with Q. Or, you can compare P with $\frac{\partial}{\partial x}\left(\int Q\,dy\right)$.

Example: Show that $\mathbf{F}(x, y) = \nabla f$ for some function f, where

$$\mathbf{F}(x, y) = (3x^2 + 4xy)\mathbf{i} + (2x^2 + 4y^3)\mathbf{j}.$$

The components of **F** are $P(x, y) = 3x^2 + 4xy$ and $Q(x, y) = 2x^2 + 4y^3$.

$\frac{\partial P}{\partial y} = 4x$ and $\frac{\partial Q}{\partial x} = 4x$. Since $\frac{\partial P}{\partial y} = \frac{\partial Q}{\partial x}$, $\mathbf{F}(x, y) = \nabla f$ for some function f.

To find f we integrate $P(x, y)$ with respect to x:

$$f(x, y) = \int (3x^2 + 4xy)\,dx = x^3 + 2x^2 y + C(y),$$

where $C(y)$ is a function only of y.

Then differentiate the result with respect to y:

$$\frac{\partial}{\partial y}(x^3 + 2x^2 y + C(y)) = 2x^2 + C'(y).$$

Comparing this to $Q(x, y) = 2x^2 + 4y^3$, $C'(y) = 4y^3$, so $C(y) = y^4 + K$. Therefore, $f(x, y) = x^3 + 2x^2 y + y^4 + K$.

5) If **F** is conservative and C is a smooth closed curve, then $\int_C \mathbf{F} \cdot d\mathbf{r} = $ _____.

0. For some f, $\mathbf{F} = \nabla f$, so $\int_C \mathbf{F} \cdot d\mathbf{r} = \int_C \nabla f \cdot d\mathbf{r} = 0$ since the curve is closed.

6) Suppose C_1 and C_2 are two distinct paths from point A to point B in a region D and $\displaystyle\int_{C_1} \mathbf{F} \cdot d\mathbf{r} = \int_{C_2} \mathbf{F} \cdot d\mathbf{r}$. Can we conclude the line integral is independent of path?

No, independence of path requires the line integrals along *all* paths in D from A to B have the same value.

7) Is $\mathbf{F}(x, y) = (2x + y)\mathbf{i} + (x + 8y)\mathbf{j}$ the gradient of some function?

Yes. For $P(x, y) = 2x + y$ and
$$Q(x,y) = x + 8y, \frac{\partial P}{\partial y} = 1 \text{ and } \frac{\partial Q}{\partial x} = 1.$$
Since $\dfrac{\partial P}{\partial y} = \dfrac{\partial Q}{\partial x}$, $\mathbf{F} = \nabla f$ for some f.

8) Find a function f whose gradient is $\nabla f = (y^2 + 1)\mathbf{i} + (2xy + 3y^2)\mathbf{j}$.

Let $P(x, y) = y^2 + 1$ and
$Q(x, y) = 2xy + 3y^2$.
$$f(x,y) = \int P\,dx = \int (y^2 + 1)\,dx$$
$$= xy^2 + x + h(y),$$
where h is function of y.

$\dfrac{\partial f}{\partial y} = 2xy + h'(y)$. Comparing to Q,

$h'(y) = 3y^2$ so $h(y) = y^3 + K$.
$f(x, y) = xy^2 + x + y^3 + K$.
For $K = 0, f(x, y) = xy^2 + x + y^3$.

9) Let C be the parabola $x = y^2$ from $(1, 1)$ to $(4, 2)$. Evaluate:
$$\int_C (2x + 3y^2)\,dx + (6xy + 10)\,dy.$$

Let $P(x, y) = 2x + 3y^2$ and
$Q(x, y) = 6xy + 10$.
Since $\dfrac{\partial P}{\partial y} = 6y = \dfrac{\partial Q}{\partial x}$, $P\mathbf{i} + Q\mathbf{j}$ is a gradient of some function $f(x, y)$.
$$f(x, y) = \int (2x + 3y^2)\,dx$$
$$= x^2 + 3xy^2 + h(y).$$
$\dfrac{\partial f}{\partial y} = 6xy + h'(y)$.

Comparing with $Q(x, y)$, $h'(y) = 10$, so $h(y) = 10y + c$. We choose $c = 0$. Thus $f(x, y) = x^2 + 3xy^2 + 10y$ and
$$\int_C P\,dx + Q\,dy = f(4, 2) - f(1, 1)$$
$$= 84 - 14 = 70.$$
Note that it did not matter that C was a parabola. C could have been any smooth curve from $(1, 1)$ to $(4, 2)$.

10) Evaluate $\int_C dx + z\,dy + y\,dz$, where C is any path from $(0, 1, 1)$ to $(3, 1, 6)$.

Let $P(x, y, z) = 1$, $Q(x, y, z) = z$, and $R(x, y, z) = y$.

$$f(x, y, z) = \int P\,dx = x + C(y, z).$$

$f_y = C_y(y, z) = Q = z.$

So, $C(y, z) = yz + D(z)$.

$f(x, y, z) = x + yz + D(z)$.

$f_z = y + D'(z) = R = y.$

$D'(z) = 0$, so $D(z) = K$. We use $K = 0$.

Therefore, $f(x, y, z) = x + yz$ and

$$\int_C dx + z\,dy + y\,dz = (x + yz)\Big|_{(0,1,1)}^{(3,1,6)}$$

$$= (3 + 6) - (0 + 1) = 8.$$

D. Technology Plus. Use a computer algebra system or a graphing calculator to solve.

T-1) Draw the vector field
$\mathbf{F}(x, y) = \langle 1 + x^2 + y^2, 2y(x + 1) \rangle$ for the region $D: 0 \le x \le 1$, $-1 \le y \le 1$. Does the field appear to be conservative?

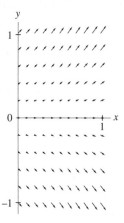

There is a chance that $\mathbf{F}$ is conservative because it appears that $\int_C \mathbf{F} \cdot d\mathbf{r}$ will be small (near zero) for a closed curve C in D. (The Riemann sum will contain positive and negative terms.)

For $P(x, y) = 1 + x^2 + y^2$ and $Q(x, y) = 2y(x + 1)$,

$$\frac{\partial P}{\partial y} = 2y \text{ and } \frac{\partial Q}{\partial x} = 2y.$$

Thus, $\mathbf{F}$ is conservative.

Section 16.4   Green's Theorem

This section presents Green's Theorem, a very useful relationship between a double integral over a plane region D and a line integral around the boundary curve for D. You should think of it as the double integral version of the Fundamental Theorem of Calculus. One of several applications of Green's Theorem provides an alternative method for calculating the area of a region D.

Concepts to Master

A. Positive orientation of a simple closed curve; Green's Theorem

B. Evaluating areas with line integrals

Summary and Focus Questions

Page 1108
(ET Page 1084)

A. Let C be a simple closed curve in the plane given by $\mathbf{r}(t)$, for $a \le t \le b$. We say C has a **positive orientation** if the traversal of C from $\mathbf{r}(a)$ to $\mathbf{r}(b)$ is counterclockwise.

The line integral for a vector field $\mathbf{F}(x, y) = P(x, y)\mathbf{i} + Q(x, y)\mathbf{j}$ about C is sometimes denoted $\oint_C \mathbf{F} \cdot d\mathbf{r}$ when the boundary C is a simple closed curve with positive orientation.

Green's Theorem   Let C be a piecewise smooth, simple closed curve with positive orientation about a region D. If $P(x, y)$ and $Q(x, y)$ have continuous partials in an open region containing D, then:

$$\oint_C P\,dx + Q\,dy = \iint_D \left(\frac{\partial Q}{\partial x} - \frac{\partial P}{\partial y} \right) dA.$$

Since $\oint_C P\,dx + Q\,dy = \int_C \mathbf{F} \cdot d\mathbf{r}$, Green's Theorem says

$$\int_C \mathbf{F} \cdot d\mathbf{r} = \iint_D \left(\frac{\partial Q}{\partial x} - \frac{\partial P}{\partial y} \right) dA.$$

Green's Theorem is the double integral version of the Fundamental Theorem of Calculus for Integrals. Recall for one-dimensional domains $[a, b]$, the integral $\int_a^b f'(x)\,dx$ is evaluated at the boundary (endpoints) of $[a, b]$ by $f(b) - f(a)$. In two-dimensional domains, Green's Theorem says the double integral $\iint_D \left(\frac{\partial Q}{\partial x} - \frac{\partial P}{\partial y} \right) dA$ may be evaluated at the boundary of D (the curve C) by $\oint_C P\,dx + Q\,dy = \int_C \mathbf{F} \cdot d\mathbf{r}$.

Example: Let **F** be the vector field
$\mathbf{F}(x,y) = (xy^2)\mathbf{i} + (2x^2 y)\mathbf{j}$ and D be the region at
the right with simple closed boundary C composed
of the parabola $y = x^2$ (curve C_1) and line $y = x$
(curve C_2).

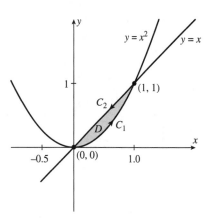

Here $P(x,y) = xy^2$ and $Q(x,y) = 2x^2 y$. We can
calculate $\oint_C \mathbf{F} \cdot d\mathbf{r}$ two ways—directly and using
Green's Theorem:

a) Since C is composed of two curves,

$$\oint_C \mathbf{F} \cdot d\mathbf{r} = \int_{C_1} \mathbf{F} \cdot d\mathbf{r} + \int_{C_2} \mathbf{F} \cdot d\mathbf{r}.$$

C_1 is parameterized by $x = t$, $y = t^2$, $0 \le t \le 1$. $dx = dt$ and $dy = 2t\, dt$.

$$\int_{C_1} \mathbf{F} \cdot d\mathbf{r} = \int_{C_1} P\, dx + Q\, dy = \int_0^1 \left[t(t^2)^2(1) + 2(t)^2(t^2)(2t) \right] dt$$
$$= \int_0^1 5t^5 dt = \frac{5}{6}.$$

C_2 is parameterized by $x = 1 - t$, $y = 1 - t$, $0 \le t \le 1$. (We use $x = 1 - t$
and $y = 1 - t$ because the curve C_2 is traversed from $(1, 1)$ to $(0, 0)$.)
$dx = (-1)dt$ and $dy = (-1)dt$.

$$\int_{C_2} \mathbf{F} \cdot d\mathbf{r} = \int_{C_2} P\, dx + Q\, dy = \int_0^1 \left[(1 - t)(1 - t)^2(-1) + 2(1 - t)^2(1 - t)(-1) \right] dt$$
$$= \int_0^1 -3(1 - t)^3 dt = -\frac{3}{4}.$$

Thus, $\oint_C \mathbf{F} \cdot d\mathbf{r} = \dfrac{5}{6} + \left(-\dfrac{3}{4} \right) = \dfrac{1}{12}.$

b) The region D may be described as $0 \le x \le 1$, $x^2 \le y \le x$.

Since $\dfrac{\partial P}{\partial y} = 2xy$ and $\dfrac{\partial Q}{\partial x} = 4xy$, by Green's Theorem

$$\oint_C \mathbf{F} \cdot d\mathbf{r} = \int_C P\, dx + Q\, dy = \iint_D \left(\frac{\partial Q}{\partial x} - \frac{\partial P}{\partial y} \right) dA$$
$$= \int_0^1 \int_{x^2}^x (4xy - 2xy)\, dy\, dx = \int_0^1 \int_{x^2}^x 2xy\, dy\, dx$$
$$= \int_0^1 xy^2 \Big]_{x^2}^x dx = \int_0^1 (x^3 - x^5)\, dx = \left(\frac{x^4}{4} - \frac{x^6}{6} \right) \Big]_0^1 = \frac{1}{12}.$$

1) Place arrows on the curve C near the point P to indicate a positive orientation to the curve

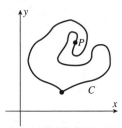

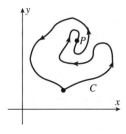

2) Evaluate using Green's Theorem:

a) $\oint_C xy\,dx + y^2\,dy$, where C is the triangle formed by $(0, 0)$, $(1, 1)$ and $(0, 1)$.

Let $P(x, y) = xy$, $Q(x, y) = y^2$.

Then $\dfrac{\partial P}{\partial y} = x$ and $\dfrac{\partial Q}{\partial x} = 0$.

C encloses the region

$D: 0 \le x \le 1,\ x \le y \le 1$.

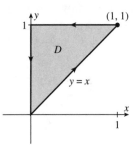

By Green's Theorem, we rewrite the line integral as a double integral:

$$\iint_D (0 - x)\,dA = \int_0^1 \int_x^1 -x\,dy\,dx$$

$$= \int_0^1 (x^2 - x)\,dx = -\frac{1}{6}.$$

b) $\oint_C (6xy^2)\,dx + (6x^2 y)\,dy$, where C is the triangle in part a).

For $P(x, y) = 6xy^2$ and $Q(x, y) = 6x^2 y$,

$$\frac{\partial P}{\partial y} = 12xy = \frac{\partial Q}{\partial x}.$$

Thus, $\dfrac{\partial Q}{\partial x} - \dfrac{\partial P}{\partial y} = 0$,

so $\oint_C P\,dx + Q\,dy = \iint_D 0\,dA = 0$.

Page 1111 (ET Page 1087)

B. Let C be the simple closed boundary of a region D. One way to find the area of D is to evaluate $\iint_D 1\,dA$.

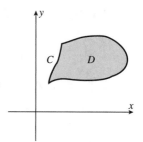

In order to use Green's Theorem to evaluate this double integral, we need to write the integrand 1 as $1 = \dfrac{\partial Q}{\partial x} - \dfrac{\partial P}{\partial y}$. There are many choices for P and Q, each resulting in a different line integral integrand. Three of the simplest combinations of P and Q for which $1 = \dfrac{\partial Q}{\partial x} - \dfrac{\partial P}{\partial y}$ are:

$P(x, y)$	$Q(x, y)$	Line integral to calculate for area of D
0	x	$\oint_C x\,dy$
$-y$	0	$-\oint_C y\,dy$
$-\dfrac{1}{2}y$	$\dfrac{1}{2}x$	$\dfrac{1}{2}\oint_C x\,dy - y\,dx$

Example: Let D be the region between the parabola $y = 1 - x^2$ and the line $y = 1 - x$. From single-variable calculus, the area of D is:

$$\int_0^1 \left[(1 - x^2) - (1 - x)\right] dx = \int_0^1 (-x^2 + x)\,dx$$

$$= \left(-\frac{x^3}{3} + \frac{x^2}{2}\right)\Bigg]_0^1 = \frac{1}{6}.$$

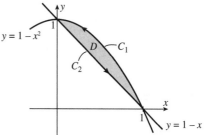

We can also find the area using Green's Theorem. Let C be the closed curve consisting of C_1 (the parabola $y = 1 - x^2$) followed by C_2 (the line $y = 1 - x$.) We parameterize the curves and find the line integrals using the form $\oint_C x\,dy$.

C_1: $x = 1 - t$
$\qquad y = 1 - (1 - t)^2 = 2t - t^2$
$\qquad 0 \le t \le 1$.

C_2: $x = t$
$\qquad y = 1 - t$
$\qquad 0 \le t \le 1$.

For the curve C_1, $dx = (-1)\, dt$, $dy = (2 - 2t)\, dt$, and

$$\int_{C_1} x\, dy = \int_0^1 (1 - t)(2 - 2t)\, dt = 2\int_0^1 (1 - t)^2\, dt = \frac{-2}{3}(1 - t)^3 \Big]_0^1 = \frac{2}{3}.$$

For the curve C_2, $dx = dt$, $dy = (-1)dt$, and

$$\int_{C_2} x\, dy = \int_0^1 (t)(-1)\, dt = -\int_0^1 t\, dt = -\frac{t^2}{2} \Big]_0^1 = -\frac{1}{2}.$$

The area of D is $\oint_C x\, dy = \int_{C_1} x\, dy + \int_{C_2} x\, dy = \frac{2}{3} - \frac{1}{2} = \frac{1}{6}.$

---

3) Use a line integral to find the area of the region D:

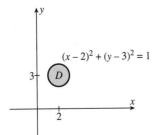

We know the answer will be π since D is a circle of radius 1. The boundary C may be described as

$x = 2 + \cos\theta$,
$y = 3 + \sin\theta$, $0 \le \theta \le 2\pi$.

Then:
$dx = -\sin\theta\, d\theta$,
$dy = \cos\theta\, d\theta$.

The area of D is $\dfrac{1}{2}\oint_C x\, dy - y\, dx$

$$= \frac{1}{2}\int_0^{2\pi} \big[(2 + \cos\theta)\cos\theta\, d\theta - (3 + \sin\theta)(-\sin\theta)\big]d\theta$$

$$= \frac{1}{2}\int_0^{2\pi} [2\cos\theta + 3\sin\theta + 1]\, d\theta$$

$$= \frac{1}{2}(2\sin\theta - 3\cos\theta + \theta)\Big]_0^{2\pi} = \pi.$$

Section 16.5 Curl and Divergence

This section describes two operations in three dimensions that resemble differentiation. Applications of these operations include the modeling of fluid flow. The curl at a point is a vector that measures the rotation of fluid about a point in the flow. The div at a point is a number that measures the rate of change of fluid flowing from the point (how fast is the fluid *diverging* from the point).

Concepts to Master

A. Del operator; Curl of a vector field

B. Divergence of a vector field

C. Vector forms of Green's Theorem

Summary and Focus Questions

Page 1115 (ET Page 1091)

A. Let **F** be a three-dimensional vector field. The operations described in this section result in either a vector field (curl **F**) or a scalar field (div **F**).

The ∇ (**del**) operator is the expression $\nabla = \mathbf{i}\dfrac{\partial}{\partial x} + \mathbf{j}\dfrac{\partial}{\partial y} + \mathbf{k}\dfrac{\partial}{\partial z}$.

The **curl** of $\mathbf{F} = P\mathbf{i} + Q\mathbf{j} + R\mathbf{k}$ is

$$\text{curl } \mathbf{F} = \nabla \times \mathbf{F} = \begin{vmatrix} \mathbf{i} & \mathbf{j} & \mathbf{k} \\ \dfrac{\partial}{\partial x} & \dfrac{\partial}{\partial y} & \dfrac{\partial}{\partial z} \\ P & Q & R \end{vmatrix}$$

$$= \left(\frac{\partial R}{\partial y} - \frac{\partial Q}{\partial z} \right)\mathbf{i} + \left(\frac{\partial P}{\partial z} - \frac{\partial R}{\partial x} \right)\mathbf{j} + \left(\frac{\partial Q}{\partial x} - \frac{\partial P}{\partial y} \right)\mathbf{k}.$$

You should think of curl **F** as a type of derivative for the vector field **F**.

Example: Let $\mathbf{F}(x, y, z) = x^2 z^3 \mathbf{i} + 2xy\mathbf{j} + yz^2\mathbf{k}$. Then:

$$P(x, y, z) = x^2 z^3, \frac{\partial P}{\partial y} = 0, \frac{\partial P}{\partial z} = 3x^2 z^2;$$

$$Q(x, y, z) = 2xy, \frac{\partial Q}{\partial x} = 2y, \frac{\partial Q}{\partial z} = 0;$$

$$R(x, y, z) = yz^2, \frac{\partial R}{\partial x} = 0, \frac{\partial R}{\partial y} = z^2.$$

$$\text{curl } \mathbf{F} = (z^2 - 0)\mathbf{i} + (3x^2 z^2 - 0)\mathbf{j} + (2y - 0)\mathbf{k} = z^2\mathbf{i} + 3x^2 z^2\mathbf{j} + 2y\mathbf{k}.$$

If **F** is the velocity of a fluid flow, then curl **F**(x, y, z) measures the circulation per unit area orthogonal to **F**(x, y, z) at (x, y, z). If curl **F** $= 0$ (irrotation) at a point, there is no circulation about the point as fluid flows through the point with velocity **F**. The direction of curl **F** is the axis of rotation around the point (use the Right Hand Rule to determine the direction of rotation.) Indeed, curl is sometimes called rotor (for rotation) and the notation used is rot **F**.

Example: In the preceding example, for the point $(1, 2, 1)$ in the fluid:

$$\mathbf{F}(1, 2, 1) = (1^2 1^3)\mathbf{i} + 2(1)(2)\mathbf{j} + 2(1^2)\mathbf{k} = \mathbf{i} + 4\mathbf{j} + 2\mathbf{k}$$

describes the direction and velocity of the flow of the fluid at $(1, 2, 1)$, and:

$$\text{curl } \mathbf{F}(1, 2, 1) = (1^2)\mathbf{i} + 3(1^2)(1^2)\mathbf{j} + 2(2)\mathbf{k} = \mathbf{i} + 3\mathbf{j} + 4\mathbf{k}$$

describes the direction and amount of circulation at the point corresponding to $(1, 2, 1)$.

There are two important results about curl:

a) If $f(x, y, z)$ has continuous second partials, curl $(\nabla f) = \mathbf{0}$, the zero vector.

b) Conversely, if **F**(x, y, z) has components with continuous partial derivatives and curl **F** $= \mathbf{0}$, then **F** is conservative.

---

1) Find the curl of
$$\mathbf{F}(x, y, z) = xy^2\mathbf{i} + x^2z\mathbf{j} + yz^2\mathbf{k}.$$

Let $P(x, y, z) = xy^2$, $Q(x, y, z) = x^2z$, and $R(x, y, z) = yz^2$.

$$\frac{\partial P}{\partial y} = 2xy, \ \frac{\partial P}{\partial z} = 0, \ \frac{\partial Q}{\partial x} = 2xz, \ \frac{\partial Q}{\partial z} = x^2,$$

$$\frac{\partial R}{\partial x} = 0, \ \frac{\partial R}{\partial y} = z^2.$$

$$\text{curl } \mathbf{F} = (z^2 - x^2)\mathbf{i} + (0 - 0)\mathbf{j} + (2xz - 2xy)\mathbf{k}$$
$$= (z^2 - x^2)\mathbf{i} + (2xz - 2xy)\mathbf{k}.$$

2) Use the curl **F** to determine whether
$\mathbf{F}(x, y, z) = xy\mathbf{i} + z^2\mathbf{j} + xz\mathbf{k}$ is conservative.

$$\frac{\partial P}{\partial y} = x, \ \frac{\partial P}{\partial z} = 0, \ \frac{\partial Q}{\partial x} = 0, \ \frac{\partial Q}{\partial z} = 2z,$$

$$\frac{\partial R}{\partial x} = z, \ \frac{\partial R}{\partial y} = 0.$$

$$\text{curl } \mathbf{F} = (0 - 2z)\mathbf{i} + (0 - z)\mathbf{j} + (0 - x)\mathbf{k}$$
$$= -2z\mathbf{i} - z\mathbf{j} - x\mathbf{k}, \text{ not the zero vector.}$$

Therefore, **F** is not conservative.

3) Show that $\mathbf{F}(x, y, z)$
$= (2x + y)\mathbf{i} + (x + 2yz^2)\mathbf{j} + 2y^2z\mathbf{k}$ is
conservative and find any f such that $\nabla f = \mathbf{F}$.

$\dfrac{\partial P}{\partial y} = 1, \ \dfrac{\partial P}{\partial z} = 0, \ \dfrac{\partial Q}{\partial x} = 1, \ \dfrac{\partial Q}{\partial z} = 4yz,$

$\dfrac{\partial R}{\partial x} = 0, \ \dfrac{\partial R}{\partial y} = 4yz.$

$\text{curl } \mathbf{F} = (4yz - 4yz)\mathbf{i} + (0 - 0)\mathbf{j} + (1 - 1)\mathbf{k}$
$= \mathbf{0}.$

Therefore, $\mathbf{F}$ is conservative and $\mathbf{F} = \nabla f$ for some f.

$f_x = 2x + y, \ f_y = x + 2yz^2, \ f_z = 2y^2z.$
From $f_x = 2x + y,$
$f(x, y, z) = x^2 + xy + C(y, z).$
$f_y = x + C_y(y, z) = x + 2yz^2.$
$C_y(y, z) = 2yz^2,$
so $C(y, z) = y^2z^2 + K(z).$
Hence, $f(x, y, z) = x^2 + xy + y^2z^2 + K(z).$
Now $f_z = 2y^2z + K_z(z) = 2y^2z,$
so $K_z(z) = 0.$ Choose $K(z) = 0.$
Therefore, $f(x, y, z) = x^2 + xy + y^2z^2.$

Page 1118 (ET Page 1094)

B. For $\mathbf{F} = P\mathbf{i} + Q\mathbf{j} + R\mathbf{k}$, the **divergence** of $\mathbf{F}$ is

$$\text{div } \mathbf{F} = \nabla \cdot \mathbf{F} = \frac{\partial P}{\partial x} + \frac{\partial Q}{\partial y} + \frac{\partial R}{\partial z}.$$

Example: Let $\mathbf{F}(x, y, z) = x^2z^3\mathbf{i} + 2xy\mathbf{j} + yz^2\mathbf{k}$ as in the previous example.
$\dfrac{\partial P}{\partial x} = 2xz^3, \ \dfrac{\partial Q}{\partial y} = 2x,$ and $\dfrac{\partial R}{\partial z} = 2yz.$ Then div $\mathbf{F} = 2xz^3 + 2x + 2yz.$

If $\mathbf{F}$ represents a fluid flow, then div $\mathbf{F}(x, y, z)$ may be interpreted as the (instantaneous) rate of change of the mass of the fluid per unit volume at (x, y, z)—the tendency of the fluid to diverge ("thin out").

If div $\mathbf{F}(x, y, z) > 0$, there is a net flow out of the point (x, y, z). (The fluid is diverging.)

Example: Again, using the preceding example and the point $(1, 2, 1)$ in the fluid:

$$\text{div } \mathbf{F}(1, 2, 1) = 2(1)(1^3) + 2(1) + 2(2)(1) = 8.$$

At the point $(1, 2, 1)$, the fluid is diverging (spreading out from the point).

For $\mathbf{F} = P\mathbf{i} + Q\mathbf{j} + R\mathbf{k}$, with continuous second partial derivatives, div curl $\mathbf{F} = 0$.

Since div curl of a vector field is always zero, this means that whenever div $\mathbf{F} \neq 0$, we can conclude that $\mathbf{F}$ is not the curl of any vector field. In our preceding example, div $\mathbf{F} = 8$, so $\mathbf{F}$ cannot be the curl of any other vector field.

4) Find div **F** for $\mathbf{F}(x, y, z)$
$= (x^2 + y)\mathbf{i} + (y^2 + z)\mathbf{j} + xyz\mathbf{j}.$

$\dfrac{\partial P}{\partial x} = 2x, \dfrac{\partial Q}{\partial y} = 2y, \dfrac{\partial R}{\partial z} = xy.$

div $\mathbf{F} = 2x + 2y + xy.$

5) True or False:
curl div $\mathbf{F} = 0.$

False. For a vector field **F**, div **F** is a scalar. Curl can be computed only for a vector field, so curl div **F** makes no sense.

6) For a fluid with velocity
$\mathbf{F} = 2\mathbf{i} + y^2\mathbf{j} + (x + z)\mathbf{k}$, is the fluid diverging at (l, 2, 1)?

Yes. div $\mathbf{F} = \dfrac{\partial P}{\partial x} + \dfrac{\partial Q}{\partial y} + \dfrac{\partial R}{\partial z} = 0 + 2y + 1$
$= 2y + 1.$

At (1, 2, 1), div $\mathbf{F} = 2(2) + 1 = 5.$ Since div $\mathbf{F} > 0$, the fluid is diverging.

7) Let **F** be given by
$\mathbf{F}(x, y, z) = (x^2 + z^3)\mathbf{i} + xyz^2\mathbf{j} + (y^2 + z)\mathbf{k}.$
Calculate div curl **F**.

We know the result must be 0.

$$\text{curl } \mathbf{F} = \begin{vmatrix} \mathbf{i} & \mathbf{j} & \mathbf{k} \\ \dfrac{\partial}{\partial x} & \dfrac{\partial}{\partial y} & \dfrac{\partial}{\partial z} \\ P & Q & R \end{vmatrix}$$

$= (2y - 2xyz)\mathbf{i} - (0 - 3z^2)\mathbf{j} + (yz^2 - 0)\mathbf{k}.$
$= (2y - 2xyz)\mathbf{i} + 3z^2\mathbf{j} + yz^2\mathbf{k}.$

Therefore,
div curl $\mathbf{F} = (0 - 2yz) + 0 + (2yz) = 0.$

8) Is there a vector field **F** such that
curl $\mathbf{F}(x, y, z) = 2x^2 y\mathbf{i} - 2xy^2\mathbf{j} + z^2\mathbf{k}$?

No. For such an **F**, div curl $\mathbf{F} = 0.$ But div curl $\mathbf{F} = 4xy - 4xy + 2z = 2z \neq 0.$

C. Suppose C is a smooth, simple close curve in the plane with a positive orientation around a region D. For a two-dimensional vector field $\mathbf{F}(x, y) = P(x, y)\mathbf{i} + Q(x, y)\mathbf{j}$, we can think of **F** as being a three-dimensional vector field by adding the variable z to P and to Q and adding the component $0\,\mathbf{k}$ to **F**. We can then use the curl and div operations to give two vector forms of Green's Theorem.

Since $\mathbf{k} \cdot \mathbf{k} = 1, \dfrac{\partial P}{\partial z} = 0,$ and $\dfrac{\partial Q}{\partial z} = 0$, the curl is simply curl $\mathbf{F} \cdot \mathbf{k} = \dfrac{\partial Q}{\partial x} - \dfrac{\partial P}{\partial y}.$

Thus we have the first vector form of Green's Theorem:

i) $\displaystyle \int_C \mathbf{F} \cdot d\mathbf{r} = \iint_D \text{curl } \mathbf{F} \cdot \mathbf{k}\, dA = \iint_D \left(\dfrac{\partial Q}{\partial x} - \dfrac{\partial P}{\partial y} \right) dA.$

This form says the line integral of the tangential component of **F** along *C* is the double integral of the vertical component (**k**-direction) of curl **F** over the region *D*.

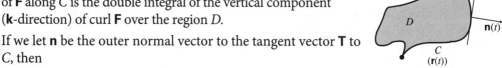

If we let **n** be the outer normal vector to the tangent vector **T** to *C*, then

$$n(t) = \frac{y'(t)}{|r'(t)|}i - \frac{x'(t)}{|r'(t)|}j \text{ and the second form of Green's Theorem is:}$$

ii) $\displaystyle\int_C F \cdot n \, ds = \iint_D \operatorname{div} F(x, y) \, dA,$

This form says the line integral of the normal component of **F** along *C* is the double integral of the divergence of **F** over the region *D*.

Example: Let $F(x, y) = (x^2 + xy)i + (y^2 + xy)j$. Let *D* be the rectangular region given by $0 \le x \le 1, 0 \le y \le 2$ and *C* be the rectangular boundary of *D*. Use the vector forms of Green's Theorem to find **a)** $\displaystyle\int_C F \cdot dr$ and **b)** $\displaystyle\int_C F \cdot n \, ds.$

Remember to think of **F** as $F(x, y, z) = (x^2 + xy)i + (y^2 + xy)j + 0k$.

a) In this case, $\dfrac{\partial P}{\partial y} = x, \dfrac{\partial P}{\partial z} = 0, \dfrac{\partial Q}{\partial x} = y, \dfrac{\partial Q}{\partial z} = 0, \dfrac{\partial R}{\partial x} = 0,$ and $\dfrac{\partial R}{\partial y} = 0.$

curl $F = (0 - 0)i - (0 - 0)j + (y - x)k$. Thus, curl $F \cdot k = y - x$ and

$$\int_C F \cdot dr = \iint_D \operatorname{curl} F \cdot k \, dA = \iint_D (y - x) \, dA = \int_0^1 \int_0^2 (y - x) \, dy \, dx$$

$$= \int_0^1 \left(\frac{y^2}{2} - xy \right) \Big|_0^2 \, dx = \int_0^1 (2 - 2x) \, dx = (2x - x^2) \Big|_0^1 = 1.$$

b) Here $\dfrac{\partial P}{\partial x} = 2x + y, \dfrac{\partial Q}{\partial y} = 2y + y,$ and $\dfrac{\partial R}{\partial z} = 0.$

div $F = (2x + y) + (2y + x) + (0) = 3x + 3y$. Therefore:

$$\int_C F \cdot n \, ds = \iint_D \operatorname{div} F \, ds = \iint_D (3x + 3y) \, dA = \int_0^1 \int_0^2 (3x + 3y) \, dy \, dx$$

$$= \int_0^1 \left(3xy + \frac{3}{2}y^2 \right) \Big|_0^2 \, dx = \int_0^1 (6x + 6) \, dx = (3x^2 + 6x) \Big|_0^1 = 9.$$

9) Let $\mathbf{F}(x, y, z) = (2xy)\mathbf{i} + (4x + 3y)\mathbf{j}$. Let D be the triangular region bounded by $x = 0$, $y = 2$, and $y = 2x$, and C be the boundary of D. Use the vector forms of Green's Theorem to find:

a) $\displaystyle\int_C \mathbf{F} \cdot d\mathbf{r}$.

The region D may be described by $0 \le x \le 1, 2x \le y \le 2$.

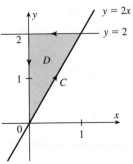

$$\text{curl } \mathbf{F} \cdot \mathbf{k} = \frac{\partial Q}{\partial x} - \frac{\partial P}{\partial y} = 4 - 2x.$$

$$\int_C \mathbf{F} \cdot d\mathbf{r} = \iint_D \text{curl } \mathbf{F} \cdot \mathbf{k}\, dA$$

$$= \iint_D (4 - 2x)\, dA$$

$$= \int_0^1 \int_{2x}^2 (4 - 2x)\, dy\, dx$$

$$= \int_0^1 (4y - 2xy)\Big]_{2x}^2 dx$$

$$= \int_0^1 (8 - 12x + 4x^2)\, dx$$

$$= \left(8x - 6x^2 + \frac{4}{3}x^3\right)\Big]_0^1 = \frac{10}{3}.$$

b) $\displaystyle\int_C \mathbf{F} \cdot \mathbf{n}\, ds$.

$$\text{div } \mathbf{F} = 2y + 3 + 0 = 2y + 3.$$

$$\int_C \mathbf{F} \cdot \mathbf{n}\, ds = \iint_D \text{div } \mathbf{F}\, ds$$

$$= \iint_D (2y + 3)\, dA$$

$$= \int_0^1 \int_{2x}^2 (2y + 3)\, dy\, dx$$

$$= \int_0^1 (y^2 + 3y)\Big]_{2x}^2 dx$$

$$= \int_0^1 (10 - 4x^2 - 6x)\, dx$$

$$= \left(10x - \frac{4}{3}x^3 - 3x^2\right)\Big]_0^1 = \frac{17}{3}.$$

Section 16.6 Parametric Surfaces and Their Areas

Surfaces in space are two-dimensional objects. This section shows how to parameterize a surface in a manner similar to the way curves are parameterized. We need two parametric variables instead of one. You will also learn how to calculate tangent planes and areas for parametric surfaces.

Concepts to Master

A. Parameterization of a surface; Grid curve

B. Tangent plane to a parametric surface

C. Area of a parametric surface

D. Surface area of the graph of the function

E. Technology Plus

Summary and Focus Questions

Page
1123
(ET Page
1099)

A. Let $\mathbf{r}(u,v) = x(u,v)\mathbf{i} + y(u,v)\mathbf{j} + z(u,v)\mathbf{k}$ be a vector function with domain D, a subset of the uv-plane. The set of all points $\mathbf{r}(u, v)$, where (u, v) varies throughout D, is a **parametric surface**, S, in three dimensions. The equations:

$$x = x(u,v) \qquad y = y(u,v) \qquad z = z(u,v)$$

parameterize the surface S.

Example: Let D be the triangular region below in the uv-plane and let $\mathbf{r}(u, v) = (3u)\mathbf{i} + (2v)\mathbf{j} + (6 - 6u - 6v)\mathbf{k}$. Then $\mathbf{r}(0, 0) = (0, 0, 6)$, $\mathbf{r}(1, 0) = (3, 0, 0)$, and $\mathbf{r}(0, 1) = (0, 2, 0)$.
The resulting parameterized surface, S, is that portion of the plane $z = 6 - 2x - 3y$ in the first octant of xyz-space.

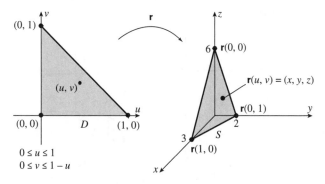

By keeping u constant, $u = u_0$, the function $\mathbf{r}(u_0, v)$ becomes a vector function of the variable v and determines a **grid curve** on the surface S. Similarly, by keeping $v = v_0$ constant, we get another grid curve determined by $\mathbf{r}(u, v_0)$.

Grid curves help visualize the parameterization. For the previous example, $u = \dfrac{1}{3}$ and $u = \dfrac{2}{3}$ produce two grid curves (lines because $\mathbf{r}$ is linear) on the surface S.

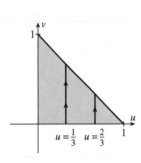
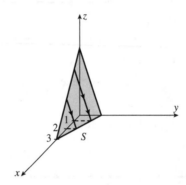

1) Parameterize the cone $z^2 = x^2 + y^2$, $0 \le z \le 1$, using polar coordinates.

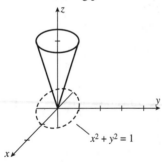

Using polar coordinates (r, θ), let D be the region in the $r\theta$-plane
$0 \le \theta \le 2\pi$,
$0 \le r \le 1$.
Then $x = r \cos \theta$,
$y = r \sin \theta$,
$z = r, 0 \le \theta \le 2\pi, 0 \le r \le 1$
parameterizes the cone.

2) Parameterize the surface of revolution obtained by revolving $y = x^2, 0 \le x \le 2$, about the x-axis.

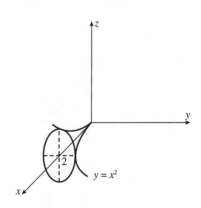

The surface is about the x-axis, so we use x as one parameter: $x = x$. For any constant x, the surface is a circle with radius x^2, so we use θ as the other parameter: $y = x^2 \cos \theta$, $z = x^2 \sin \theta$. Our parameterization is
$$x = x,$$
$$y = x^2 \cos \theta,$$
$$z = x^2 \sin \theta$$
with $0 \le x \le 2, 0 \le \theta \le 2\pi$.

3) Describe the grid curves for the surface S parameterized by
$$x = u$$
$$y = u^2 + v^2$$
$$z = v$$
where u and v are real numbers.

The surface S is a paraboloid along the y-axis with vertex $(0, 0, 0)$ because $y = u^2 + v^2 = x^2 + z^2$.
If we fix $u = u_0$, then $x = u_0$ is a constant and $y = u_0^2 + v^2$ is a parabola in the plane $x = u_0$ parallel to the yz-plane.

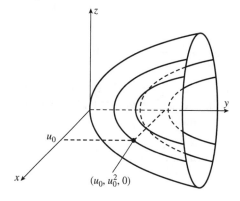

If we fix $v = v_0$, then $z = v_0$ is a constant and $y = u^2 + v_0^2$ is a parabola in the plane $z = v_0$ parallel to the xy-plane.

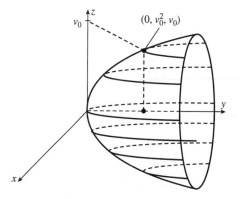

Page
1127
(ET Page
1104)

B. The tangent plane at (x_0, y_0, z_0) to a surface S parameterized by $\mathbf{r}(u, v)$ has as a normal vector $\mathbf{r}_u \times \mathbf{r}_v$. The equation of the tangent plane to S is

$$(\mathbf{r}_u \times \mathbf{r}_v) \cdot \langle x - x_0, y - y_0, z - z_0 \rangle = 0.$$

The surface S is **smooth** if the normal vector is not **0**.

Example: Let S be a surface in three dimensions parameterized by $x = u$, $y = u^2 + v^2$, $z = v$, where $u \geq 0$ and $v \geq 0$. Find the normal vector and equation of the tangent plane at the point $(1, 5, 2)$.

$\mathbf{r}(u, v) = \langle u, u^2 + v^2, v \rangle$, so $\mathbf{r}_u = \langle 1, 2u, 0 \rangle$ and $\mathbf{r}_v = \langle 0, 2v, 1 \rangle$. The normal vector is:

$$\mathbf{n} = \mathbf{r}_u \times \mathbf{r}_v = \begin{vmatrix} \mathbf{i} & \mathbf{j} & \mathbf{k} \\ 1 & 2u & 0 \\ 0 & 2v & 1 \end{vmatrix} = (2u - 0)\mathbf{i} - (1 - 0)\mathbf{j} + (2v - 0)\mathbf{k} = 2u\mathbf{i} - \mathbf{j} + 2v\mathbf{k}.$$

At the point $(1, 5, 2)$, where $u = 1$ and $v = 2$, $\mathbf{n} = \langle 2(1), -1, 2(2) \rangle = \langle 2, -1, 4 \rangle$. The tangent plane to S at $(1, 5, 2)$ has equation:

$$2(x - 1) - 1(y - 5) + 4(z - 2) = 0$$
$$2x - y + 4z = 5.$$

Since $\langle 2, -1, 4 \rangle \neq \mathbf{0}$, the surface is smooth at $(1, 5, 2)$.

---

4) Find the equation of the plane tangent to the surface S at $(u, v) = (1, 1)$, where S is parameterized by

$x = 2u^2v^2$
$y = u + 2v$
$z = 2u + v.$

$\mathbf{r}(u, v) = x\mathbf{i} + y\mathbf{j} + z\mathbf{k}.$

$\mathbf{r}_u = \dfrac{\partial x}{\partial u}\mathbf{i} + \dfrac{\partial y}{\partial u}\mathbf{j} + \dfrac{\partial z}{\partial u}\mathbf{k} = 4uv^2\mathbf{i} + 1\mathbf{j} + 2\mathbf{k}.$

$\mathbf{r}_v = 4u^2v\mathbf{i} + 2\mathbf{j} + 1\mathbf{k}.$

$$\mathbf{r}_u \times \mathbf{r}_v = \begin{vmatrix} \mathbf{i} & \mathbf{j} & \mathbf{k} \\ 4uv^2 & 1 & 2 \\ 4u^2v & 2 & 1 \end{vmatrix}$$
$$= -3\mathbf{i} + (8u^2v - 4uv^2)\mathbf{j} + (8uv^2 - 4u^2v)\mathbf{k}.$$

At $(u, v) = (1, 1)$, $\mathbf{r}_u \times \mathbf{r}_v = -3\mathbf{i} + 4\mathbf{j} + 4\mathbf{k}$ and $(x, y, z) = (2, 3, 3)$ is a point on the plane.

The tangent plane is
$-3(x - 2) + 4(y - 3) + 4(z - 3) = 0$
$3x - 4y - 4z + 18 = 0.$

Page
1128
(ET Page
1104)

C. The surface area of a surface S parameterized by $\mathbf{r}(u, v)$ for $(u, v) \in D$ is

$$A(S) = \iint_D |\mathbf{r}_u \times \mathbf{r}_v| \, dA.$$

Example: Let S be a surface parameterized by $x = uv$, $y = e^u$, $z = e^v$. Find an iterated integral for the surface area of S above the region $D = \{(u,v) \mid 0 \le u \le 1, \, 0 \le v \le 1\}$. For $\mathbf{r}(u, v) = \langle uv, e^u, e^v \rangle$, $\mathbf{r}_u = \langle v, e^u, 0 \rangle$ and $\mathbf{r}_v = \langle u, 0, e^v \rangle$.

$$\mathbf{r}_u \times \mathbf{r}_v = \begin{vmatrix} \mathbf{i} & \mathbf{j} & \mathbf{k} \\ v & e^u & 0 \\ u & 0 & e^v \end{vmatrix} = e^{u+v}\mathbf{i} - ve^v\mathbf{j} - ue^u\mathbf{k}$$

$$|\mathbf{r}_u \times \mathbf{r}_v| = \sqrt{(e^{u+v})^2 + (-ve^v)^2 + (-ue^u)^2} = \sqrt{e^{2u+2v} + v^2e^{2v} + u^2e^{2u}}$$

$$A(S) = \iint_D \sqrt{e^{2u+2v} + v^2e^{2v} + u^2e^{2u}}\, dA = \int_0^1\int_0^1 \sqrt{e^{2u+2v} + v^2e^{2v} + u^2e^{2u}}\, dv\, du.$$

5) Set up an iterated integral for the area of each surface S.

a) S is parameterized by
$$x = 2uv$$
$$y = u + v$$
$$z = u - v$$
$$0 \le u \le 2, \, 0 \le v \le 1.$$

Let D be the region $0 \le u \le 2$, $0 \le v \le 1$.
$\mathbf{r}(u, v) = \langle 2uv, u + v, u - v \rangle$, so $\mathbf{r}_u = \langle 2v, 1, 1 \rangle$
and $\mathbf{r}_v = \langle 2u, 1, -1 \rangle$.

$$\mathbf{r}_u \times \mathbf{r}_v = \begin{vmatrix} \mathbf{i} & \mathbf{j} & \mathbf{k} \\ 2v & 1 & 1 \\ 2u & 1 & -1 \end{vmatrix}$$
$$= (-2)\mathbf{i} - (-2v - 2u)\mathbf{j} + (2v - 2u)\mathbf{k}$$
$$|\mathbf{r}_u \times \mathbf{r}_v| = \sqrt{(-2)^2 + (2v + 2u)^2 + (2v - 2u)^2}$$
$$= 2\sqrt{1 + 2u^2 + 2v^2}.$$
This area is
$$A(S) = \iint_D 2\sqrt{1 + 2u^2 + 2v^2}\, dA$$
$$= \int_0^2\int_0^1 2\sqrt{2u^2 + 2v^2 + 1}\, dv\, du.$$

b) S is the surface
$$z = x^2 - y^2,$$
$$0 \le x \le 1, \, 0 \le y \le 1.$$

Parameterize S with $x = u$, $y = v$, $z = u^2 - v^2$. Since $x = u$ and $y = v$ we have $0 \le u \le 1$, $0 \le v \le 1$.
$\mathbf{r}(u, v) = \langle u, v, u^2 - v^2 \rangle$, so $\mathbf{r}_u = \langle 1, 0, 2u \rangle$
and $\mathbf{r}_v = \langle 0, 1, -2v \rangle$.

$$\mathbf{r}_u \times \mathbf{r}_v = \begin{vmatrix} \mathbf{i} & \mathbf{j} & \mathbf{k} \\ 1 & 0 & 2u \\ 0 & 1 & -2v \end{vmatrix}$$
$$= (-2u)\mathbf{i} - (-2v)\mathbf{j} + (1)\mathbf{k}.$$
$$|\mathbf{r}_u \times \mathbf{r}_v| = \sqrt{(-2u)^2 + (2v)^2 + (1)^2}$$
$$= \sqrt{1 + 4u^2 + 4v^2}.$$

The area is $\int_0^1\int_0^1 \sqrt{1 + 4u^2 + 4v^2}\, dv\, du.$

Page
1130
(ET Page
1106)

D. The surface of the graph of a function $z = f(x, y)$ with domain D and continuous partial derivatives is a special case of the parameterized surfaces just considered. In this case $\mathbf{r}(u, v) = (u, v, f(x, y))$, $\mathbf{r}_x = \langle 1, 0, f_x \rangle$, $\mathbf{r}_y = \langle 0, 1, f_y \rangle$, and $\mathbf{r}_x \times \mathbf{r}_y = (-f_x)\mathbf{i} - (f_y)\mathbf{j} + (1)\mathbf{k}$. Therefore, the surface area of the graph of f over the region D is:

$$\iint_D \sqrt{[f_x(x, y)]^2 + [f_y(x, y)]^2 + 1}\, dA.$$

Example: In Exercise 5b) the surface is given by the function $z = x^2 - y^2$. Then $f_x = 2x$ and $f_y = -2y$, so we may write directly that:

$$A(S) = \iint_D \sqrt{(2x)^2 + (-2y)^2 + 1}\, dA = \iint_D \sqrt{1 + 4x^2 + 4y^2}\, dA,$$

which is the same integral as in 5b) except for the names of the variables.

6) Find the surface area of the paraboloid $z = x^2 + 2y^2$ above the triangular region D:

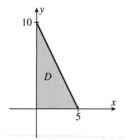

$f_x = 2x, f_y = 4y.$

The region D is $0 \le x \le 5$, $0 \le y \le 10 - 2x$.

The surface area is

$$A(S) = \iint_D \sqrt{(2x)^2 + (4y)^2 + 1}\, dA$$
$$= \int_0^5 \int_0^{10-2x} \sqrt{4x^2 + 16y^2 + 1}\, dy\, dx.$$
$$\approx 355.8.$$

7) Find the surface area of the top portion of the sphere $x^2 + y^2 + z^2 = 4$ inside the cylinder $x^2 + y^2 = 1$.

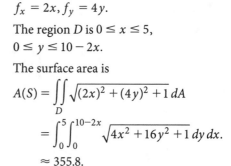

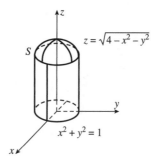

For $z = \sqrt{4 - x^2 - y^2}$,

$$\frac{\partial z}{\partial x} = \frac{-x}{\sqrt{4 - x^2 - y^2}} \quad \text{and}$$

$$\frac{\partial z}{\partial y} = \frac{-y}{\sqrt{4 - x^2 - y^2}}.$$

Thus, $\left(\dfrac{\partial z}{\partial x}\right)^2 + \left(\dfrac{\partial z}{\partial y}\right)^2 + 1$

$$= \frac{x^2}{4 - x^2 - y^2} + \frac{y^2}{4 - x^2 - y^2} + 1$$

$$= \frac{4}{4 - x^2 - y^2}.$$

The surface area of S is

$$\iint_D \frac{2}{\sqrt{4 - x^2 - y^2}}\, dA, \text{ where } D$$

is the area inside the circle $x^2 + y^2 = 1$.
Switch to polar coordinates, $x = r\cos\theta$,
$y = r\sin\theta$, $0 \le \theta \le 2\pi$, $0 \le r \le 1$.
This integral is

$$\int_0^{2\pi}\int_0^1 \frac{2}{\sqrt{4 - r^2}} r\, dr\, d\theta$$

$$= \int_0^{2\pi}\left(-2\sqrt{4 - r^2}\,\Big]_0^1\right) d\theta$$

$$= \int_0^{2\pi}(-2\sqrt{3} - (-2)(2))\, d\theta$$

$$= \int_0^{2\pi}(4 - 2\sqrt{3})\, d\theta = (4 - 2\sqrt{3})\theta\,\Big]_0^{2\pi}$$

$$= (4 - 2\sqrt{3})2\pi = 8\pi - 4\pi\sqrt{3}.$$

E. Technology Plus. Use a computer algebra system or a graphing calculator to solve.

T-1) a) Sketch the graph of the surface

$$x = 4 + u\cos v + \frac{u^2}{4}$$

$$y = 4 + u\sin v + \frac{u^2}{4}$$

$$z = u$$

$$0 \le u \le 3, \quad 0 \le v \le 2\pi.$$

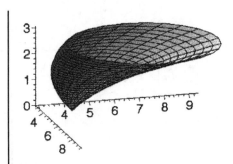

b) Set up an iterated integral for the surface area of the "tornado."

Let $\mathbf{r}(u, v) = \langle x, y, z \rangle$.

$$\mathbf{r}_u = \left\langle \cos v + \frac{u}{2}, \ \sin v + \frac{u}{2}, \ 1 \right\rangle$$

$$\mathbf{r}_v = \langle -u \sin v, \ u \cos v, \ 0 \rangle$$

$$\mathbf{r}_u \times \mathbf{r}_v = \left\langle -u \cos v, \ -u \sin v, \ u + \frac{u^2}{2}(\cos v + \sin v) \right\rangle$$

$$|\mathbf{r}_u \times \mathbf{r}_v|$$

$$= \sqrt{2u^2 + u^3(\cos v + \sin v) + \frac{u^4}{4}(1 + 2 \cos v \, \sin v)}$$

$$= \sqrt{2u^2 + u^3(\cos v + \sin v) + \frac{u^4}{4}(1 + \sin 2v)}.$$

The surface area is $\displaystyle\int_0^3 \int_0^{2\pi} |\mathbf{r}_u \times \mathbf{r}_v| \, dv \, du$

$$= \int_0^3 \int_0^{2\pi} \sqrt{2u^2 + u^3(\cos v + \sin v) + \frac{u^4}{4}(1 + \sin 2v)} \, dv \, du.$$

T-2) Sketch the surface
$z = x + y - x^2 - y^2,\ 0 \le x \le 1,$
$0 \le y \le 1$ and find its area to three decimal places.

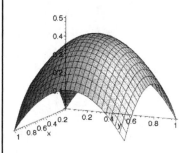

$f(x, y) = x + y - x^2 - y^2$
$f_x = 1 - 2x,\, f_y = 1 - 2y.$
The surface area is

$$\iint_D \sqrt{(1 - 2x)^2 + (1 - 2y)^2 + 1} \, dA,$$

where D is the $[0, 1] \times [0, 1]$ square.
The integral is

$$\int_0^1 \int_0^1 \sqrt{3 - 4x + 4x^2 - 4y + 4y^2} \, dy \, dx,$$

which, to three decimal places, is 1.281.

Section 16.7   Surface Integrals

In this section, we will see that a surface integral is the two-dimensional version of a line integral. Just as a line integral was used to determine the length, mass, and so on, of a curve, a surface integral may be used to calculate the area, mass, and so on, of a surface.

Concepts to Master

A. Surface integral

B. Orientation of a surface; Surface integral of a vector field

---

Summary and Focus Questions

Page 1137 (ET Page 1110)

A. Suppose S is a surface parameterized by

$$\mathbf{r}(u, v) = x(u, v)\mathbf{i} + y(u, v)\mathbf{j} + z(u, v)\mathbf{k},$$

with domain D in the uv-plane. The **surface integral** of a function $f(x, y, z)$ whose domain includes S is

$$\iint_S f(x, y, z)\, dS = \iint_D f(\mathbf{r}(u, v))|\mathbf{r}_u \times \mathbf{r}_v|\, dA.$$

In the special case where S is parameterized by a function $z = g(x, y)$,

$$\iint_S f(x, y, z)\, dS = \iint_D f(x, y, g(x, y))\sqrt{(g_x)^2 + (g_y)^2 + 1}\, dA.$$

1) Find an iterated integral for the surface integral of $f(x, y, z) = x$, where S is that portion of the plane $6x + 3y + 2z = 12$ in the first octant.

A parameterization of S is

$$\mathbf{r}(u, v) = u\mathbf{i} + v\mathbf{j} + \left(6 - 3u - \frac{3}{2}v\right)\mathbf{k}.$$

for $\mathbf{r}(u, v) \in D$, $0 \le u \le 2$, $0 \le v \le 4 - 2u$. See the figure.

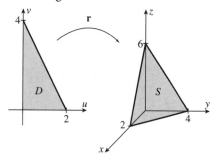

$$\mathbf{r}_u \times \mathbf{r}_v = \begin{vmatrix} \mathbf{i} & \mathbf{j} & \mathbf{k} \\ 1 & 0 & -3 \\ 0 & 1 & -\dfrac{3}{2} \end{vmatrix} = 3\mathbf{i} + \dfrac{3}{2}\mathbf{j} + \mathbf{k}.$$

$$|\mathbf{r}_u \times \mathbf{r}_v| = \sqrt{3^2 + \left(\dfrac{3}{2}\right)^2 + 1^2} = \dfrac{7}{2}.$$

Since $f(x, y, z) = x$, $f(\mathbf{r}(u, v)) = u$.

$$\iint_S x \, dS = \iint_D u\left(\dfrac{7}{2}\right) dA = \int_0^2 \int_0^{4-2u} \dfrac{7u}{2} \, dv \, du.$$

2) Evaluate $\displaystyle\iint_S y^2 \, dS$, where S is that part of the cylinder $x^2 + z^2 = 1$ between the planes $y = 0$ and $y = 3 - x$.

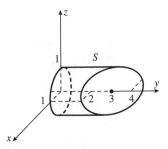

We parameterize S with $\mathbf{r}(u, v) = \langle x, y, z \rangle$ where

$x = \cos u$
$y = v$
$z = \sin u$,

and $0 \le u \le 2\pi$, $0 \le v \le 3 - \cos u$.

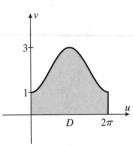

$\mathbf{r}_u = \langle -\sin u, 0, \cos u \rangle$ and $\mathbf{r}_v = \langle 0, 1, 0 \rangle$.

$$\mathbf{r}_u \times \mathbf{r}_v = \begin{vmatrix} \mathbf{i} & \mathbf{j} & \mathbf{k} \\ -\sin u & 0 & \cos u \\ 0 & 1 & 0 \end{vmatrix}$$

$$= -\cos u\mathbf{i} - \sin u\mathbf{k}.$$

$$|\mathbf{r}_u \times \mathbf{r}_v| = \sqrt{(-\cos u)^2 + 0^2 + (-\sin u)^2}$$

$$= 1.$$

$$\left|\iint_S = \int_0^{2\pi} \int_0^{3-\cos u} v^2(1)\, dv\, du\right.$$

$$= \int_0^{2\pi} \left(\frac{1}{3}v^3\right]_0^{3-\cos u}\, du$$

$$= \frac{1}{3}\int_0^{2\pi} (3-\cos u)^3\, du$$

$$= \frac{1}{3}\int_0^{2\pi} (27 - 27\cos u + 9\cos^2 u - \cos^3 u)\, du$$

$$= \frac{1}{3}\left(27u - 27\sin u + 9\left(\frac{1}{2}u + \frac{1}{4}\sin 2u\right)\right.$$

$$\left. -\frac{1}{3}\left((2+\cos^2 u)\sin u\right)\right]_0^{2\pi}$$

$$= 21\pi.$$

Page 1139 (ET Page 1115)

B. A surface such as the paraboloid $z = 4 - x^2 - y^2$ is two-sided,[*] meaning it has a top side (in the positive z-direction) and an underside. Thus, at any point on the paraboloid we can draw an outward-pointing normal vector that varies continuously as we move around the surface.

For a given surface S, if it is possible to choose a normal vector **n** at each point on the surface that varies continuously, then we say that S is **oriented** (with orientation provided by the given choices of **n**). For a closed surface (which forms the boundary of a solid), a **positive orientation** is the orientation that points outward from the solid.

If $\mathbf{F} = P\mathbf{i} + Q\mathbf{j} + R\mathbf{k}$ is a continuous vector field whose domain includes an oriented surface S (with unit normal **n**), the **surface integral**, or **flux**, of **F** across S is defined as

$$\iint_S \mathbf{F}\cdot d\mathbf{S} = \iint_S \mathbf{F}\cdot\mathbf{n}\, dS.$$

There are two special cases for surface integrals:

1) When S is given by $\mathbf{r}(u, v) = x(u, v)\mathbf{i} + y(u, v)\mathbf{j} + z(u, v)\mathbf{k}$, for (u, v) in a domain D:

$$\iint_S \mathbf{F}\cdot d\mathbf{S} = \iint_D \mathbf{F}\cdot(\mathbf{r}_u \times \mathbf{r}_v)\, dA.$$

2) When S is the graph of a function $z = g(x, y)$:

$$\iint_S \mathbf{F}\cdot d\mathbf{S} = \iint_D \left(-P\frac{\partial g}{\partial x} - Q\frac{\partial g}{\partial y} + R\right) dA.$$

[*]Not all surfaces in three dimensions have two sides. See, for example, the Möbius strip.

Example: Let S be that part of the parabolic cylinder $z = 1 - x^2$ that is above the square $0 \le x \le 1, 0 \le y \le 1$. Let $\mathbf{F}$ be the vector field $\mathbf{F}(x, y, z) = 2y\mathbf{i} + 3z\mathbf{j} + 4x\mathbf{k}$.

Find $\displaystyle\iint_S \mathbf{F} \cdot d\mathbf{S}$ two ways.

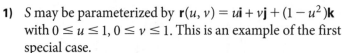

1) S may be parameterized by $\mathbf{r}(u, v) = u\mathbf{i} + v\mathbf{j} + (1 - u^2)\mathbf{k}$ with $0 \le u \le 1, 0 \le v \le 1$. This is an example of the first special case.

$$\mathbf{r}_u = \mathbf{i} + 0\mathbf{j} + (-2u)\mathbf{k} \text{ and } \mathbf{r}_v = 0\mathbf{i} + \mathbf{j} + 0\mathbf{k}$$

$$\mathbf{r}_u \times \mathbf{r}_v = \begin{vmatrix} \mathbf{i} & \mathbf{j} & \mathbf{k} \\ 1 & 0 & -2u \\ 0 & 1 & 0 \end{vmatrix} = 2u\mathbf{i} - 0\mathbf{j} + \mathbf{k}$$

$$\mathbf{F}(\mathbf{r}(u, v)) = 2v\mathbf{i} + 3(1 - u^2)\mathbf{j} + 4u\mathbf{k}$$

$$\mathbf{F}(\mathbf{r}(u, v)) \cdot (\mathbf{r}_u \times \mathbf{r}_v) = (2v)(2u) + 3(1 - u^2)(0) + (4u)(1) = 4uv + 4u$$

$$\iint_S \mathbf{F} \cdot d\mathbf{S} = \int_0^1 \int_0^1 (4uv + 4u) \, dv \, du = \int_0^1 (2uv^2 + 4uv) \Big]_0^1 du$$

$$= \int_0^1 6u \, du = 3u^2 \Big]_0^1 = 3.$$

2) For the vector field $\mathbf{F}$, $P(x, y, z) = 2y$, $Q(x, y, z) = 3z$, and $R(x, y, z) = 4x$. Since S is given by the function $z = g(x, y) = 1 - x^2$, this is an example of the second special case.

$$g_x = -2x \text{ and } g_y = 0$$

$$-P\frac{\partial g}{\partial x} - Q\frac{\partial g}{\partial y} + R = (-2y)(-2x) - (3z)(0) + (4x) = 4xy + 4x.$$

$$\iint_S \mathbf{F} \cdot d\mathbf{S} = \int_0^1 \int_0^1 (4xy + 4x) \, dy \, dx.$$

This is the same integral as in 1) except for the names of the variables.

Therefore, in this case, too: $\displaystyle\iint_S \mathbf{F} \cdot d\mathbf{S} = 3$.

3) Evaluate $\iint\limits_{S} \mathbf{F} \cdot d\mathbf{S}$, where S is the top half of the sphere $x^2 + y^2 + z^2 = 36$ and $\mathbf{F}(x, y, z) = z\mathbf{k}$.

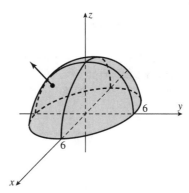

In spherical coordinates, the hemisphere is $\rho = 6$ with $0 \leq \theta \leq 2\pi$ and $0 \leq \phi \leq \dfrac{\pi}{2}$.

A parameterization of S is

$x = 6 \sin \phi \cos \theta,$

$y = 6 \sin \phi \sin \theta,$
$z = 6 \cos \phi,$

$0 \leq \phi \leq \dfrac{\pi}{2}, 0 \leq \theta \leq 2\pi.$

$\mathbf{r}_{\phi} \times \mathbf{r}_{\theta}$

$$= \begin{vmatrix} \mathbf{i} & \mathbf{j} & \mathbf{k} \\ 6 \cos \phi \cos \theta & 6 \cos \phi \sin \theta & -6 \sin \phi \\ -6 \sin \phi \sin \theta & 6 \sin \phi \cos \theta & 0 \end{vmatrix}$$

$$= 36(\sin^2 \phi \cos \theta \mathbf{i} + \sin^2 \phi \sin \theta \mathbf{j} + \sin \phi \cos \phi \mathbf{k}).$$

$\mathbf{F} \cdot (\mathbf{r}_{\phi} \times \mathbf{r}_{\theta}) = z(36 \sin \phi \cos \phi)$
$$= 6 \cos \phi 36 \sin \phi \cos \phi$$
$$= 216 \sin \phi \cos^2 \phi.$$

$$\iint\limits_{S} \mathbf{F} \cdot d\mathbf{S} = \int_0^{\pi/2} \int_0^{2\pi} 216 \sin \phi \cos^2 \phi \, d\theta \, d\phi$$

$$= 216 \int_0^{2\pi} \sin \phi \cos^2 \phi \, \theta \Big]_0^{2\pi} d\phi$$

$$= 512\pi \int_0^{\pi/2} \sin \phi \cos^2 \phi \, d\phi$$

$$= \frac{512\pi}{-3} \cos^3 \phi \Big]_0^{\pi/2}$$

$$= \frac{512\pi}{3}.$$

4) Evaluate $\displaystyle\iint_S \mathbf{F}\cdot d\mathbf{S}$, where

$\mathbf{F}(x, y, z) = (x + y + z)\mathbf{i} + (x - y)\mathbf{j} + (xy + z)\mathbf{k}$ and S is the graph of the function $z = xy$ over the region D: $0 \le x \le 1, 0 \le y \le 1$.

From $z = xy$,

$P(x,y,z) = x + y + z = x + y + xy$
$Q(x,y,z) = x - y$
$R(x,y,z) = xy + z = xy + xy = 2xy.$

$$\frac{\partial z}{\partial x} = y \text{ and } \frac{\partial z}{\partial y} = x.$$

$$-P\frac{\partial z}{\partial x} - Q\frac{\partial z}{\partial y} + R$$
$$= -(x + y + xy)y - (x - y)x + 2xy$$
$$= 2xy - y^2 - xy^2 - x^2.$$

$$\iint_S \mathbf{F}\cdot d\mathbf{S} = \iint_D (2xy - y^2 - xy^2 - x^2)\,dA$$

$$= \int_0^1\int_0^1 (2xy - y^2 - xy^2 - x^2)\,dy\,dx$$

$$= \int_0^1 \left(xy^2 - \frac{y^3}{3} - x\frac{y^3}{3} - x^2 y \right)\Bigg]_0^1 dx$$

$$= \int_0^1 \left(\left(x - \frac{1}{3} - \frac{1}{3}x - x^2 \right) - (0) \right) dx$$

$$= \int_0^1 \left(-x^2 + \frac{2}{3}x - \frac{1}{3} \right) dx$$

$$= \left(-\frac{x^3}{3} + \frac{x^2}{3} - \frac{x}{3} \right)\Bigg]_0^1 = -\frac{1}{3}.$$

Section 16.8 Stokes' Theorem

This section and the next provide extensions of Green's Theorem to three dimensions. Green's Theorem relates a double over a (flat) region in the plane to a line integral around the closed boundary curve. Stokes' Theorem relates a surface integral for a surface S in three dimensions to a line integral around the closed boundary curve of S.

Concept to Master

Stokes' Theorem

Summary and Focus Questions

Stokes' Theorem provides a relationship between a surface integral and a line integral around the boundary of the surface.

Page 1146 (ET Page 1122)

Stokes' Theorem Let S be a smooth, simply connected, orientable surface bounded by a simple closed curve C. Let $\mathbf{F} = P\mathbf{i} + Q\mathbf{j} + R\mathbf{k}$, where P, Q, and R are component functions with continuous first partial derivatives. Finally, suppose C has a positive orientation (meaning that as you look down on the surface from any outward normal to the surface, the orientation of C is counterclockwise). Then:

$$\int_C \mathbf{F} \cdot d\mathbf{r} = \iint_S (\text{curl } \mathbf{F}) \cdot d\mathbf{S}.$$

1) Use Stokes' Theorem to rewrite $\int_C \mathbf{F} \cdot d\mathbf{r}$ where $\mathbf{F}(x, y, z) = xy\mathbf{i} + yz\mathbf{j} + xz\mathbf{k}$ and C is the counterclockwise-oriented triangular boundary of the intersection of the plane $x + 2y + 4z = 8$ with the first octant.

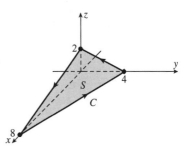

Let S be to plane surface bounded by C. By Stokes' Theorem:

$$\int_C \mathbf{F} \cdot d\mathbf{r} = \iint_S \text{curl } \mathbf{F} \cdot d\mathbf{S}$$

From the components of **F**, $\dfrac{\partial P}{\partial y} = x$, $\dfrac{\partial P}{\partial z} = 0$,

$\dfrac{\partial Q}{\partial x} = 0$, $\dfrac{\partial Q}{\partial z} = y$, $\dfrac{\partial R}{\partial x} = z$, and $\dfrac{\partial R}{\partial y} = 0$.

$$\text{curl } \mathbf{F} = (0 - y)\mathbf{i} + (0 - z)\mathbf{j} + (0 - x)\mathbf{k}$$
$$= -y\mathbf{i} - z\mathbf{j} - x\mathbf{k}.$$

The surface S is $\mathbf{r}(u, v) = <x, y, z>$, where
$x = 8 - 2u - 4v,\ y = u,\ z = v,$
$0 \le u \le 4, 0 \le v \le 2 - \dfrac{u}{2}.$
$\mathbf{r}_u = \langle -2, 1, 0 \rangle, \mathbf{r}_v = \langle -4, 0, 1 \rangle.$

$$\mathbf{r}_u \times \mathbf{r}_v = \begin{vmatrix} \mathbf{i} & \mathbf{j} & \mathbf{k} \\ -2 & 1 & 0 \\ -4 & 0 & 1 \end{vmatrix} = \mathbf{i} + 2\mathbf{j} + 4\mathbf{k}$$

is the normal vector to the plane S.

$\text{curl } \mathbf{F} \cdot (\mathbf{r}_u \times \mathbf{r}_v)$
$\quad = -y(1) + (-z)2 + (-x)4$
$\quad = -y - 2z - 4x$
$\quad = -u - 2v - 4(8 - 2u - 4v)$
$\quad = 7u + 14v - 32.$

$$\iint\limits_{S} \text{curl } \mathbf{F} \cdot d\mathbf{S}$$
$$= \int_0^4 \int_0^{2-(u/2)} (7u + 14v - 32)\, dv\, du.$$

2) Use Stokes' Theorem to rewrite
$\displaystyle\iint\limits_{S} \text{curl } \mathbf{F} \cdot d\mathbf{S}$, where:
$\mathbf{F}(x, y, z) = (x - y)\mathbf{i} + (y - z)\mathbf{j} + (x - z)\mathbf{k}$
and S is the paraboloid
$z = 4 - x^2 - y^2,\ z \ge 0$, oriented upward.

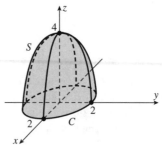

The boundary of S is the circle C in the
xy plane: $x^2 + y^2 = 4,\ z = 0$. Parameterize
C with

$\mathbf{r}(t) = 2\cos t\, \mathbf{i} + 2\sin t\, \mathbf{j} + 0\mathbf{k},\ 0 \le t \le 2\pi.$
$\mathbf{r}'(t) = -2\sin t\, \mathbf{i} + 2\cos t\, \mathbf{j}.$
$\mathbf{F}(\mathbf{r}(t)) = 2((\cos t - \sin t)\mathbf{i} + \sin t\, \mathbf{j} + \cos t\, \mathbf{k}).$
$\mathbf{F}(\mathbf{r}(t)) \cdot \mathbf{r}'(t) = 4(-\cos t \sin t + \sin^2 t + \cos t \sin t)$
$\quad\quad\quad\quad\quad = 4\sin^2 t.$

$$\iint\limits_{S} \text{curl } \mathbf{F} \cdot d\mathbf{S} = \iint\limits_{C} \mathbf{F} \cdot d\mathbf{r} = \int_0^{2\pi} 4\sin^2 t\, dt.$$

Section 16.9 The Divergence Theorem

Stokes' Theorem (from the previous section) moves Green's Theorem up to three dimensions by increasing the domains to three dimensions—replacing the double integral of Green's Theorem with a surface integral and the line integral in the plane with a line integral in three dimensions. In this section the Divergence Theorem may be thought of as boosting the integrals up by one dimension— replacing the double integral of Green's Theorem by a triple integral of a solid and replacing the line integral around a region by a surface integral around the boundary of the solid.

Concept to Master

Divergence Theorem

Summary and Focus Questions

Page 1152 (ET Page 1129)

Divergence Theorem   Let E be a simple solid whose boundary surface S has positive orientation. Let $\mathbf{F} = P\mathbf{i} + Q\mathbf{j} + R\mathbf{k}$, where P, Q, and R are component functions with continuous first partial derivatives on an open region containing E. Then

$$\iint_S \mathbf{F} \cdot d\mathbf{S} = \iiint_E \operatorname{div} \mathbf{F} \, dV.$$

1) Write a triple iterated integral for $\iint_S \mathbf{F} \cdot d\mathbf{S}$, where $\mathbf{F}(x, y, z) = xy\mathbf{i} + y^2\mathbf{k} + xz\mathbf{k}$ and S is the triangular surface bounded by the coordinate planes and $6x + 4y + 3z = 12$ in the first octant.

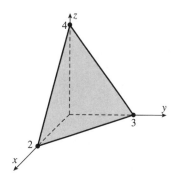

Let E be the solid region enclosed by S:

$0 \le x \le 2$

$0 \le y \le 3 - \dfrac{3}{2}x$

$0 \le z \le 4 - 2x - \dfrac{4}{3}y.$

$P(x, y, z) = xy, Q(x, y, z) = 2y,$

$R(x, y, z) = xz. \dfrac{\partial P}{\partial x} = y, \dfrac{\partial Q}{\partial y} = 2y, \dfrac{\partial R}{\partial z} = x.$

$\operatorname{div} \mathbf{F} = y + 2y + x = x + 3y.$

$$\iint_S \mathbf{F} \cdot d\mathbf{S} = \iiint_E (x + 3y)\, dV$$

$$= \int_0^2 \int_0^{3-(3/2)x} \int_0^{4-2x-(4/3)y} (x + 3y)\, dz\, dy\, dx.$$

2) Let $\mathbf{F} = (x, y, z) = \left\langle x, y^2, z^3 \right\rangle.$

   a) Find div $\mathbf{F}$.

$\operatorname{div} \mathbf{F} = 1 + 2y + 3z^2.$

   b) Write a surface integral for
$$\iiint_E (1 + 2y + 3z^2)\, dV,$$
     where E is the solid ball
     $x^2 + y^2 + z^2 \le 1$ in the first octant.

In the first octant, the surface S of E may be parameterized by $\mathbf{r}(u,v) = \langle x, y, z \rangle$:

$x = \sin u \cos v$

$y = \sin u \sin v$

$z = \cos u$

$0 \le u \le \dfrac{\pi}{2}, 0 \le v \le \dfrac{\pi}{2}.$

$$\iiint_E (1 + 2y + 3z^2)\, dV = \iint_S \mathbf{F} \cdot d\mathbf{S}$$

   c) Write an iterated integral for your
     answer to part b).

$\mathbf{r}_u = \langle \cos u \cos v,\ \cos u \sin v,\ -\sin u \rangle$ and
$\mathbf{r}_v = \langle -\sin u \sin v,\ \sin u \cos v,\ 0 \rangle.$

$$\mathbf{r}_u \times \mathbf{r}_v = \begin{vmatrix} \mathbf{i} & \mathbf{j} & \mathbf{k} \\ \cos u \cos v & \cos u \sin v & -\sin u \\ -\sin u \sin v & \sin u \cos v & 0 \end{vmatrix}$$

$= (\sin^2 u \cos v)\mathbf{i} + (\sin^2 u \sin v)\mathbf{j} + (\sin u \cos u)\mathbf{k}$

$\left| \mathbf{r}_u \times \mathbf{r}_v \right|$

$= \sqrt{\sin^4 u \cos^2 v + \sin^4 u \sin^2 v + \sin^2 u \cos^2 u}$

$= \sin u$ since $0 \le u \le \dfrac{\pi}{2}.$

$$\iint_S \mathbf{F} \cdot d\mathbf{S}$$

$$= \int_0^{\pi/2} \int_0^{\pi/2} (1 + 2\sin u \sin v + 3\cos^2 u)\sin u\, dv\, du.$$

Chapter 17 — Second-Order Differential Equations

Section 17.1   Second-Order Linear Equations

This chapter is a continuation of the chapter on differential equations, where you studied first-order linear differential equations. This section begins the study of second-order linear equations.

Concepts to Master

A. Second-order linear differential equations; Homogeneous; Linear combination; Linearly independent; Auxiliary equation

B. Initial values; Boundary values

C. Technology Plus

Summary and Focus Questions

Page 1166 (ET Page 1142)*

A. A **second-order linear differential equation** has the form:
$$P(x)y'' + Q(x)y' + R(x)y = G(x).$$
This section considers the **homogeneous** case, where $G(x) = 0$.

Examples: a) $x^2 y'' + (2x + 1)y' + 10y = 0$ is homogeneous.
b) $x^2 y'' + (2x + y)y' + 10xy = 0$ is not homogeneous because of the term yy'.
c) $5y''' + xy' + 3y = 0$ is not second order.

If y_1 and y_2 are two solutions to the homogeneous equation, then all linear combinations $c_1 y_1 + c_2 y_2$ are also solutions (c_1, c_2 real numbers).

Solutions y_1 and y_2 are **linearly independent** means neither y_1 nor y_2 is a constant multiple of the other.

If y_1 and y_2 are linearly independent solutions, then *all* solutions may be written as linear combinations of y_1 and y_2: $c_1 y_1 + c_2 y_2$.

In the special case where P, Q, and R are constants, the differential equation has the form:
$$ay'' + by' + cy = 0.$$

*When using the Early Transcendentals text, use the page number in parentheses.

The corresponding **auxiliary equation** (with variable r) is $ar^2 + br + c = 0$.
The type of roots (real or imaginary) of the auxiliary equation determine the
form of the general solution to $ay'' + by' + cy = 0$:

$ar^2 + br + c = 0$		
Discriminant	**Roots**	**Solution to $ay'' + by' + cy = 0$**
$b^2 - 4ac > 0$	r_1, r_2 real, $r_1 \neq r_2$	$y = c_1 e^{r_1 x} + c_2 e^{r_2 x}$
$b^2 - 4ac = 0$	r real	$y = c_1 e^{rx} + c_2 x e^{rx}$
$b^2 - 4ac < 0$	r_1, r_2 complex, $r_1 \neq r_2$ $r_1 = \alpha + \beta i$ $r_2 = \alpha - \beta i$	$y = e^{\alpha x}(c_1 \cos \beta x + c_2 \sin \beta x)$

Example: Solve the equation $y'' - 8y' + 16 = 0$.
The discriminant is $(-8)^2 - 4(1)(16) = 0$. There is one solution to the
auxiliary equation $r^2 - 8r + 16 = 0$. Find the solution by factoring:

$$(r - 4)^2 = 0$$
$$r = 4.$$

The general solution to $y'' - 8y' + 16 = 0$ is $y = c_1 e^{4x} + c_2 x e^{4x}$, where c_1 and
c_2 are real numbers.

1) True or False:
$xy' + y'' + 5x + 10y = 0$ is homogeneous.

False (because of the term $5x$).

2) True or False:
$y_1 = 2xy - 3y^2$ and $y_2 = 12y^2 - 8xy$ are
linearly independent.

False. Since $-4y_1 = y_2$, they are dependent.

3) Sometimes, Always, or Never:
If y_1 and y_2 are solutions to
$ay'' + by' + cy = 0$, then all solutions are
of the form $c_1 y_1 + c_2 y_2$.

Sometimes. Solutions y_2 and y_1 must be
linearly independent for this to be true.

4) Find the general solution to each:
  a) $y'' + 6y' + 8y = 0$.

The auxiliary equation is $r^2 + 6r + 8 = 0$.
$(r + 2)(r + 4) = 0$.
$r = -2, -4$.
The general solution is $y = c_1 e^{-2x} + c_2 e^{-4x}$.

b) $y'' - 6y' + 13y = 0.$

The auxiliary equation is $r^2 - 6r + 13 = 0.$
$$r = \frac{-(-6) \pm \sqrt{36 - 4(1)(13)}}{2} = \frac{6 \pm \sqrt{-16}}{2}$$
$$= 3 \pm 2i.$$
The general solution is
$$y = e^{3x}(c_1 \cos 2x + c_2 \sin 2x).$$

c) $4y'' - 12y' + 9y = 0.$

The auxiliary equation is
$$4r^2 - 12r + 9 = 0.$$
$$(2r - 3)^2 = 0.$$
$$r = \frac{3}{2}.$$
The general solution is
$$y = c_1 e^{(3/2)x} + c_2 x e^{(3/2)x}.$$

Page 1170 (ET Page 1146)

B. Because the general solution has two parameters, c_1 and c_2, two conditions must be given to the general solution of a second-order differential equation to specify a particular solution. Depending on what is specified, we have two types of problems.

1) If $y(x_0) = y_0$ and $y'(x_0) = y_1$ are specified, we have an **initial-value problem** (initial values for y and y' for x_0 are given).

2) If $y(x_0) = y_0$ and $y(x_1) = y_1$ are specified, we have a **boundary-value problem** (values for y for two different points x_0 and x_1 are given).

For continuous functions P, Q, R, and G with $P(x) \neq 0$, initial-value problems will always have a solution, but boundary-value problems need not necessarily have a solution. The method for determining the particular solution involves substituting the conditions in y and y' and solving two equations in the two unknowns c_1 and c_2.

Example: Solve the equation $y'' - 8y' + 16 = 0$ with initial conditions $y(0) = 2$ and $y'(0) = 5$.

In a previous example, we saw that the general solution to $y'' - 8y' + 16 = 0$ is $y = c_1 e^{4x} + c_2 x e^{4x}$.

$y(0) = 2$. Also $y(0) = c_1 e^{4(0)} + c_2(0)e^{4(0)} = c_1$. Therefore, $c_1 = 2$.

Next use $y'(0) = 5$ to find c_2.

$y'(x) = 4c_1 e^{4x} + c_2 x(4 e^{4x}) + c_2 e^{4x}(1).$

$y'(0) = 4c_1 e^{4(0)} + c_2(0)(4e^{4(0)}) + c_2 e^{4(0)} = 4c_1 + c_2.$

Since $c_1 = 2$, $5 = 4(2) + c_2$. Therefore, $c_2 = -3$.

The particular solution is $y = 2e^{4x} - 3xe^{4x}.$

5) Solve $y'' + 2y' - 8y = 0$ with initial conditions $y(0) = 28$ and $y'(0) = 2$.

<After finding the general solution, use the initial conditions to find two equations in the two variables c_1 and c_2.>

$r^2 + 2r - 8 = 0$ has solutions $r = -4, 2$.
The general solution is $y = c_1 e^{2x} + c_2 e^{-4x}$.
Since $y(0) = 28$, $c_1 + c_2 = 28$.

$y' = 2c_1 e^{2x} - 4c_2 e^{-4x}$.

Since $y'(0) = 2$, $2c_1 - 4c_2 = 2$ or

$c_1 - 2c_2 = 1$.
The system $c_1 + c_2 = 28$
$c_1 - 2c_2 = 1$

has solution $3c_2 = 27$
$c_2 = 9$.

$c_1 - 2(9) = 1$, $c_1 = 19$.
The particular solution is

$y = 19e^{2x} + 9e^{-4x}$.

6) Solve $y'' - 4y' + 3y = 0$ with boundary conditions $y(0) = e^2$, $y(1) = e$.

$r^2 - 4r + 3 = 0$
$(r - 3)(r - 1) = 0$
$r = 1, 3$.
The general solution is $y = c_1 e^x + c_2 e^{3x}$.
$y(0) = e^2$ implies $c_1 + c_2 = e^2$.
$y(1) = e$ implies $c_1 e + c_2 e^3 = e$, or
$c_1 + c_2 e^2 = 1$.

Subtract the second equation from the first:
$c_2 - c_2 e^2 = e^2 - 1$
$c_2(1 - e^2) = e^2 - 1 = -(1 - e^2)$
$c_2 = -1$.
Then $c_1 - 1 = e^2$, so $c_1 = e^2 + 1$.
The solution is $y = (e^2 + 1)e^x - e^{3x}$.

7) Solve $y'' - 6y' + 9y = 0$ with boundary conditions $y(0) = 2$, $y(1) = e^3$.

$r^2 - 6r + 9 = 0$ has solution $r = 3$.
The general solution is $y = c_1 e^{3x} + c_2 x e^{3x}$.
$y(0) = 2$ implies $c_1 = 2$.
$y(1) = 1$ implies $c_1 e^3 + c_2 e^3 = e^3$.
$c_1 + c_2 = 1$.
Since $c_1 = 2$, $c_2 = -1$.
The particular solution is $y = 2e^{3x} - x e^{3x}$.

C. Technology Plus. Use a computer algebra system or a graphing calculator to solve.

T-1) **a)** Find and graph two linearly independent solutions $y_1 = e^{r_1 x}$ and $y_2 = e^{r_2 x}$ for $12y'' + y' - y = 0$.

$12r^2 + r - 1 = 0.$
$(3r + 1)(4r - 1) = 0$

$r = -\dfrac{1}{3}$ and $r = \dfrac{1}{4}$.

The general solution is $y = c_1 e^{-x/3} + c_2 e^{x/4}$.
For $c_1 = 1$ and $c_2 = 0$, $y_1 = e^{-x/3}$.
For $c_1 = 0$ and $c_2 = 1$, $y_2 = e^{x/4}$.

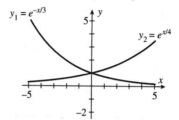

b) Now, on the same screen, graph $y = c_1 e^{-x/3} + c_2 e^{x/4}$ for several values of c_1 and c_2, such as $c_1 = \pm 1, \pm 2$ and $c_2 = \pm 1, \pm 2$.

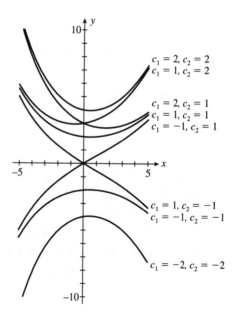

c) What features do all the solutions have in common?

For any c_1 and c_2, not both zero,

$$\lim_{x \to \infty} y = \pm\infty \text{ and } \lim_{x \to -\infty} y = \pm\infty.$$

All solutions cross the y-axis at $c_1 + c_2$.

Section 17.2 Nonhomogeneous Linear Equations

The previous section discussed methods for solving second-order homogeneous equations of the form $ay'' + by' + cy = 0$. This section considers nonhomogeneous equations of the form $ay'' + by' + cy = G(x)$. The first step in finding solutions to an equation of this form is solving the corresponding homogeneous case.

Concepts to Master

A. Nonhomogeneous linear differential equation; Method of Undetermined Coefficients

B. Method of Variation of Parameters

Summary and Focus Questions

Page
1172
(ET Page
1148)

A. A second-order **linear nonhomogeneous differential equation** with constant differential coefficients has the form:

$$ay'' + by' + cy = G(x),$$

where $G(x)$ is continuous. The corresponding **complementary differential equation** is:

$$ay'' + by' + cy = 0.$$

To solve $ay'' + by' + cy' = G(x)$:

Step 1. Find the general solution y_c to $ay'' + by' + cy = 0$.
Step 2. Find a particular solution y_p to $ay'' + by' + cy = G(x)$.
Step 3. The general solution is $y(x) = y_c(x) + y_p(x)$.

The particular solution y_p in Step 2 can sometimes be found using the method of **undetermined coefficients**:

i) Choose a trial form for the particular solution $y_p(x)$ from the following table.

$G(x)$	$y_p(x)$
$G(x) = e^{kx}P(x)$, where $P(x)$ is a polynomial of a degree n.	$y_p(x) = e^{kx}Q(x)$, where $Q(x)$ is polynomial of a degree n: $Q(x) = A_n x^n + \cdots + A_1 x + A_0$.
$G(x) = e^{kx}P(x) \cos mx$ or $G(x) = e^{kx}P(x) \sin mx$, where $P(x)$ is a polynomial of degree n.	$y_p(x) = e^{kx}Q(x) \cos mx + e^{kx}R(x) \sin mx$, where $Q(x)$ and $R(x)$ are polynomials of degree n: $Q(x) = A_n x^n + \cdots + A_1 x + A_0$ $R(x) = B_n x^n + \cdots + B_1 x + B_0$.

ii) Substitute $y_p''(x)$, $y_p'(x)$, and $y_p(x)$ into $ay'' + by' + cy = G(x)$ and solve for the coefficients of $Q(x)$ and, if appropriate, $R(x)$.

iii) Check to see whether $y_p(x)$ solves the complementary equation $ay'' + by' + cy = 0$.

iv) If any term of y_p is not a solution to the complementary equation, use y_p as the particular solution. If any term of y_p is a solution to the complementary equation and xy_p is not, xy_p is the particular solution. If any terms of y_p and xy_p are solutions to the complementary equation, use $x^2 y_p$ as the particular solution.

Example: Find the general solution to $y'' - 4y' + 3y = 10\cos x$.

Step 1. To find the general solution to $y'' - 4y' + 3y = 0$, we see that $r^2 - 4r + 3 = 0$ has solution $r = 1$ and $r = 3$. Therefore, $y_c(x) = c_1 e^x + c_2 e^{3x}$.

Step 2. Find a particular solution to the nonhomogeneous equation.

i) $G(x) = 10\cos x$ has the form $e^{kx} P(x) \cos mx$ where $k = 0$, $P(x) = 10$, and $m = 1$. Therefore, Q and R have degree $n = 0$ and have the form $Q(x) = A$ and $R(x) = B$. Thus, $y_p(x) = A\cos x + B\sin x$.

ii) $y_p'(x) = -A\sin x + B\cos x$
$y_p''(x) = -A\cos x - B\sin x$
Substitute these results into $y'' - 4y' + 3y = 10\cos x$:
$(-A\cos x - B\sin x) - 4(-A\sin x + B\cos x) + 3(A\cos x + B\sin x) = 10\cos x$
and collect terms:

$$(2A - 4B)\cos x + (4A + 2B)\sin x = 10\cos x$$

Compare coefficients:

$$2A - 4B = 10 \text{ and } 4A + 2B = 0.$$

These equations have solutions $A = 1$ and $B = -2$.
Therefore, $y_p(x) = \cos x - 2\sin x$.

iii) No term of $y_p(x) = \cos x - 2\sin x$ is a solution to $y'' - 4y' + 3y = 0$.

iv) Our particular solution is $y_p(x) = \cos x - 2\sin x$.

Step 3. The general solution to $y'' - 4y' + 3y = 10\cos x$ is

$$y_c(x) = c_1 e^x + c_2 e^{3x} + \cos x - 2\sin x.$$

When $G(x)$ is the sum of terms, find particular solutions for each term and add them. For example, if $G(x) = G_1(x) + G_2(x)$:

Find $y_{p_1}(x)$, a particular solution to $ay'' + by' + cy = G_1(x)$.
Find $y_{p_2}(x)$, a particular solution to $ay'' + by' + cy = G_2(x)$.
Then $y_p(x) = y_{p_1}(x) + y_{p_2}(x)$ is a particular solution to $ay'' + by' + cy = G(x)$.

1) Solve each equation.

    a) $y'' + 3y' - 10y = 16e^{3x}$.

Step 1. The complementary equation
$y' + 3y' - 10y = 0$ has auxiliary equation
$r^2 + 3r - 10 = 0$.
$(r - 2)(r + 5) = 0$,
so $r = 2, -5$.
The general solution to the homogeneous
equation is $y(x) = c_1 e^{2x} + c_2 e^{-5x}$.

Therefore, the general solution to
the nonhomogeneous equation is
$y_c(x) = c_1 e^{2x} + c_2 e^{-5x} + y_p(x)$.

Step 2. Find $y_p(x)$.
    i) $G(x) = 16e^{3x}$, so $k = 3$, $P(x) = 16$, $n = 0$.
       Therefore $Q(x) = A$.
       So, $y_p(x) = Ae^{3x}$.

    ii) $y_p'(x) = 3Ae^{3x}$, $y_p''(x) = 9Ae^{3x}$.

       Substituting these into the original
       equation yields
       $9Ae^{3x} + 3(3Ae^{3x}) - 10(Ae^{3x}) = 16e^{3x}$,

       $8Ae^{3x} = 16e^{3x}$, so $A = 2$.
       Hence $y_p = 2e^{3x}$.
    iii) We note $y = 2e^{3x}$ is not a solution to the
       complementary equation.
    iv) Therefore $y_p(x) = 2e^{3x}$.

Step 3. The general solution is
$y(x) = c_1 e^{2x} + c_2 e^{-5x} + 2e^{3x}$.

    b) $y'' - 7y' + 12y = 24x - 2$.

Step 1. The complementary equation
$y'' - 7y' + 12y = 0$ has auxiliary equation

$r^2 - 7r + 12 = 0$.
$r = 3$ and $r = 4$.
The general solution to $y'' - 7y' + 12y = 0$ is
$y_c(x) = c_1 e^{3x} + c_2 e^{4x}$.

Step 2. Find $y_p(x)$.
    i) $G(x) = 24x - 2$ is of the form $e^{kx} P(x)$,
       where $k = 0$, $n = 1$, and $P(x) = 24x - 2$.
       So, $Q(x) = A_1 x + A_0$ and
       $y_p(x) = (A_1 x + A_0)e^0 = A_1 x + A_0$.

ii) $y_p'(x) = A_1$ and $y_p''(x) = 0$.

Substituting into $y'' - 7y' + 12y = 24x - 2$,

$0 - 7(A_1) + 12(A_1 x + A_0) = 24x - 2$

$12A_1 x + (-7A_1 + 12A_0) = 24x - 2$

$12A_1 = 24$ and $-7A_1 + 12A_0 = -2$.

Hence, $A_1 = 2$ and $-7(2) + 12A_0 = -2$,

$12A_0 = 12$, so $A_0 = 1$.

Therefore, $y_p(x) = 2x + 1$.

iii) No term of $y_p(x) = 2x + 1$ is a solution to $y'' - 7y' + 12y = 0$.

iv) $y_p(x) = 2x + 1$.

Step 3. The general solution to the nonhomogeneous equation is

$y(x) = c_1 e^{3x} + c_2 e^{4x} + 2x + 1$.

Page 1178
(ET Page 1153)

B. There is an alternate method for Step 2 that, unlike the undetermined coefficients method, does not rely on the form of $G(x)$.

To find a particular solution $y_p(x)$ to $ay'' + by' + cy = G(x)$, use the method of **variation of parameters**:

i) Find any pair of linearly independent solutions y_1 and y_2 to the complementary equation $ay'' + by' + cy = 0$.

ii) Solve the following system for u_1' and u_2':

$$u_1' y_1 + u_2' y_2 = 0$$

$$a(u_1' y_1' + u_2' y_2') = G(x).$$

The solution is $u_1' = \dfrac{-G(x)y_2}{a(y_1 y_2' - y_2 y_1')}$ and $u_2' = \dfrac{G(x)y_1}{a(y_1 y_2' - y_2 y_1')}$.

iii) Integrate u_1' and u_2' to get $u_1(x)$ and $u_2(x)$.

iv) A particular solution is $y_p(x) = u_1(x)y_1(x) + u_2(x)y_2(x)$.

Example: Solve the equation $y'' + 3y' + 2y = \cos x$ using the method of variation of parameters.

Step 1. The auxiliary equation for the complementary differential equation $y'' + 3y' + 2y = 0$ is $r^2 + 3r + 2 = 0$ and has solution $r = -1$ and $r = -2$. Thus, the complementary equation has general solution $y_c(x) = c_1 e^{-x} + c_2 e^{-2x}$.

Step 2.

i) Choose $c_1 = 1, c_2 = 0$ and $c_1 = 0, c_2 = 1$ to give the pair of independent solutions to $y'' + 3y' + 2y = 0$:

$$y_1 = e^{-x} \text{ and } y_2 = e^{-2x}.$$

The solutions y_1 and y_2 are linearly independent because neither is a constant multiple of the other.

ii) $y_1' = -e^{-x}$ and $y_2' = -2e^{-2x}$.

We solve the following system for u_1' and u_2':

$$u_1' \, (e^{-x}) + u_2' \, (e^{-2x}) = 0$$
$$1(u_1' \, (-e^{-x}) + u_2' \, (-2e^{-2x}) = \cos x.$$

First determine the denominator of the solution:

$$a(y_1y_2' - y_2y_1') = 1\big((e^{-x})(-2e^{-2x}) - (e^{-2x})(-e^{-x})\big) = -e^{-3x}.$$

Now find u_1' and u_2':

$$u_1' = \frac{-(\cos x)e^{-2x}}{-e^{-3x}} = (\cos x)e^x \text{ and } u_2' = \frac{(\cos x)e^{-x}}{-e^{-3x}} = -(\cos x)e^{2x}.$$

iii) Use integration by parts to find u_1 and u_2:

$$u_1 = \int (\cos x)e^x \, dx = \frac{e^x}{2}(\cos x + \sin x)$$

$$u_2 = \int -(\cos x)e^{2x} \, dx = -\frac{e^{2x}}{5}(2\cos x + \sin x).$$

There is no need to include the constants of integration.

iv) A particular solution to $y'' + 3y' + 2y = \cos x$ is:

$$y_p = u_1y_1 + u_2y_2 = \left(\frac{e^x}{2}(\cos x + \sin x) \right)e^{-x} + \left(-\frac{e^{2x}}{5}(2\cos x + \sin x) \right)e^{-2x}$$

$$= \frac{1}{2}(\cos x + \sin x) + \left(-\frac{1}{5} \right)(2\cos x + \sin x) = \frac{1}{10}\cos x + \frac{3}{10}\sin x.$$

Step 3. The general solution is:

$$y(x) = c_1e^{-x} + c_2e^{-2x} + \frac{1}{10}\cos x + \frac{3}{10}\sin x.$$

3) Solve $y'' - 5y' + 4y = 30e^{6x}$.

Step 1. The auxiliary equation is
$y'' - 5y' + 4y = 0$.
From $r^2 - 5r + 4 = 0$,
$(r-1)(r-4) = 0, r = 1, r = 4$.
Thus, the general solution to
$y'' - 5y' + 4y = 0$ is $y_c(x) = c_1 e^x + c_2 e^{4x}$.
Step 2.

i) Let $c_1 = 1, c_2 = 0$; then $y_1 = e^x$.
   Let $c_1 = 0, c_2 = 1$; then $y_2 = e^{4x}$.
   y_1 and y_2 are linearly independent and
   $y_1' = e^x$, $y_2' = 4e^{4x}$.

ii) We solve the system
   $u_1' e^x + u_2' e^{4x} = 0$.

   $u_1' e^x + u_2' (4e^{4x}) = 30e^{6x}$.

   $a(y_1 y_2' - y_2 y_1') = 1((e^x)(4e^{4x}) - (e^{4x})(e^x))$
   $\qquad\qquad\qquad = 3e^{5x}$.

   $u_1' = \dfrac{-(30e^{6x})e^{4x}}{3e^{5x}} = -10e^{5x}$ and

   $u_2' = \dfrac{(30e^{6x})e^x}{3e^{5x}} = 10e^{2x}$.

iii) From $u_1' = -10e^{5x}, u_1 = -2e^{5x}$.
   From $u_2' = 10e^{2x}, u_2 = 5e^{2x}$.

iv) The particular solution is
   $y_p(x) = u_1 y_1 + u_2 y_2$
   $\qquad = (-2e^{5x})e^x + (5e^{2x})e^{4x} = 3e^{6x}$.
Step 3. The general solution to the
nonhomogeneous equation is
$y(x) = c_1 e^x + c_2 e^{4x} + 3e^{6x}$.

Section 17.3 Applications of Second-Order Differential Equations

Second-order differential equations have applications in many diverse areas. This section presents two from science and engineering—vibrating springs and electrical circuits.

Concepts to Master

Phenomena modeled by $ay'' + by' + cy = G(x)$

Summary and Focus Questions

Page 1180 (ET Page 1156)

Each of the following applications is modeled by $ay'' + by' + cy = G(x)$, whose solution methods are found in Section 17.2. The interpretation of the constants and functions differ from application to application, but the mathematics is the same.

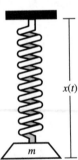

1) Vibrating spring: $mx'' + kx = 0$.

   Interpretation: Stretching or compressing a spring by x units from its natural length:

   $\quad x$ = displacement from equilibrium at time t
   $\quad x', x''$ = velocity and acceleration
   $\quad\quad m$ = mass of an object at the end of the spring
   $\quad\quad k$ = spring constant

2) Damped oscillating motion: $mx'' + cx' + kx = 0$.

   Interpretation: Stretching or compressing a spring that is also subject to a damping force (such as friction or and movement through a thick liquid) by x units:

   $\quad x$ = displacement from equilibrium at time t
   $\quad x', x''$ = velocity and acceleration
   $\quad\quad m$ = mass of an object at the end of the spring
   $\quad\quad k$ = damping constant (such as fluid resistance)
   $\quad\quad c$ = spring constant

3) Damped oscillating motion: $mx'' + cx' + kx = F(t)$.

   Interpretation: Same as before, but with the added external force F.

4) Electric circuits: $LQ'' + RQ' + \dfrac{1}{C}Q = E(t)$.

Interpretation: An electric charge E (such as from a battery) applied to a circuit with components inducing various voltage drops:

Q = charge on the capacitor at time t
Q' = current
L = inductance constant for an inductor
R = resistance constant for a resistor
$\dfrac{1}{C}$ = elastance constant for a capacitor
$E(t)$ = electromotive force applied to a circuit

1) A spring with a 10-kg mass can be stretched 0.5 m beyond its equilibrium by a force of 85 N (newtons). Suppose the spring is held in a fluid with damping constant 20. If the mass starts at equilibrium with an initial velocity of 1 m/s, find the position of the mass after t seconds.

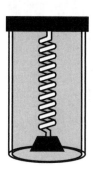

We are given $m = 10$, $F(t) = 0$, $c = 20$, $x(0) = 0$, and $x'(0) = 1$.
To find the spring constant, k, use Hooke's Law: $k(0.5) = 85$, $k = 170$.
The differential equation model is
$10x'' + 20x' + 170x = 0$, or
$x'' + 2x' + 17x = 0$, with initial values
$x(0) = 0$, $x'(0) = 1$.
From $r^2 + 2r + 17 = 0$, $r = -1 \pm 4i$.
Thus, the general solution is
$x = e^{-t}(c_1 \cos 4t + c_2 \sin 4t)$.
At $t = 0$, $x = 0$, so
$0 = 1(c_1 \cos 0 + c_2 \sin 0)$.
Therefore, $c_1 = 0$.
Thus, $x = e^{-t}(c_2 \sin 4t)$.
$x' = e^{-t}(4c_2 \cos 4t) - e^{-t}(c_2 \sin 4t)$.
At $t = 0$, $x' = 1$, so
$1 = 1(4c_2) - 1(c_2(0))$
$c_2 = 0.25$.
Therefore, $x = 0.25e^{-t} \sin 4t$ is the position of the mass at time t seconds.

2) A series circuit consists of a resistor with $R = 20\ \Omega$ (ohms), an inductor with $L = 1$ H (henries), a capacitor with $C = 0.01$ F (farads), and a generator producing a voltage of $360 - 200t$. Find the general solution for the differential equation for determining the charge at time t.

We are given
$L = 1, R = 20, C = 0.01, E(t) = 360 - 200t$.
The equation is
$Q'' + 20Q' + 100Q = 360 - 200t$.

Step 1. The auxiliary equation is
$r^2 + 20r + 100 = 0$ and has solution
$r = -10$. Thus, the complementary equation has general solution
$y_c = c_1 e^{-10t} + c_2 x e^{-10t}$.

Step 2. Use the method of undetermined coefficients.

i) Since $E(t) = 360 - 200t$, let $y_p = A_1 t + A_0$. Then $y_p' = A_1$ and $y_p'' = 0$.

ii) Substituting into the differential equation
$$0 + 20A_1 + 100(A_1 t + A_0) = 360 - 200t$$
$$100A_1 t + (20A_1 + 100A_0) = 360 - 200t$$
Thus,
$$100A_1 = -200$$
$$20A_1 + 100A_0 = 360.$$
The solution to the system is
$$A_1 = -2 \text{ and } A_0 = 4.$$
Thus, $y_p = 4 - 2t$.

iii) None of the terms solve the complementary equation.

iv) Therefore, the particular solution is
$$y_p(x) = 4 - 2t.$$

Step 3. The general solution is
$$y = c_1 e^{-10t} + c_2 x e^{-10t} + 4 - 2t.$$

Section 17.4 Series Solutions

Some second-order equations, even some simple-looking ones, cannot be solved explicitly using combinations of our familiar functions. This section shows how to solve some equations by representing the solution as a power series.

Concepts to Master

Power series solution to a differentiation equation

Summary and Focus Questions

Page 1188 (ET Page 1164)

The method of solving a differential equation using a power series has these steps:

i)   Assume the equation has a power series solution of the form:

$$y = \sum_{n=0}^{\infty} c_n x^n = c_0 + c_1 x + c_3 x^3 + \cdots.$$

ii) Obtain expressions for y', y'', and so on:

$$y' = \sum_{n=1}^{\infty} n c_n x^{n-1} = c_1 + 2c_2 x + 3c_3 x^2 + \cdots.$$

$$y'' = \sum_{n=2}^{\infty} n(n-1) c_n x^{n-2} = 2c_2 + 3 \cdot 2c_3 x + 4 \cdot 3c_4 x^2 + \cdots.$$

iii) Substitute these expressions into the differential equation.

iv) Equate coefficients of corresponding powers of x to obtain equations for determining $c_0, c_1, c_2, \ldots$. The general expression for c_i will often be a recursive expression in terms of previous c_j, where $j < i$.

Example: Find the general solution to $y = y'$ using a power series.

i)   Assume $y = c_0 + c_1 x + c_2 x^2 + c_3 x^3 + c_4 x^4 \cdots$.

ii) Then $y' = c_1 + c_2 2x + c_3 3x^2 + c_4 4x^3 \cdots$.

iii) Substitute into $y = y'$:

$$c_0 + c_1 x + c_2 x^2 + c_3 x^3 + c_4 x^4 \cdots = c_1 + c_2 + c_3 3x^2 + c_4 4x^3 \cdots.$$

iv) Equate coefficients:

$$c_0 = c_1.$$

$$c_1 = c_2 2, \text{ so } c_2 = \frac{c_1}{2} = \frac{c_0}{2}.$$

$$c_2 = c_3 3, \text{ so } c_3 = \frac{c_2}{3} = \frac{\frac{c_0}{2}}{3} = \frac{c_0}{3!}.$$

$$c_3 = c_4 4, \text{ so } c_4 = \frac{c_3}{4} = \frac{\frac{c_0}{3!}}{4} = \frac{c_0}{4!}.$$

In general, $c_n = \dfrac{c_0}{n!}$ for $n = 0, 1, 2, 3,\ldots$. and the solution is

$$y = \sum_{n=0}^{\infty} \frac{c_0}{n!} x^n = c_0 \sum_{n=0}^{\infty} \frac{x^n}{n!}.$$

We note this solution is the familiar power series for $y = c_0 e^x$, which satisfies $y = y'$.

1) Find a series solution to $y'' = xy'$.

i) $y = c_0 + c_1 x + c_2 x^2 + c_3 x^3 + \cdots$.

ii) $y' = c_1 + 2c_2 x + 3c_3 x^2 + 4c_4 x^3 + \cdots$ and
$y'' = 2c_2 + 3 \cdot 2c_3 x + 4 \cdot 3c_4 x^2 + 5 \cdot 4c_5 x^3 + \cdots$.

iii) $xy' = c_1 x + 2c_2 x^2 + 3c_3 x^3 + \cdots$.

iv) From the equation $y'' = xy'$, we equate coefficients.

Power of x	Relation
x^0	$2c_2 = 0$
x^1	$3 \cdot 2c_3 = c_1$
x^2	$4 \cdot 3c_4 = 2c_2$
x^3	$5 \cdot 4c_5 = 3c_3$

In general, $n(n-1)c_n = (n-2)c_{n-2}$.

$$c_n = \frac{(n-2)c_{n-2}}{n(n-1)}.$$

Thus, c_0 is arbitrary, c_1 is arbitrary, $c_2 = 0$,

$$c_3 = \frac{c_1}{3 \cdot 2}, \quad c_4 = 0,$$

$$c_5 = \frac{3c_3}{5 \cdot 4} = \frac{3c_1}{5 \cdot 4 \cdot 3 \cdot 2}, \quad c_6 = 0,$$

$$c_7 = \frac{5c_5}{7 \cdot 6} = \frac{5 \cdot 3c_1}{7 \cdot 6 \cdot 5 \cdot 4 \cdot 3 \cdot 2}, \quad \text{and so on.}$$

In general,

$$c_n = \begin{cases} 0 & n \text{ even} \\ \dfrac{(n-2)(n-4)\cdots 1}{n!} c_1 & n \text{ odd} \end{cases}$$

or, equivalently, $c_{2n} = 0$ and

$$c_{2n+1} = \frac{(2n-1)\cdots 3 \cdot 1}{n!} c_1 \text{ for all } n.$$

Therefore,

$$y = c_0 + c_1 \sum_{n=1}^{\infty} \frac{(2n-1)\cdots 3\cdot 1}{(2n+1)!} x^{2n+1}$$

is the solution to $y'' = xy'$, where c_0 and c_1 are arbitrary constants.

2) Find a series solution to $y'' + xy' - y = 0$, $y(0) = 1, y'(0) = 0$.

i) $y = \sum_{n=0}^{\infty} c_n x^n$.

ii) $y' = \sum_{n=1}^{\infty} n c_n x^{n-1}$

$$xy' = \sum_{n=1}^{\infty} n c_n x^n = \sum_{n=0}^{\infty} n c_n x^n$$

$$y'' = \sum_{n=2}^{\infty} n(n-1) c_n x^{n-2}$$

$$= \sum_{n=0}^{\infty} (n+2)(n+1) c_{n+2} x^n.$$

iii) Substitute into $y'' + xy' - y = 0$.

$$\sum_{n=0}^{\infty} (n+2)(n+1) c_{n+2} x^n$$

$$+ \sum_{n=0}^{\infty} n c_n x^n - \sum_{n=0}^{\infty} c_n x^n = 0.$$

$$\sum_{n=0}^{\infty} \left[(n+2)(n+1) c_{n+2} + (n-1) c_n\right] x^n = 0.$$

iv) $(n+2)(n+1) c_{n+2} + (n-1) c_n = 0$
for all n.

In general, $c_{n+2} = \frac{-(n-1)}{(n+2)(n+1)} c_n$.

$n = 0$: $c_2 = \frac{1}{2} c_0$.

$n = 1$: $c_3 = 0$.

$n = 2$: $c_4 = \frac{-1}{4\cdot 3} c_2 = \frac{-1}{4\cdot 3\cdot 2} c_0 = \frac{(-1)}{4!} c_0$.

$n = 3$: $c_5 = 0$.

$n = 4$: $c_6 = \frac{(-3)}{6\cdot 5} c_4 = \frac{(-3)(-1)}{6!} c_0$.

$n = 5$: $c_7 = 0$.

$n = 6$: $c_8 = \frac{(-5)}{8\cdot 7} c_4 = \frac{(-5)(-3)(-1)}{8!} c_0$.

and so on.

Since $y(0) = 1, c_0 = 1$.

Since $y'(0) = 0$, $c_1 = 0$.

For $n \geq 2$ and n odd, $c_n = 0$.

For $n \geq 2$ and n even,

$$c_n = \frac{(3-n)(5-n)\cdots(-3)(-1)(1)}{n!}.$$

Therefore,

$$y = 1 + \frac{1}{2!}x^2 - \frac{(1)(1)}{4!}x^4$$
$$+ \frac{(3)(1)(1)}{6!}x^6 - \frac{(5)(3))(1)(1)}{8!}x^8 + \cdots.$$

Since only the even-power terms of y have nonzero coefficients, we can rewrite c_n:

$$c_n = \frac{(3-n)(5-n)\cdots(-3)(-1)(1)}{n!}$$
$$= \frac{(-1)^{\frac{n}{2}+1}(n-3)(n-5)\cdots(3)(1)(1)}{n!},$$

for $n = 2, 4, 6, \cdots$.

Therefore, for $n = 1, 2, 3, \cdots$,

$$c_{2n} = \frac{(-1)^{n+1}(2n-3)(2n-5)\cdots(3)(1)(1)}{(2n)!}.$$

Then we may rewrite y:

$$y = 1 + \sum_{n=1}^{\infty} \frac{(-1)^{n+1}(2n-3)(2n-5)\cdots(3)(1)(1)}{(2n)!}x^{2n}.$$

Section 10.1

_____ **1.** The graph of $x = 2 + 3t$, $y = 4 - t$ is a:

    a) circle   b) ellipse

    c) line   d) parabola

_____ **2.** The graph of $x = \cos t$, $y = \sin^2 t$ is:

a)

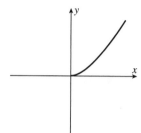

b)

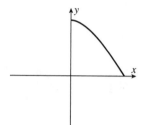

c)

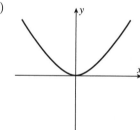

d)

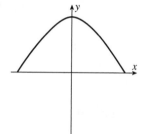

_____ **3.** Elimination of the parameter in $x = 2t^{3/2}$, $y = t^{2/3}$ gives:

    a) $x^4 = 16y^9$ b) $16x^4 = y^4$

    c) $x^3 = 8y^4$ d) $8x^3 = y^4$

On Your Own

Section 10.2

___ 1. Find $\frac{dy}{dx}$ if $x = \sqrt{t}$, $y = \sin 2t$.

    a) $\frac{4\cos t}{\sqrt{t}}$ b) $\frac{\cos 2t}{\sqrt{t}}$ c) $\frac{\cos 2t}{2\sqrt{t}}$ d) $4\sqrt{t}\cos 2t$

___ 2. Find $\frac{d^2y}{dx^2}$ for $x = 3t^2 + 1$, $y = t^6 + 6t^5$.

    a) $t^4 + 5t^3$ b) $4t^3 + 15t^2$

    c) $\frac{2}{3}t^2 + \frac{5}{2}t$ d) $t^3 + \frac{1}{2}t^2$

___ 3. The slope of the tangent line at the point where $t = \frac{\pi}{6}$ to the curve
     $y = \sin 2t$, $x = \cos 3t$ is:

    a) $\frac{1}{3}$ b) $-\frac{1}{3}$ c) 3 d) -3

___ 4. A definite integral for the area under the curve described by $x = t^2 + 1$,
     $y = 2t$, $0 \le t \le 1$ is:

    a) $\int_0^1 (2t^3 + 2t)\,dt$ b) $\int_0^1 4t^2\,dt$

    c) $\int_0^1 (2t^2 + 2)\,dt$ d) $\int_0^1 4t\,dt$

___ 5. The length of the curve given by $x = 3t^2 + 2$, $y = 2t^3$, $t \in [0, 1]$ is:

    a) $4\sqrt{2} - 2$ b) $8\sqrt{2} - 1$

    c) $\frac{2}{3}\left(2\sqrt{2} - 1\right)$ d) $\sqrt{2} - 1$

___ 6. Find a definite integral for the area of the surface of revolution about the
     x-axis obtained by rotating the curve $y = t^2$, $x = 1 + 3t$, $0 \le t \le 2$.

    a) $\int_0^2 2\pi t^2 \sqrt{t^4 + 9t^2 + 6t + 1}\,dt$

    b) $\int_0^2 2\pi t^2 \sqrt{4t^2 + 9}\,dt$

    c) $\int_0^2 2\pi (2t)\sqrt{t^4 + 9t^2 + 6t + 1}\,dt$

    d) $\int_0^2 2\pi (2t)\sqrt{4t^4 + 9}\,dt$

Section 10.3

____ **1.** The polar coordinates of the
point P in the figure at the right are:

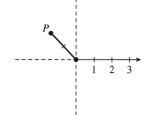

a) $\left(-2, \frac{\pi}{4}\right)$ b) $\left(-2, \frac{3\pi}{4}\right)$

c) $\left(2, -\frac{\pi}{4}\right)$ d) $\left(2, \frac{3\pi}{4}\right)$

____ **2.** Polar coordinates of the point with rectangular coordinates $(5, 5)$ are:

a) $(25, 0)$ b) $\left(5, \frac{\pi}{4}\right)$ c) $\left(5\sqrt{2}, \frac{\pi}{4}\right)$ d) $\left(50, -\frac{\pi}{4}\right)$

____ **3.** Rectangular coordinates of the point with polar coordinates $\left(-1, \frac{3\pi}{2}\right)$ are:

a) $(-1, 0)$ b) $(0, 1)$ c) $(0, -1)$ d) $(1, 0)$

____ **4.** The graph of $\theta = 2$ in polar coordinates is a:

a) circle   b) line   c) spiral   d) 3-leaved rose

____ **5.** Which is the best graph of $r = 1 - \sin \theta$ for $0 \le \theta \le \pi$?

a)
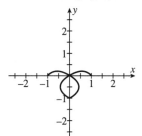

b)
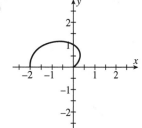

c)
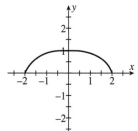

d)
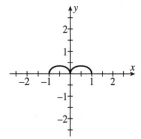

____ **6.** The slope of the tangent line to $r = \cos \theta$ at $\theta = \frac{\pi}{3}$ is:

a) $\sqrt{3}$ b) $\frac{1}{\sqrt{3}}$ c) $-\sqrt{3}$ d) $-\frac{1}{\sqrt{3}}$

Section 10.4

1. The area of the region bounded by $\theta = \frac{\pi}{3}$, $\theta = \frac{\pi}{4}$, and $r = \sec\theta$ is:

a) $\frac{1}{2}\left(\sqrt{3} - 1\right)$ b) $\sqrt{3}$

c) $2(\sqrt{3} - 1)$ d) $2\sqrt{3}$

2. The area of the shaded region is given by:

a) $\displaystyle\int_0^{\pi/2} \sin 3\theta \, d\theta$

b) $\displaystyle\int_0^{\pi/2} 2\sin^2 3\theta \, d\theta$

c) $\displaystyle\int_0^{\pi/3} \sin 3\theta \, d\theta$

d) $\displaystyle\int_0^{\pi/3} 2\sin^2 3\theta \, d\theta$

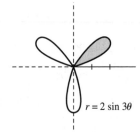

$r = 2\sin 3\theta$

3. The length of the arc $r = e^\theta$ for $0 \le \theta \le \pi$ is given by:

a) $\displaystyle\int_0^\pi \sqrt{e^\theta} \, d\theta$

b) $\displaystyle\int_0^\pi 2e^{2\theta} \, d\theta$

c) $\displaystyle\int_0^\pi \sqrt{2e^\theta} \, d\theta$

d) $\displaystyle\int_0^\pi \sqrt{2e^\theta} \, d\theta$

____ **1.** The graph of $\frac{y^2}{16} = 1 + \frac{x^2}{25}$ is:

a)

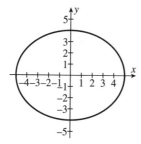

b)

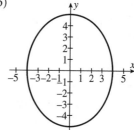

c)

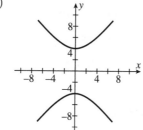

d)

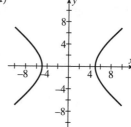

____ **2.** The center of $4x^2 + 8x - 3y^2 + 12y + 4 = 0$ is

     a) $(1, -2)$ b) $(-1, 2)$ c) $(2, -1)$ d) $(-2, 1)$

____ **3.** The conic section whose equation is $y(3 - y) + 4x^2 = 2x(1 + 2x) - y$ is a(n)

     a) parabola   b) ellipse   c) hyperbola   d) None of these

Section 10.6

____ **1.** The figure at the right shows one point P on a conic and the distances of P to the focus and the directrix of a conic. What type of conic is it?

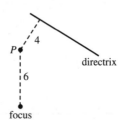

a) parabola
b) ellipse
c) hyperbola
d) There is not enough information is provided to answer the question.

____ **2.** The polar equation of the conic with eccentricity 3 and directrix $x = -7$ is:

a) $r = \dfrac{21}{1 + 3 \cos \theta}$

b) $r = \dfrac{21}{1 - 3 \cos \theta}$

c) $r = \dfrac{21}{1 + 3 \sin \theta}$

d) $r = \dfrac{21}{1 - 3 \sin \theta}$

____ **3.** The directrix of the conic given by $r = \dfrac{6}{2 + 10 \sin \theta}$ is:

a) $x = \dfrac{5}{3}$

b) $x = \dfrac{3}{5}$

c) $y = \dfrac{5}{3}$

d) $y = \dfrac{3}{5}$

____ **4.** The polar form for the graph at the right is:

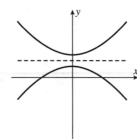

a) $r = \dfrac{ed}{1 + e \cos \theta}$

b) $r = \dfrac{ed}{1 - e \cos \theta}$

c) $r = \dfrac{ed}{1 + e \sin \theta}$

d) $r = \dfrac{ed}{1 - e \sin \theta}$

Section 11.1

_____ **1.** $\lim\limits_{n \to \infty} \dfrac{n^2 + 3n}{2n^2 + n + 1} =$

    a) 0   b) $\dfrac{1}{2}$ c) 1   d) ∞

_____ **2.** Sometimes, Always, or Never:
If $a_n \geq b_n \geq 0$ for all n and $\{b_n\}$ diverges, then $\{a_n\}$ diverges.

_____ **3.** Sometimes, Always, or Never:
If $a_n \geq b_n \geq c_n$ for all n and both $\{a_n\}$ and $\{c_n\}$ converge, then $\{b_n\}$
converges.

_____ **4.** Sometimes, Always, or Never:
If $\{a_n\}$ is increasing and bounded above, then $\{a_n\}$ converges.

_____ **5.** True or False:
$a_n = \dfrac{(-1)^n}{n^2}$ is monotonic.

_____ **6.** $\lim\limits_{n \to \infty} \dfrac{\arctan n}{2} =$

    a) $\dfrac{\pi}{4}$ b) $\dfrac{\pi}{2}$ c) π d) does not exist

_____ **7.** $\lim\limits_{n \to \infty} \dfrac{2^n}{n!} =$

    a) 2   b) 1   c) 0   d) does not exist

_____ **8.** The fourth term of $\{a_n\}$ defined by $a_1 = 3$, and $a_{n+1} = \dfrac{2}{3}a_n$
for $n = 1, 2, 3, \ldots$ is:

    a) $\dfrac{16}{81}$ b) $\dfrac{16}{27}$ c) $\dfrac{8}{27}$ d) $\dfrac{8}{9}$

Section 11.2

_____ **1.** The third partial sum of $\sum_{n=1}^{\infty} \frac{n}{n^2+1}$ is:

 a) $\frac{3}{10}$ b) $\frac{1}{2}$ c) $\frac{3}{5}$ d) $\frac{6}{5}$

_____ **2.** True or False:

$$\sum_{n=1}^{\infty} a_n \text{ converges mean } \lim_{k \to \infty} \sum_{n=1}^{k} a_n \text{ exists.}$$

_____ **3.** True or False:

$$\sum_{n=1}^{\infty} a_n \text{ converges if and only if } \lim_{n \to \infty} a_n = 0.$$

_____ **4.** True or False:

If $\sum_{n=1}^{\infty} a_n$ converges and $\sum_{n=1}^{\infty} b_n$ converges, then $\sum_{n=1}^{\infty} (a_n - b_n)$ converges.

_____ **5.** The harmonic series is:

 a) $1 + 2 + 3 + 4 + ...$ b) $1 + \frac{1}{2} + \frac{1}{3} + \frac{1}{4} + ...$

 c) $1 + \frac{1}{2} + \frac{1}{4} + \frac{1}{8} + ...$ d) $1 - \frac{1}{2} + \frac{1}{4} - \frac{1}{8} + ...$

_____ **6.** True or False:

$$\sum_{n=1}^{\infty} \frac{n}{n+1} \text{ converges.}$$

_____ **7.** $\sum_{n=1}^{\infty} 2\left(\frac{1}{4}\right)^n$ converges to:

 a) $\frac{9}{4}$ b) 2 c) $\frac{2}{3}$ d) the series diverges

_____ **8.** True or False:

$-3 + 1 - \frac{1}{3} + \frac{1}{9} - \frac{1}{27} + ...$ is a geometric series.

Section 11.3

_____ **1.** For what values of p does the series $\sum_{n=1}^{\infty} \frac{1}{(n^2)^p}$ converge?

    a) $p > -\frac{1}{2}$ b) $p < -\frac{1}{2}$ c) $p > \frac{1}{2}$ d) $p < \frac{1}{2}$

_____ **2.** True or False:

If $f(x)$ is continuous and decreasing, $f(n) = a_n$ for all $n = 1, 2, 3, \ldots,$ and $\int_{1}^{\infty} f(x)\,dx = M$, then $\sum_{n=1}^{\infty} a_n = M$.

_____ **3.** True or False:

By the Integral Test, $\sum_{n=2}^{\infty} \frac{1}{n \ln n}$ converges.

_____ **4.** True or False:

$\sum_{n=1}^{\infty} \frac{1}{n^{2/3}}$ is a convergent series.

_____ **5.** True or False:

$\sum_{n=1}^{\infty} \frac{n + 2}{(n^2 + 4n + 1)^2}$ is a convergent series.

_____ **6.** For $s = \sum_{n=1}^{\infty} \frac{1}{n^3}$, an upper bound estimate for $s - s_6$ (where s_6 is the sixth partial sum) is:

    a) $\int_{1}^{\infty} x^{-3}\,dx$ b) $\int_{6}^{\infty} x^{-3}\,dx$

    c) $\int_{7}^{\infty} x^{-3}\,dx$ d)   the series does not converge

Section 11.4

_____ **1.** Sometimes, Always, or Never:

If $0 \leq a_n \leq b_n$ for all n and $\sum_{n=1}^{\infty} a_n$ diverges, then $\sum_{n=1}^{\infty} b_n$ converges.

_____ **2.** True or False:

$\sum_{n=1}^{\infty} \dfrac{n+1}{n^3}$ is a convergent series.

_____ **3.** True or False:

$\sum_{n=1}^{\infty} \dfrac{\sqrt{n} + \sqrt[3]{n}}{n^{2/3} + n^{3/2} + 1}$ is a convergent series.

_____ **4.** True or False:

$\sum_{n=1}^{\infty} \dfrac{\cos^2(2^n)}{2^n}$ is a convergent series.

_____ **5.** $s = \sum_{n=1}^{\infty} \dfrac{1}{n \cdot 2^n}$ converges by the Comparison Test, comparing it to $\sum_{n=1}^{\infty} \dfrac{1}{2^n}$.

Using this information, make an estimate of the difference between s and its third partial sum.

a) $\dfrac{1}{16}$ b) $\dfrac{1}{8}$ c) $\dfrac{1}{32}$ d) 1

Section 11.5

_____ **1.** True or False:

$$\sum_{n=1}^{\infty} \frac{(-1)^{n+1} \ln n}{n^2}$$ is a convergent series.

_____ **2.** True or False:

$$\sum_{n=1}^{\infty} \frac{(-1)^n}{\sqrt[4]{n+1}}$$ is a convergent series.

_____ **3.** For what value of n is the nth partial sum within 0.01 of the value of $\sum_{n=1}^{\infty} \frac{(-1)^n}{2^n}$? (Choose the smallest such n.)

    a) $n = 4$ b) $n = 6$ c) $n = 8$ d) $n = 10$

_____ **4.** True or False:

The Alternating Series Test may be applied to determine the convergence of $\sum_{n=1}^{\infty} \frac{2 + (-1)^n}{2n^2}$.

Section 11.6

_____ **1.** True or False:

If $\displaystyle\sum_{n=1}^{\infty} a_n$ converges absolutely, then it converges conditionally.

_____ **2.** True or False:

If $\displaystyle\lim_{n\to\infty}\left|\frac{a_n}{a_{n+1}}\right| = 3$, then $\displaystyle\sum_{n=1}^{\infty} a_n$ converge absolutely.

_____ **3.** True or False:

Every series must do one of these: converge absolutely, converge conditionally, or diverge.

_____ **4.** Which is true about the series $\displaystyle\sum_{n=1}^{\infty} 2^{-n} n!$?

a) diverges

b) converges absolutely

c) converges conditionally

d) converges, but not absolutely and not conditionally

_____ **5.** Which is true about the series $\displaystyle\sum_{n=1}^{\infty} \frac{(-5)^{n+1}}{n^n}$?

a) diverges

b) converges absolutely

c) converges conditionally

d) converges, but not absolutely and not conditionally

_____ **6.** True or False:

The series $\displaystyle\sum_{n=1}^{\infty} \frac{(-1)^n}{n^3}$ converges to a number that we will call s.

All rearrangements of $\displaystyle\sum_{n=1}^{\infty} \frac{(-1)^n}{n^3}$ also converge to s.

Section 11.7

_____ **1.** True or False:
$$\sum_{n=2}^{\infty} \frac{1}{(\ln n)^n} \text{ converges.}$$

_____ **2.** True or False:
$$\sum_{n=1}^{\infty} \frac{6}{7n + 8} \text{ converges.}$$

_____ **3.** True or False:
$$\sum_{n=1}^{\infty} \frac{(-1)^n \sqrt{n}}{n + 3} \text{ converges.}$$

_____ **4.** True or False:
$$\sum_{n=1}^{\infty} \frac{e^n}{n!} \text{ converges.}$$

_____ **5.** True or False:
$$\sum_{n=1}^{\infty} \frac{(-1)^n}{\sqrt[4]{n}} \text{ converges conditionally.}$$

_____ **6.** True or False:
$$\sum_{n=1}^{\infty} \frac{(-1)^n}{(1.1)^n} \text{ converges absolutely.}$$

Section 11.8

____ **1.** Sometimes, Always, or Never:

The interval of convergence of a power series $\sum\limits_{n=0}^{\infty} a_n(x-c)^n$ is an open interval $(c-R, c+R)$. (When $R = 0$ we mean $\{c\}$ and when $R = \infty$ we mean $(-\infty, \infty)$.)

____ **2.** True or False:

If a number p is in the interval of convergence of $\sum\limits_{n=0}^{\infty} a_n x^n$ then so is the number $\frac{p}{2}$.

____ **3.** For $f(x) = \sum\limits_{n=0}^{\infty} \frac{(x-1)^n}{3^n}$, $f(3) =$

   a) 0   b) 2
   c) 3   d) $f(3)$ does not exist

____ **4.** The interval of convergence of $\sum\limits_{n=1}^{\infty} \frac{x^n}{\sqrt{n}}$ is:

   a) $[-1, 1]$ b) $[-1, 1)$ c) $(-1, 1]$ d) $(-1, 1)$

____ **5.** The radius of convergence of $\sum\limits_{n=0}^{\infty} \frac{n(x-5)^n}{3^n}$ is:

   a) $\frac{1}{3}$ b) 1   c) 3   d) ∞

Section 11.9

____ **1.** Given that $e^x = \sum_{n=0}^{\infty} \frac{x^n}{n!}$ for all x, a power series for xe^{x^2} is:

a) $\sum_{n=0}^{\infty} \frac{x^2}{n!}$

b) $\sum_{n=0}^{\infty} \frac{x^{2n}}{n!}$

c) $\sum_{n=0}^{\infty} \frac{x^{2n+1}}{n!}$

d) $\sum_{n=0}^{\infty} \frac{x^{2n}}{(n+1)!}$

____ **2.** For $f(x) = \sum_{n=0}^{\infty} \frac{x^{2n}}{n!}$, $f'(x) =$

a) $\sum_{n=1}^{\infty} \frac{x^{2n-1}}{n!}$

b) $\sum_{n=1}^{\infty} \frac{2x^{2n-1}}{(n-1)!}$

c) $\sum_{n=1}^{\infty} \frac{2^n x^{2n-1}}{n!}$

d) $\sum_{n=1}^{\infty} \frac{(2n-1)x^{2n-1}}{(n-1)!}$

____ **3.** Using $\frac{1}{1-x} = \sum_{n=0}^{\infty} x^n$ for $|x| < 1$, $\int \frac{x}{1-x^2}\, dx =$

a) $\sum_{n=0}^{\infty} \frac{x^{2n}}{2n}$

b) $\sum_{n=0}^{\infty} \frac{x^{2n+1}}{2n+1}$

c) $\sum_{n=0}^{\infty} \frac{x^{2n+2}}{2n+2}$

d) $\sum_{n=0}^{\infty} x^{2n+1}$

____ **4.** Using $\frac{1}{1-x} = \sum_{n=0}^{\infty} x^n$ for $|x| < 1$ and differentiation, find a power series for $\frac{1}{(1+x)^2}$.

a) $\sum_{n=1}^{\infty} \frac{x^{n-1}}{n}$

b) $\sum_{n=1}^{\infty} -n\, x^{n-1}$

c) $\sum_{n=1}^{\infty} (-n)^n x^{n-1}$

d) $\sum_{n=1}^{\infty} n(-1)^n x^{n-1}$

____ **5.** From $\frac{1}{1-x} = \sum_{n=0}^{\infty} x^n$ for $|x| < 1$ and substituting $4x^2$ for x, the resulting power series $\sum_{n=0}^{\infty} (4x^2)^n$ has interval of convergence

a) $\left(-\frac{1}{4}, \frac{1}{4}\right)$

b) $\left(-\frac{1}{2}, \frac{1}{2}\right)$

c) $(-4, 4)$

d) $(-2, 2)$

____ **1.** Given the Taylor Series $e^x = \sum\limits_{n=0}^{\infty} \frac{x^n}{n!}$, a Taylor series for $e^{x/2}$ is:

a) $\sum\limits_{n=0}^{\infty} \frac{2^n x^n}{n!}$

b) $\sum\limits_{n=0}^{\infty} \frac{2x^n}{n!}$

c) $\sum\limits_{n=0}^{\infty} \frac{x^n}{2^n n!}$

d) $\sum\limits_{n=0}^{\infty} \frac{x^n}{2n!}$

____ **2.** Using the power series for $\cos x$, the sum of the series $\sum\limits_{n=0}^{\infty} \frac{(-1)^n}{(2n)!}(0.25)^n$ is:

a) $\cos \sqrt{0.5}$ b) $\cos(0.0625)$ c) $\cos(0.25)$ d) $\cos(0.5)$

____ **3.** The nth term in the Taylor series centered at 1 for $f(x) = x^{-2}$ is:

a) $(-1)^{n+1}(n+1)!(x-1)^n$ b) $(-1)^n n!(x-1)^n$
c) $(-1)^n (n+1)(x-1)^n$ d) $(-1)^n n(x-1)^n$

____ **4.** The Taylor polynomial of degree 3 for $f(x) = x(\ln x - 1)$ about 1 is $T_3(x) =$

a) $-1 + \frac{(x-1)^2}{2} - \frac{(x-1)^3}{6}$

b) $-2 + x + \frac{(x-1)^2}{2} - \frac{(x-1)^3}{6}$

c) $-1 + x - x^2 + x^3$

d) $-1 - x + x^2 - x^3$

____ **5.** Use the Taylor polynomial of degree 2 for $f(x) = \sqrt{x}$ centered at 1 to estimate $\sqrt{1.6}$.

a) 1.250 b) 1.255 c) 1.265 d) 1.270

____ **6.** True or False:

If $T_n(x)$ is the nth Taylor polynomial for $f(x)$ centered at c, then $T_n^{(k)}(c) = f^{(k)}(c)$ for $k = 0, 1, ..., n$.

____ **7.** If $f(x) = \sin 2x$, an upper bound for $|f^{(n+1)}(x)|$ is

a) 2 b) 2^n c) 2^{n+1} d) 2^{2n}

____ **8.** If we know that $\left|f^{(n+1)}(x)\right| \le M$ for all $|x - a| \le d$, then the absolute value of the nth remainder of the Taylor series for $y = f(x)$ about a for $|x - a| \le d$ is less than or equal to

a) $\frac{M}{n!} |x - a|^n$

b) $\frac{M}{(n+1)!} |x - a|^{n+1}$

c) $\frac{M}{n!} |x - a|^{n+1}$

d) $\frac{M}{(n+1)!} |x - a|^n$

(continued on next page)

_____ **9.** $\dbinom{\frac{1}{2}}{3} =$

     a) $\dfrac{5}{16}$ b) $\dfrac{1}{16}$ c) $\dfrac{5}{8}$ d) $-\dfrac{5}{8}$

_____ **10.** $\displaystyle\sum_{n=0}^{\infty} \dbinom{\frac{2}{3}}{n} x^n$ is the binomial series for:

     a) $(1 + x)^{2/3}$ b) $(1 + x)^{-2/3}$

     c) $(1 + x)^{3/2}$ d) $(1 + x)^{-3/2}$

_____ **11.** Using a binomial series, the Maclaurin series for $\dfrac{1}{1 + x^2}$ is:

     a) $\displaystyle\sum_{n=0}^{\infty} \dbinom{1}{n} x^n$ b) $\displaystyle\sum_{n=0}^{\infty} \dbinom{-1}{n} x^n$

     c) $\displaystyle\sum_{n=0}^{\infty} \dbinom{1}{n} x^{2n}$ d) $\displaystyle\sum_{n=0}^{\infty} \dbinom{-1}{n} x^{2n}$

_____ **12.** From the Maclaurin series for $\ln(1 + x)$ and $\sin x$, the first three terms of the Maclaurin series for $(\ln(1 + x))(\sin x)$ is

     a) $x^2 - \dfrac{x^3}{2} - \dfrac{x^4}{6}$ b) $x^2 - \dfrac{x^3}{2} + \dfrac{x^4}{6}$

     c) $x - \dfrac{x^2}{2} - \dfrac{x^3}{6}$ d) $x - \dfrac{x^2}{2} + \dfrac{x^3}{6}$

Section 11.11

_____ **1.** The quadratic approximation for $f(x) = x^6$ at $a = 1$ is:

   a) $P(x) = 6(x - 1) + 30(x - 1)^2$

   b) $P(x) = 1 + 6(x - 1) + 30(x - 1)^2$

   c) $P(x) = 6(x - 1) + 15(x - 1)^2$

   d) $P(x) = 1 + 6(x - 1) + 15(x - 1)^2$

_____ **2.** If the Maclaurin polynomial of degree 2 for $f(x) = e^x$ is used to approximate $e^{0.2}$ then the best estimate for the error with $0 < x < 1$ is:

   a) $\frac{1}{24}e$ b) $\frac{1}{6}e$ c) $\frac{1}{2}e$ d) e

_____ **3.** What degree Taylor polynomial about $a = 1$ is needed to approximate $e^{1.05}$ accurate to within 0.0001?

   a) $n = 2$ b) $n = 3$ c) $n = 4$ d) $n = 5$

Section 12.1

_____ **1.** The point plotted at the right has coordinates:

a) $(2, 5, 4)$ b) $(5, 2, 4)$

c) $(4, 2, 5)$ d) $(2, 4, 5)$

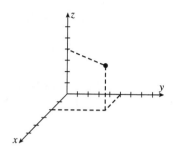

_____ **2.** The equation of the plane partially drawn in the figure is:

a) $x = 3$

b) $y = 3$

c) $z = 3$

d) $x + y = 3$

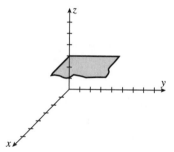

_____ **3.** The distance between $(7, 4, -3)$ and $(-1, 2, 3)$ is:

a) 4   b) 16   c) 104   d) $\sqrt{104}$

_____ **4.** The sphere graphed at the right has equation:

a) $(x - 2)^2 + (y - 3)^2 + (z - 4)^2 = 1$

b) $(x - 2)^2 + (y - 4)^2 + (z - 4)^2 = 4$

c) $(x - 2)^2 + (y - 5)^2 + (z - 4)^2 = 4$

d) $(x - 2)^2 + (y - 4)^2 + (z - 4)^2 = 1$

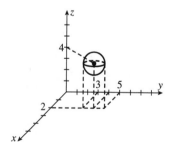

_____ **5.** The region described by $y^2 + z^2 = 4$ is:

a) a circle, center at the origin   b) a sphere, center at the origin

c) a cone, along the x-axis   d) a cylinder, along the x-axis

Section 12.2

_____ **1.** The vector represented by $\overrightarrow{AB}$ where $A(4, 8)$ and $B(6, 6)$ is:

    a) $\langle -2, 2 \rangle$ b) $\langle 2, -2 \rangle$ c) $\langle 10, 14 \rangle$ d) $\langle 14, 10 \rangle$

_____ **2.** The length of $\mathbf{a} = 4\mathbf{i} - \mathbf{j} - 2\mathbf{k}$ is:

    a) $\sqrt{11}$ b) 11 c) $\sqrt{21}$ d) 21

_____ **3.** For $\mathbf{a} = 6\mathbf{i} - \mathbf{j}$ and $\mathbf{b} = 2\mathbf{i} + 3\mathbf{j}$, $\mathbf{a} + 2\mathbf{b} =$

    a) $2\mathbf{i} - 4\mathbf{j}$ b) $8\mathbf{i} - 2\mathbf{j}$ c) $16\mathbf{i} - 4\mathbf{j}$ d) $10\mathbf{i} + 5\mathbf{j}$

_____ **4.** True or False:

    $\mathbf{i} + \mathbf{j}$ is a unit vector in the direction of $5\mathbf{i} + 5\mathbf{j}$.

_____ **5.** The vector **c** in the figure at the right is:

    a) $\mathbf{a} - \mathbf{b}$
    b) $\mathbf{b} - \mathbf{a}$
    c) $\mathbf{a} + \mathbf{b}$
    d) $-\mathbf{a} - \mathbf{b}$

_____ **6.** The unit vector in the direction of $\mathbf{u} = \langle 1, 3, -1 \rangle$ is:

    a) $\left\langle \frac{1}{3}, 1, -\frac{1}{3} \right\rangle$

    b) $\left\langle \frac{1}{11}, 1, -\frac{1}{11} \right\rangle$

    c) $\left\langle \frac{1}{11}, \frac{3}{11}, -\frac{1}{11} \right\rangle$

    d) $\left\langle \frac{1}{\sqrt{11}}, \frac{3}{\sqrt{11}}, -\frac{1}{\sqrt{11}} \right\rangle$

Section 12.3

_____ **1.** For $\mathbf{a} = 2\mathbf{i} + 3\mathbf{j} - 4\mathbf{k}$ and $\mathbf{b} = -\mathbf{i} + 2\mathbf{j} - \mathbf{k}$, $\mathbf{a} \cdot \mathbf{b} =$

    a) 0   b) 8   c) 10   d) 12

_____ **2.** The angle θ at the right has cosine:

    a) $\dfrac{9}{5\sqrt{10}}$ b) $\dfrac{9}{250}$

    c) $\dfrac{9}{\sqrt{50}}$ d) $\dfrac{9}{50}$

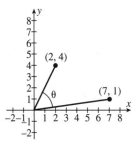

_____ **3.** $(\mathbf{i} \cdot \mathbf{j})\mathbf{k} =$

    a) $\mathbf{k}$ b) $-\mathbf{k}$ c) 0 (the number 0)   d) $\mathbf{0}$ (the zero vector)

_____ **4.** The scalar projection of $\mathbf{a} = \langle 1, 4 \rangle$ on $\mathbf{b} = \langle 6, 3 \rangle$ is:

    a) $\dfrac{18}{\sqrt{17}}$ b) $\dfrac{6}{\sqrt{5}}$ c) $\dfrac{6}{\sqrt{17}\sqrt{5}}$ d) $\dfrac{18}{\sqrt{17}\sqrt{5}}$

_____ **5.** The amount of work done by a horizontal force of 10 N moving an object 5 m up a 45° incline (ignoring gravity) is:

    a) $5\sqrt{2}$ J   b) $\dfrac{25\sqrt{2}}{2}$ J   c) $50\sqrt{2}$ J   d) $25\sqrt{2}$ J

_____ **6.** The direction cosines for the angles α, β, γ in the figure are:

    a) $\alpha: \dfrac{2}{9}$, $\beta: \dfrac{1}{9}$, $\gamma: \dfrac{2}{9}$

    b) $\alpha: \dfrac{2}{3}$, $\beta: \dfrac{1}{3}$, $\gamma: \dfrac{2}{3}$

    c) $\alpha: \sqrt{\dfrac{2}{9}}$, $\beta: \sqrt{\dfrac{1}{9}}$, $\gamma: \sqrt{\dfrac{2}{9}}$

    d) $\alpha: \sqrt{\dfrac{2}{3}}$, $\beta: \sqrt{\dfrac{1}{3}}$, $\gamma: \sqrt{\dfrac{2}{3}}$

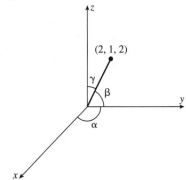

Section 12.4

____ **1.** For **a** = $\langle 1, 2, 1 \rangle$ and **b** = $\langle 3, 0, 2 \rangle$, **a** × **b** =

    a) $\langle -6, -1, 4 \rangle$ b) $\langle -6, 1, 4 \rangle$

    c) $\langle 4, -1, -6 \rangle$ d) $\langle 4, 1, -6 \rangle$

____ **2.** True or False:

    a × **b** is a vector in the plane formed by vectors **a** and **b**.

____ **3.** **j** × (−**k**) =

    a) **i**   b) −**i**   c) **j** + **k**   d) −**j** − **k**

____ **4.** True or False:

    If the vector **c** is orthogonal to both vectors **a** and **b**, then c is a scalar multiple of **a** × **b**.

____ **5.** The area of the parallelogram at the right is:

    a) 26

    b) $\sqrt{14}$

    c) $\sqrt{86}$

    d) $\sqrt{300}$

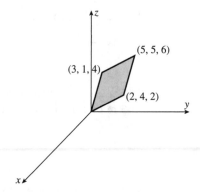

____ **6.** For **a** = $\langle 6, 2, 3 \rangle$, **b** = $\langle 4, 7, 9 \rangle$, and **c** = $\langle 8, 1, 5 \rangle$, **c** · (**a** × **b**) =

    a) $\begin{vmatrix} 6 & 2 & 3 \\ 4 & 7 & 9 \\ 8 & 1 & 5 \end{vmatrix}$ b) $\begin{vmatrix} 6 & 2 & 3 \\ 8 & 1 & 5 \\ 4 & 7 & 9 \end{vmatrix}$ c) $\begin{vmatrix} 8 & 1 & 5 \\ 6 & 2 & 3 \\ 4 & 7 & 9 \end{vmatrix}$ d) $\begin{vmatrix} 8 & 1 & 5 \\ 4 & 7 & 9 \\ 6 & 2 & 3 \end{vmatrix}$

Section 12.5

_____ **1.** The equation of the line through $(4, 10, 8)$ and $(3, 5, 1)$ is:

    a) $\dfrac{x-4}{3} = \dfrac{y-10}{5} = \dfrac{z-8}{1}$
          b) $\dfrac{x-3}{4} = \dfrac{y-5}{10} = \dfrac{z-1}{8}$

    c) $\dfrac{x-3}{7} = \dfrac{y-5}{15} = \dfrac{z-1}{9}$
          d) $\dfrac{x-4}{1} = \dfrac{y-10}{5} = \dfrac{z-8}{7}$

_____ **2.** True or False:

    These lines are skew: $\begin{aligned} x &= 1 + 5t \\ y &= 3 - 6t \\ z &= 1 - 2t \end{aligned}$ $\begin{aligned} x &= 4 + 7s \\ y &= 5 + s \\ z &= 1 - 4s \end{aligned}$

_____ **3.** The equation of the plane through $\langle -7, 2, 3 \rangle$ with normal vector $\langle 6, -4, 1 \rangle$ is:

    a) $6x - 4y + z = -47$
          b) $6x - 4y + z = 0$

    c) $-7x + 2y + 3z = -20$
          d) $-7x + 2y + 3z = -12$

_____ **4.** The equation of the plane formed by the two lines

    $\begin{aligned} x &= 3 + 2t \\ y &= 1 - 4t \\ z &= 5 + t \end{aligned}$ and $\begin{aligned} x &= 3 + t \\ y &= 1 + 2t \\ z &= 5 + 2t \end{aligned}$

    is:

    a) $2(x - 3) - 4(y - 1) + (z - 5) = 0$
    b) $(x - 3) + 2(y - 1) + 2(z - 5) = 0$
    c) $(x - 3) - 6(y - 1) + (z - 5) = 0$
    d) $-10(x - 3) - 3(y - 1) + 8(z - 5) = 0$

_____ **5.** The distance from $(1, 2, 1)$ to the plane $6x + 5y + 8z = 34$ is:

    a) $\sqrt{5}$
        b) $2\sqrt{5}$
        c) $\dfrac{2}{\sqrt{5}}$
        d) $\dfrac{1}{\sqrt{5}}$

_____ **6.** A normal vector to the plane $2x + 3y + 6z = -20$ is:

    a) $\langle 2, 3, 6 \rangle$
        b) $\langle 15, 10, 5 \rangle$
        c) $\langle 1, 1, 1 \rangle$
        d) $\langle 6, 3, 2 \rangle$

Section 12.6

_____ **1.** $\frac{x^2}{25} + 1 = \frac{z^2}{16} - \frac{y^2}{9}$ is a(n):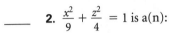

     a) hyperboloid of one sheet   b) hyperboloid of two sheets

     c) ellipsoid   d) hyperbolic cone

_____ **2.** $\frac{x^2}{9} + \frac{z^2}{4} = 1$ is a(n):

     a) ellipsoid   b) elliptic cone

     c) elliptic paraboloid   d) elliptic cylinder

_____ **3.** The graph at the right has equation:

     a) $\frac{x^2}{a^2} + \frac{y^2}{b^2} = \frac{z^2}{c^2}$

     b) $\frac{x^2}{a^2} + \frac{z^2}{c^2} = \frac{y^2}{b^2}$

     c) $\frac{y^2}{b^2} + \frac{z^2}{c^2} = \frac{x^2}{a^2}$

     d) $\frac{x^2}{a^2} + \frac{y^2}{b^2} + \frac{z^2}{c^2} = 0$

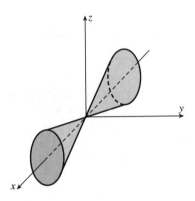

_____ **4.** The graph at the right has equation:

     a) $x = \frac{y^2}{b^2} + \frac{z^2}{c^2}$

     b) $y = \frac{x^2}{a^2} + \frac{z^2}{c^2}$

     c) $z = \frac{x^2}{a^2} + \frac{y^2}{b^2}$

     d) $y^2 = \frac{x^2}{a^2} + \frac{z^2}{c^2}$

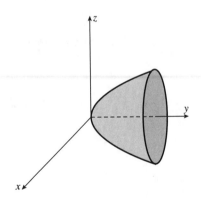

_____ **5.** $x^2 + 6x + y^2 - 10y + z = 0$ is a(n):

     a) ellipsoid

     b) paraboloid

     c) hyperbolic cone

     d) elliptic cone

_____ **6.** True or False:

     The graph of $x = 5$ is a cylinder.

Section 13.1

_____ **1.** For $\mathbf{r}(t) = \sin\frac{\pi}{t}\mathbf{i} - \cos\frac{t\pi}{4}\mathbf{j}$, $\mathbf{r}(3) =$

a) $\frac{\sqrt{3}}{2}\mathbf{i} + \frac{\sqrt{2}}{2}\mathbf{j}$ b) $\frac{\sqrt{2}}{2}\mathbf{i} + \frac{\sqrt{3}}{2}\mathbf{j}$

c) $\frac{\sqrt{3}}{2}\mathbf{i} - \frac{\sqrt{2}}{2}\mathbf{j}$ d) $\frac{\sqrt{2}}{2}\mathbf{i} - \frac{\sqrt{3}}{2}\mathbf{j}$

_____ **2.** The curve given by $\mathbf{r}(t) = 2\mathbf{i} + t\mathbf{j} + 2t\mathbf{k}$ is a:

a) line   b) plane   c) spiral   d) circle

_____ **3.** $\lim\limits_{t\to\infty} \left\langle e^{-2t}, \cos\frac{1}{t} \right\rangle =$

a) $\langle 1, 1 \rangle$ b) $\langle 0, 1 \rangle$ c) $\langle 0, 0 \rangle$ d) It does not exist.

_____ **4.** The equation that best describes the curve at the right is:

a) $\mathbf{r}(t) = \langle t, \sin t, \cos t \rangle$
b) $\mathbf{r}(t) = \langle \cos t, t, \sin t \rangle$
c) $\mathbf{r}(t) = \langle \cos t, \sin t, t \rangle$
d) $\mathbf{r}(t) = \langle \cos t, \sin t + \cos t, \sin t \rangle$

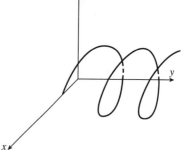

_____ **5.** True or False:
The vector function $\mathbf{r}(t) = \left\langle \frac{t-3}{|t-3|}, t^2, (t-3)^2 \right\rangle$ is continuous at $t = 3$.

Section 13.2

_____ **1.** For $\mathbf{r}(t) = t^3\mathbf{i} + \sin t\mathbf{j} - (t^2 + 2t)\mathbf{k}$, $\mathbf{r}'(0) =$

    a) $\mathbf{j} - 2\mathbf{k}$ b) $3\mathbf{i} - 4\mathbf{k}$ c) $3\mathbf{i} + \mathbf{j} - 2\mathbf{k}$ d) $\mathbf{0}$ (zero vector)

_____ **2.** True or False:
$[\mathbf{r}(t) \times \mathbf{s}(t)]' = \mathbf{r}(t) \times \mathbf{s}'(t) + \mathbf{s}(t) \times \mathbf{r}'(t)$.

_____ **3.** Find the unit tangent vector for $\mathbf{r}(t) = 4t^3\mathbf{i} + (2t + t^2)\mathbf{j} + (7t + t^2)\mathbf{k}$ at $t = -1$.

    a) $\langle 12, 0, 5 \rangle$ b) $\langle 12, 5, 0 \rangle$ c) $\left\langle \frac{12}{13}, 0, \frac{5}{13} \right\rangle$ d) $\left\langle \frac{12}{13}, \frac{5}{13}, 0 \right\rangle$

_____ **4.** $\displaystyle\int_0^1 (e^t\mathbf{i} + 3\sqrt{t}\mathbf{j} + 2t\mathbf{k})\, dt =$

    a) $e\mathbf{i} + 2\mathbf{j} + \mathbf{k}$

    b) $e\mathbf{i} + \frac{3}{2}\mathbf{j} + 2\mathbf{k}$

    c) $(e - 1)\mathbf{i} + 2\mathbf{j} + \mathbf{k}$

    d) $(e - 1)\mathbf{i} + \frac{3}{2}\mathbf{j} + 2\mathbf{k}$

_____ **1.** A definite integral for the length of the curve given by $\mathbf{r}(t) = \langle t, 3 + t, t^2 \rangle$ for $1 \le t \le 2$ is:

a) $\displaystyle\int_1^2 \sqrt{2 + 4t^2}\, dt$

b) $\displaystyle\int_1^2 \sqrt{9 + 6t + 2t^2 + t^4}\, dt$

c) $\displaystyle\int_1^2 \sqrt{9 + 4t}\, dt$

d) $\displaystyle\int_1^2 2t\, dt$

_____ **2.** A parameterization of the curve at the right is:

a) $x = 4t, y = 4 - t^2, 0 \le t \le 2$
b) $x = 4 - t^2, y = 4t, 0 \le t \le 2$
c) $x = t, y = 4 - t^2, 0 \le t \le 2$
d) $x = 4 - t^2, y = t, 0 \le t \le 2$

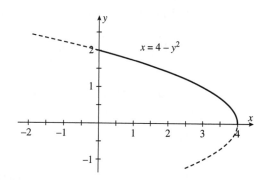

_____ **3.** For the vector function $\mathbf{r}(t)$ with unit tangent $\mathbf{T}$, the principal unit normal vector is:

a) $\dfrac{\mathbf{T}}{|\mathbf{T}|}$

b) $\dfrac{\mathbf{r}'}{|\mathbf{r}'|}$

c) $\dfrac{\mathbf{r}''}{|\mathbf{r}''|}$

d) $\dfrac{\mathbf{T}'}{|\mathbf{T}'|}$

_____ **4.** True or False:
$\mathbf{r}(t) = \langle t^2 - 2t, \sin t \rangle$, for $-\pi \le t \le \pi$, is smooth.

_____ **5.** The curvature of $\mathbf{r}(t) = -e^t\mathbf{i} + t\mathbf{j} + e^t\mathbf{k}$ at $t = 0$ is:

a) $\dfrac{\sqrt{2}}{3\sqrt{3}}$

b) $\dfrac{\sqrt{2}e}{(2e + 1)^{3/2}}$

c) $\dfrac{1}{(2e)^{3/2}}$

d) $\dfrac{\sqrt{2}e}{(2e^2 + 1)^{3/2}}$

_____ **6.** True or False:
If f is twice differentiable and x_0 is an inflection point for f, then the curvature of f at x_0 is 0.

_____ **7.** The normal plane to $\mathbf{r}(t) = t^2\mathbf{i} - t^3\mathbf{j} + t^4\mathbf{k}$ at $t = 1$ has equation:

a) $2(x - 1) - 3(y + 1) + 4(z - 1) = 9$
b) $2(x - 1) - 3(y + 1) + 4(z - 1) = 20$
c) $(x - 2) - (y + 3) + (z - 4) = 9$
d) $(x - 2) - (y + 3) + (z - 4) = 20$

Section 13.4

____ 1. The acceleration for a particle whose position at time t is
$3t^3\mathbf{i} + \ln t\mathbf{j} - \sin 2t\mathbf{k}$ is:

    a) $9t^2\mathbf{i} + \frac{1}{t}\mathbf{j} - 2\cos 2t\mathbf{k}$ b) $18t\mathbf{i} - \frac{1}{t^2}\mathbf{j} + 4\sin 2t\mathbf{k}$

    c) $18\mathbf{i} + \frac{2}{t^3}\mathbf{j} - 8\cos 2t\mathbf{k}$ d) $\frac{3}{4}t^4\mathbf{i} + (t\ln t - t)\mathbf{j} + \frac{1}{2}\cos 2t\mathbf{k}$

____ 2. Find the position function for which (at $t = 0$) the initial position is $\mathbf{i}$, the
initial velocity is $2\mathbf{j}$, and the initial acceleration is $30t\mathbf{i} + 60t^2\mathbf{k}$.

    a) $5t^3\mathbf{i} + 2\mathbf{j} + 5t^4\mathbf{k}$ b) $(5t^3 + 1)\mathbf{i} + (2t + 1)\mathbf{j} + 5t^4\mathbf{k}$

    c) $(5t^3 + 1)\mathbf{i} + 2t\mathbf{j} + 5t^4\mathbf{k}$ d) $5t^3\mathbf{i} + 2\mathbf{j} + (5t^4 + 1)\mathbf{k}$

____ 3. The force needed for a 10-kg object to attain velocity $6t^3\mathbf{i} + 10t\mathbf{j}$ is:

    a) $20t^3\mathbf{i} + 50t^2\mathbf{j}$ b) $5t^4\mathbf{i} + \frac{50}{3}t^3\mathbf{j}$

    c) $180t^2\mathbf{i} + 100\mathbf{j}$ d) $120\mathbf{i}$

____ 4. The tangential component of acceleration for $\mathbf{r}(t) = e^t\mathbf{i} - e^{-t}\mathbf{j}$ at $t = 0$ is:

    a) 0 b) $\frac{e^4 - 1}{\sqrt{e^6 - e^2}}$ c) $\frac{e^4}{\sqrt{e^2 - 1}}$ d) $\frac{e^4 - 1}{e}$

Section 14.1

_____ **1.** Which of these points is not in the domain of $f(x, y) = \sqrt{8 - x - 2y^2}$?

    a) $(0, 0)$ b) $(-6, 2)$

    c) $(3, 3)$ d) All these points are in the domain.

_____ **2.** For $f(x, y) = \ln(x + y), f(e^2, e^3) =$

    a) $2 \ln(1 + e)$ b) $2 + \ln(1 + e)$ c) 5 d) $\dfrac{3}{2}$

_____ **3.** Each level curve (for $k \neq 0$) of $f(x, y) = xy$ is a(n):

    a) ellipse   b) hyperbola   c) parabola   d) pair of lines

_____ **4.** Which function best fits the graph at the right?

    a) $f(x, y) = 4 - x^2$

    b) $f(x, y) = 4 - y^2$

    c) $f(x, y) = 4 - xy$

    d) $f(x, y) = 4 - x^2 - y^2$

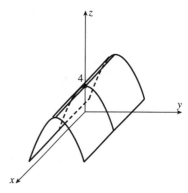

_____ **5.** The range of $f(x, y) = \sqrt{x} + \dfrac{1}{\sqrt{y}}$ is:

    a) $(-\infty, \infty)$ b) $[0, \infty)$ c) $(0, \infty)$ d) $[1, \infty]$

Section 14.2

_____ **1.** True or False:

If $\lim_{(x,y)\to(a,b)} f(x, y) = L$ **and** $\lim_{(x,y)\to(a,b)} g(x, y) = M \neq 0,$ **then**

$\lim_{(x,y)\to(a,b)} \dfrac{f(x, y)}{g(x, y)} = \dfrac{L}{M}.$

_____ **2.** $\lim_{(x,y)\to(-1,2)} \dfrac{xy}{x^2 - y^2} =$

     a) $\dfrac{2}{3}$ b) $-\dfrac{2}{3}$ c) $-\dfrac{2}{5}$ d) It does not exist.

_____ **3.** $\lim_{(x,y)\to(0,0)} \dfrac{\sin(xy)}{y^2} =$

     a) 0   b) 1   c) π d) It does not exist.

_____ **4.** $f(x, y) = \sqrt{x + y}$ is continuous for all (x, y) such that:

     a) $x \geq 0$ and $y \geq 0$ b) $y > x$

     c) $y > -x$ d) $-y \leq x$

_____ **5.** True or False:

The function f is continuous at $(0, 0)$, where $f(x, y) = \begin{cases} \dfrac{\sin x}{y} & y \neq 0 \\ 1 & y = 0 \end{cases}.$

Section 14.3

_____ **1.** $f_x(a, b)$ is the slope of the tangent line to the curve formed by the intersection of the surface $z = f(x, y)$ at (a, b) and:

    a) the trace through $x = a$

    b) the trace through $y = b$

    c) the trace through $x = y$

    d) the intersection of the two traces $x = a$ and $y = b$

_____ **2.** For $f(x, y) = x \sin y + y \cos x + x^2 y^2$, $\dfrac{\partial f}{\partial x} =$

    a) $\cos y - x \sin y + y \sin x + 4xy$

    b) $\sin y + x \cos y - y \sin x + 2x$

    c) $\sin y - y \sin x + 2xy^2$

    d) $\cos y - y \sin x + 2xy^2$

_____ **3.** For $f(x, y) = e^{x^2 y^3}$, $f_y(x, y) =$

    a) $3y^2 e^{x^2 y^3}$ b) $3x^2 y^2 e^{x^2 y^3}$

    c) $(2xy^3 + 3x^2 y^3)e^{x^2 y^3}$ d) $e^{3x^2 y^3}$

_____ **4.** For $f(x, y) = \ln(2x + 3y)$, $f_{yx} =$

    a) $\dfrac{-6}{(2x + 3y)^2}$ b) $\dfrac{3}{2x + 3y}$

    c) $\dfrac{6}{2x + 3y}$ d) $\dfrac{2}{x} + \dfrac{3}{y}$

_____ **5.** Sometimes, Always, or Never:

    $f_{xy} = f_{yx}$.

_____ **6.** How many second partial derivatives will there be for $z = f(r, s, t, x, y)$? (Do not assume that they are continuous.)

    a) 5   b) 10   c) 20   d) 25

Section 14.4

1. True or False:
  If dz exists at (x, y), then $z = f(x, y)$ is differentiable at (x, y).

2. The equation of the tangent plane to $z = x^2 + xy + y^3$ for $(x, y) = (2, -1)$ is:
  a) $3x + 5y - z = 0$ b) $3x + 5y - z = 2$
  c) $5x + 3y - z = 0$ d) $5x + 3y - z = 2$

3. For $z = \sin(xy)$, $dz =$
  a) $xy \cos(xy)dx + xy \cos(xy)dy$
  b) $x \cos(xy)dx + y \cos(xy)dy$
  c) $y \cos(xy)dx + x \cos(xy)dy$
  d) 0

4. The volume of a pyramid is $\frac{1}{3}$ times the area of its base times its altitude. Using differentials, an estimate for the volume of the pyramid at the right is:

  a) 15.10
  b) 15.12
  c) 15.13
  d) 15.14

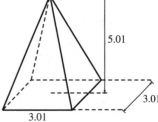

5.01

3.01

3.01

_____ **1.** For $z = xy^2$, $x = t + v^2$, $y = t^2 + v$, $\frac{\partial z}{\partial v} =$

    a) $2[(t^2 + v)^2 + (t^2 + v)(t + v^2)]$
    b) $2[(t^2 + v)^2 + (t + v^2)^2]$
    c) $2(t^2 + v)(2t^2v + v^2)$
    d) $2(t^2 + v)(t^2v + t + 2v^2)$

_____ **2.** Find $\frac{\partial f}{\partial r}$ for $f(x, y) = x^2 + 2y^2$, $x = 2rs$, $y = 4r^2s^2$.

    a) $8rs^2 + 128r^3s^4$ b) $2r^2s^2 + 64r^3s^4$
    c) $4rs^2 + 64r^3s^4$ d) $2r^2s^2 + 128r^3s^4$

_____ **3.** Find $\frac{dy}{dx}$ for y defined implicitly by $x^2 - 6xy + y^2 = 20$.

    a) $\frac{x - 3y}{3x - y}$ b) $\frac{x - 3y}{y - 3x}$ c) $\frac{y - 3x}{x - 3y}$ d) $\frac{3x - y}{x - 3y}$

_____ **4.** Find $\frac{\partial z}{\partial x}$ for $z = f(x, y)$ defined implicitly by $\frac{x^2}{4} - \frac{y^2}{9} + \frac{z^2}{16} = 0$.

    a) $\frac{-x}{4z}$ b) $\frac{x}{4z}$ c) $\frac{-4x}{z}$ d) $\frac{4x}{z}$

____ **1.** For $f(x, y) = x^3y + xy^2$ and $\mathbf{u} = \left\langle \frac{3}{5}, -\frac{4}{5} \right\rangle$, $D_{\mathbf{u}}f(-1, 1) =$

     a) $\frac{12}{5}$ b) $-\frac{12}{5}$ c) $\frac{24}{5}$ d) $-\frac{24}{5}$

____ **2.** True or False:
$D_{\mathbf{u}}f(a, b) = \nabla f(a, b) \cdot \mathbf{u}$.

____ **3.** For $z = f(x, y)$ and $\mathbf{u} = -\mathbf{j}$, $D_{\mathbf{u}}f(a, b) =$
     a) $f_x(a, b)$ (b) $f_y(a, b)$ (c) $-f_x(a, b)$ (d) $-f_y(a, b)$

____ **4.** In what direction $\mathbf{u}$ is $D_{\mathbf{u}}f(-1, 1)$ maximum for $f(x, y) = x^3y^4$?

     a) $\langle 3, -4 \rangle$ b) $\left\langle \frac{3}{5}, -\frac{4}{5} \right\rangle$ c) $\langle 4, -3 \rangle$ d) $\left\langle \frac{4}{5}, -\frac{3}{5} \right\rangle$

____ **5.** The equation of the plane tangent to $x^2 + 2y^2 + 2z^2 = 5$ at $(-1, -1, 1)$ is:
     a) $-(x + 1) - (y + 1) + (z - 1) = 0$
     b) $-(x - 1) - (y - 1) + (z + 1) = 0$
     c) $-2(x + 1) - 4(y + 1) + 4(z - 1) = 0$
     d) $-4(x + 1) - 4(y + 1) - 2(z - 1) = 0$

____ **6.** Which vector at the right could be a gradient for $f(x, y)$?
     a) **a**
     b) **b**
     c) **c**
     d) **d**

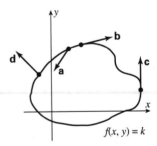

On Your Own

Section 14.7

_____ **1.** A critical point of $f(x, y) = 60 - 6x + x^2 + 8y - y^2$ is:

    a) $(3, 4)$ b) $(-3, -4)$ c) $(-3, 4)$ d) $(3, -4)$

_____ **2.** If (a, b) is a critical point of $f(x, y)$, all second derivatives of f are continuous, and $f_{xx}(a, b) = 10, f_{yy}(a, b) = 3, f_{xy}(a, b) = 5$, then (a, b) is a:

    a) local maximum
    b) local minimum
    c) saddle point
    d) You cannot tell from the information given.

_____ **3.** The absolute minimum of $f(x, y) = 40 - 3x^2 - 2y^2$ with domain $D = \{(x, y) \mid -2 \le x \le 3, -1 \le y \le 2\}$ is:

    a) 0   b) 5   c) 12   d) 40

_____ **4.** A box with half a lid is to hold 48 cm^3. See the figure at the right. If the box is to have minimum surface area, then the value of y must be:

    a) 4   b) $\sqrt[3]{48}$
    c) 6   d) 8

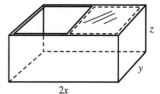

_____ **5.** True or False:

If $f_{xx}f_{yy} - (f_{xy})^2 = 0$ at (x_0, y_0), then (x_0, y_0) cannot be a saddle point.

Section 14.8

_____ **1.** The absolute maximum of $f(x, y) = 12x - 8y$ subject to $x^2 + 2y^2 = 11$ occurs at:

 a) $(3, -1)$ b) $(-3, 1)$ c) $(\sqrt{11}, 0)$ d) $\left(0, -\sqrt{\dfrac{11}{2}}\right)$

_____ **2.** Suppose we have a rectangular box whose volume is 100 cm^3 and total edge length is 50 cm. Suppose we which to solve the problem of finding the minimum surface area of such a box. Which set of equations results when finding a solution by Lagrange multipliers?

 a) $2y + 2z = \lambda_1 yz + 4\lambda_2$
 $2x + 2z = \lambda_1 xz + 4\lambda_2$
 $2x + 2y = \lambda_1 xy + 4\lambda_2$
 $xyz = 100$
 $4x + 4y + 4z = 50$

 b) $yz = \lambda_1(2y + 2z) + 4\lambda_2$
 $xz = \lambda_1(2x + 2z) + 4\lambda_2$
 $xy = \lambda_1(2x + 2y) + 4\lambda_2$
 $xyz = 100$
 $4x + 4y + 4z = 50$

 c) $4y + 4z = \lambda_1 yz + \lambda_2(2y + 2z)$
 $4x + 4z = \lambda_1 xz + \lambda_2(2x + 2z)$
 $4x + 4y = \lambda_1 xy + \lambda_2(2x + 2y)$
 $xyz = 100$
 $4x + 4y + 4z = 50$

 d) $4 = \lambda_1 yz + \lambda_2(2y + 2z)$
 $4 = \lambda_1 xz + \lambda_2(2x + 2z)$
 $4 = \lambda_1 xy + \lambda_2(2x + 2y)$
 $xyz = 100$
 $4x + 4y + 4z = 50$

Section 15.1

___ 1. How many subrectangles of $[0, 4] \times [3, 8]$ are determined by the partition with $\Delta x = 2$ and $\Delta y = 1$?

     a) 6   b) 8   c) 10   d) 20

___ 2. Let R be the rectangle $[2, 10] \times [1, 7]$. Find the Riemann sum for $\iint_R (x + y) \, dA$ using midpoints of the subrectangles for the partition of R given at the right.

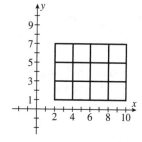

     a) 120

     b) 240

     c) 360

     d) 480

___ 3. Let $R = \{(x, y) \mid 3 \le x \le 5, 2 \le y \le 4\}$ and $f(x, y) = y^2 - x^2$. For every partition of R, the double Riemann sum for $\iint_R (y^2 - x^2) \, dA$ will be largest if, for each subrectangle, we choose (x_i^*, y_j^*) so that:

     a) x_i^* is the left (lower) endpoint and y_j^* is the left (lower) endpoint.

     b) x_i^* is the left (lower) endpoint and y_j^* is the right (upper) endpoint.

     c) x_i^* is the right (upper) endpoint and y_j^* is the left (lower) endpoint.

     d) x_i^* is the right (upper) endpoint and y_j^* is the right (upper) endpoint.

___ 4. Sometimes, Always, or Never:

$\iint_R f(x, y) \, dA$ is the volume above the rectangular region R and under $z = f(x, y)$.

___ 5. True or False:

If R is a rectangular region in the plane, then $\iint_R x^2 y^3 \, dA = \left(\iint_R x^2 \, dA \right)\left(\iint_R y^3 \, dA \right)$.

_____ **1.** Evaluate $\int_0^2 \int_0^1 6xy^2 \, dx \, dy$.

    a) 6   b) 8   c) 24   d) 30

_____ **2.** True or False:

$$\int_1^5 \int_2^4 (x^2 + y^2) \, dx \, dy = \int_2^4 \int_1^5 (x^2 + y^2) \, dx \, dy.$$

_____ **3.** An iterated integral for the volume of the solid shown is:

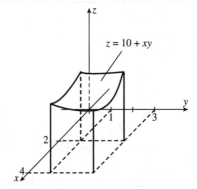

    a) $\int_2^4 \int_1^3 (10x + xy) \, dx \, dy$

    b) $\int_2^4 \int_1^3 (10x + xy) \, dy \, dx$

    c) $\int_1^2 \int_3^4 (10x + xy) \, dx \, dy$

    d) $\int_1^2 \int_3^4 (10x + xy) \, dy \, dx$

Section 15.3

___ **1.** Write $\iint_D x\,dA$ as an iterated integral, where D is shown at the right.

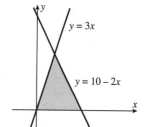

a) $\displaystyle\int_0^2 \int_{10-2x}^{3x} x\,dx\,dy$

b) $\displaystyle\int_0^5 \int_{3x}^{10-2x} x\,dy\,dx$

c) $\displaystyle\int_0^2 \int_{(x+10)/2}^{x/3} x\,dy\,dx$

d) $\displaystyle\int_0^6 \int_{y/3}^{(10-y)/2} x\,dx\,dy$

___ **2.** True or false:

$$\int_1^3 \int_2^5 xy^2\,dy\,dx = \int_2^5 \int_1^3 xy^2\,dy\,dx.$$

___ **3.** Rewritten in reverse order, $\displaystyle\int_0^1 \int_{y-1}^0 x^2y^2\,dx\,dy =$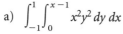

a) $\displaystyle\int_{-1}^1 \int_0^{x-1} x^2y^2\,dy\,dx$

b) $\displaystyle\int_{-1}^0 \int_0^{x+1} x^2y^2\,dy\,dx$

c) $\displaystyle\int_0^1 \int_0^{x+1} x^2y^2\,dy\,dx$

d) $\displaystyle\int_0^1 \int_0^{x-1} x^2y^2\,dy\,dx$

___ **4.** The area of the region at the right is:

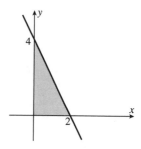

a) $\displaystyle\int_0^2 \int_0^4 1\,dy\,dx$

b) $\displaystyle\int_0^2 \int_0^{4-2x} 1\,dy\,dx$

c) $\displaystyle\int_0^4 \int_0^{(y-4)/2} 1\,dx\,dy$

d) $\displaystyle\int_0^4 \int_0^2 1\,dx\,dy$

_____ **1.** Rewritten as an iterated integral in polar coordinates, $\int_0^2 \int_0^{\sqrt{4-x^2}} y \, dy \, dx =$

a) $\int_0^2 \int_0^{\pi/2} r \sin \theta \, d\theta \, dr$

b) $\int_0^2 \int_0^{\pi/2} r^2 \sin \theta \, d\theta \, dr$

c) $\int_0^{\pi/2} \int_0^2 r^2 \cos \theta \, dr \, d\theta$

d) $\int_0^{\pi/2} \int_0^2 r \cos \theta \, dr \, d\theta$

_____ **2.** An iterated integral for the volume of the solid under $z = x^2 + y^2$ and above the shaded region D is:

a) $\int_0^5 \int_0^r 5r^2 \cos 2\theta \, d\theta \, dr$

b) $\int_0^5 \int_0^r 5r^2 \cos 2\theta \, d\theta \, dr$

c) $\int_0^{\pi/4} \int_0^{5 \cos 2\theta} r^2 \, dr \, d\theta$

d) $\int_0^{\pi/4} \int_0^{5 \cos 2\theta} r^3 \, dr \, d\theta$

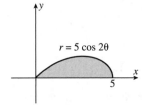

$r = 5 \cos 2\theta$

1. The mass of the lamina at the right, which has a density of $x^2 + y^2$ at each point (x, y), is given by:

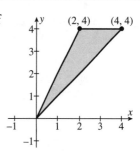

a) $\int_0^4 \int_{y/2}^y (x^2 + y^2)\, dx\, dy$

b) $\int_0^4 \int_x^{2x} (x^2 + y^2)\, dy\, dx$

c) $\int_0^4 \int_y^{2y} (x^2 + y^2)\, dx\, dy$

d) $\int_2^4 \int_{y/2}^y (x^2 + y^2)\, dx\, dy$

2. The y-coordinate of the center of mass of a lamina D with density $\rho(x, y)$ and mass m is:

a) $m \iint_D y\rho(x, y)\, dA$

b) $m \iint_D x\rho(x, y)\, dA$

c) $\dfrac{\iint_D y\rho(x, y)\, dA}{m}$

d) $\dfrac{\iint_D x\rho(x, y)\, dA}{m}$

3. True or False:

Let $f(x, y) = \begin{cases} x + y^2 & \text{if } 0 \leq x \leq 1, 0 \leq y \leq x \\ 0 & \text{otherwise} \end{cases}$. Then $f(x, y)$ is a joint probability function.

4. True or False:

For $f(x, y) = x^2(x + y)$, x and y are independent variables.

1. Find the surface area of that portion of the graph of $z = x^2 + \sqrt{3}y$ lying above the region at the right.

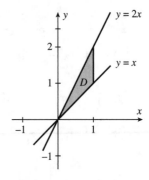

 a) $\dfrac{2\sqrt{2} - 1}{3}$

 b) $2\sqrt{2} - 1$

 c) $\dfrac{4\sqrt{2} - 2}{3}$

 d) $4\sqrt{2} - 2$

Section 15.7

____ **1.** True or False:

$$\int_3^6 \int_1^5 \int_2^4 xyz\, dx\, dz\, dy = \int_1^4 \int_3^6 \int_2^5 xyz\, dy\, dx\, dz.$$

____ **2.** $\displaystyle \int_1^2 \int_0^x \int_y^{x+y} 12x\, dz\, dy\, dx =$

    a) 45   b) 15   c) 63   d) 127

____ **3.** $\displaystyle \iiint_E z\, dV$, where E is the wedge-shaped solid shown at the right, equals:

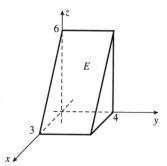

    a) $\displaystyle \int_0^4 \int_0^3 \int_0^{6-2x} z\, dz\, dx\, dy$

    b) $\displaystyle \int_0^4 \int_0^3 \int_0^6 z\, dz\, dx\, dy$

    c) $\displaystyle \int_0^4 \int_0^3 \int_0^{6-z} z\, dx\, dz\, dy$

    d) $\displaystyle \int_0^4 \int_0^3 \int_0^{6-x-y} z\, dz\, dx\, dy$

____ **4.** The moment about the xz-plane of solid E whose density is $\rho(x, y, z) = x$ is:

    a) $\displaystyle \iiint_E x^2z\, dV$ b) $\displaystyle \iiint_E xy\, dV$ c) $\displaystyle \iiint_E x\, dV$ d) $\displaystyle \iiint_E xyz\, dV$

_____ **1.** The cylindrical coordinates of the point with rectangular coordinates $(5, 5, 4)$ are:

a) $\left(5\sqrt{2}, \frac{\pi}{4}, 4\right)$ b) $\left(5\sqrt{2}, \frac{\pi}{2}, 4\right)$ c) $\left(5, \frac{\pi}{4}, 4\right)$ d) $\left(5, \frac{\pi}{2}, 4\right)$

_____ **2.** The rectangular coordinates of the point with cylindrical coordinates $\left(1, \frac{\pi}{6}, 1\right)$ are:

a) $(1, 0, 1)$ b) $\left(\frac{1}{2}, \frac{\sqrt{3}}{2}, 1\right)$ c) $\left(\frac{\sqrt{3}}{2}, \frac{1}{2}, 1\right)$ d) $(0, 1, 1)$

_____ **3.** The surface given by $z = -r^2$ in cylindrical coordinates is a(n):

a) elliptic cone   b) hyperboloid of one sheet
c) hyperbolic cylinder   d) elliptic paraboloid

_____ **4.** Write $\iiint\limits_E (x^2 + y^2 + z^2)\, dV$ in cylindrical coordinates, where E is the solid at the right.

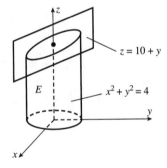

a) $\int_0^{2\pi} \int_0^2 \int_0^{10+r\sin\theta} (r^2 + z^2) r\, dz\, dr\, d\theta$

b) $\int_0^{2\pi} \int_0^4 \int_0^{r^2} (r^2 + z^2) r\, dz\, dr\, d\theta$

c) $\int_0^{2\pi} \int_0^2 \int_0^{10+r\sin\theta} (r^2 + z^2)\, dz\, dr\, d\theta$

d) $\int_0^{2\pi} \int_0^4 \int_0^{r^2} (r^2 + z^2)\, dz\, dr\, d\theta$

Section 15.9

_____ **1.** Spherical coordinates of the point with rectangular coordinates $(-2, 2\sqrt{3}, 4\sqrt{3})$ are:

a) $\left(8, \frac{2\pi}{3}, \frac{\pi}{6}\right)$ b) $\left(8, \frac{\pi}{3}, -\frac{\pi}{6}\right)$ c) $\left(8, \frac{\pi}{3}, \frac{\pi}{6}\right)$ d) $\left(8, \frac{2\pi}{3}, -\frac{\pi}{6}\right)$

_____ **2.** Rectangular coordinates of the point with spherical coordinates $\left(2, \frac{\pi}{2}, \frac{\pi}{4}\right)$ are:

a) $(2, \sqrt{2}, \sqrt{2})$ b) $(\sqrt{2}, \sqrt{2}, 0)$ c) $(\sqrt{2}, 0, \sqrt{2})$ d) $(0, \sqrt{2}, \sqrt{2})$

_____ **3.** The point P graphed at the right has spherical coordinates:

a) $\left(2, \frac{\pi}{6}, \frac{\pi}{3}\right)$

b) $\left(2, \frac{\pi}{3}, \frac{\pi}{6}\right)$

c) $\left(4, \frac{\pi}{6}, \frac{\pi}{3}\right)$

d) $\left(4, \frac{\pi}{3}, \frac{\pi}{6}\right)$

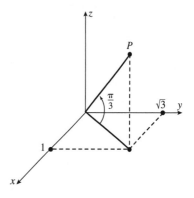

_____ **4.** Write $\iiint\limits_{E} 2\, dV$ in spherical coordinates, where E is the bottom half of the sphere at the right.

a) $\int_{\pi/2}^{\pi}\int_{0}^{2\pi}\int_{0}^{2} 2\rho^2 \sin\phi\, d\rho\, d\theta\, d\phi$

b) $\int_{\pi/2}^{\pi}\int_{0}^{2\pi}\int_{0}^{4} 2\rho^2 \sin\phi\, d\rho\, d\theta\, d\phi$

c) $\int_{\pi/2}^{\pi}\int_{0}^{2\pi}\int_{0}^{2} 2\, d\rho\, d\theta\, d\phi$

d) $\int_{\pi/2}^{\pi}\int_{0}^{2\pi}\int_{0}^{4} 2\, d\rho\, d\theta\, d\phi$

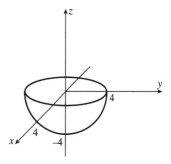

Section 15.10

_____ **1.** Find the Jacobian for $x = u^2v^2$, $y = u^2 + v^2$.

    a) $2uv^3 - 2u^3v$ b) $4u^2v - 4uv^2$

    c) $2u^2v - 2uv^2$ d) $4uv^3 - 4u^3v$

_____ **2.** Find the iterated integral for $\iint\limits_R dA$, where R is the first quadrant of the

ellipse $4x^2 + 9y^2 = 36$ and the transformation is $x = 3u\cos v$, $y = 2u\sin v$.

    a) $\displaystyle\int_0^1 \int_0^{\pi/2} 1\, dv\, du$ b) $\displaystyle\int_0^1 \int_0^{\pi/2} 6u\, dv\, du$

    c) $\displaystyle\int_0^1 \int_0^{\pi/2} 3u^2\, dv\, du$ d) $\displaystyle\int_0^1 \int_0^{\pi/2} u\, dv\, du$

Section 16.1

_____ **1.** For $\mathbf{F}(x, y) = (x + y)\mathbf{i} + (2x - 4y)\mathbf{j}$, which vector, **a**, **b**, **c**, or **d**, best represents $\mathbf{F}(1, 1)$?

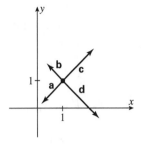

- a) **a**
- b) **b**
- c) **c**
- d) **d**

_____ **2.** The gradient vector field for $f(x, y) = x^2 y^3$ is:

- a) $x^2\mathbf{i} + y^3\mathbf{j}$
- b) $2x\mathbf{i} + 3y^2\mathbf{j}$
- c) $2xy^3\mathbf{i} + 3x^2 y^2\mathbf{j}$
- d) $y^3\mathbf{i} + x^2\mathbf{j}$

_____ **3.** If $\nabla f = \mathbf{F}$, then the vector field $\mathbf{F}$ is:

- a) linear
- b) conservative
- c) gradient
- d) tangent

_____ **4.** True or False:

$\mathbf{F}(x, y) = 2y\mathbf{i} + 2x\mathbf{j}$ is conservative.

Section 16.2

____ **1.** A definite integral for $\int_C (x +y)\, ds$, where C is the curve given by $x = 3t$, $y = t, t \in [0, 1]$ is:

a) $\int_0^1 4t\, dt$ b) $\int_0^1 4\sqrt{10}t\, dt$

c) $\int_0^1 4t\sqrt{10t}\, dt$ d) $\int_0^1 4t\sqrt{t}\, dt$

____ **2.** True or False:
The curve at the right appears to be piecewise smooth.

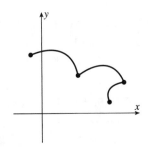

____ **3.** A definite integral for $\int_C \mathbf{F} \cdot d\mathbf{r}$, where $\mathbf{F}(x, y) = x^2\mathbf{i} + y^2\mathbf{j}$ and C is the line from $(1, 0)$ to $(0, 1)$ given by $x = 1 - t, y = t, t \in [0, 1]$, is:

a) $\int_0^1 (2t - 1)\, dt$ b) $\int_0^1 \left(t^2 + (1 - t)^2\right) dt$

c) $\int_0^1 \left(t^3 + (1 - t)^2 t\right) dt$ d) $\int_0^1 2\, dt$

____ **4.** True or False:
If C is a curve given by $\mathbf{r}(t)$ when $a \le t \le b$, then
$$\int_C f(x, y)\, dy = \int_a^b f(x(t), y(t))\, dt.$$

____ **5.** The mass of a thin wire shaped like the curve $y = x^2$ from $(0, 0)$ to $(1, 1)$ with density $1 + x^2 + y^2$ at (x, y) is:

a) $\int_0^1 (1 + t^2 + t^4)\sqrt{1 + 4t^2}\, dt$ b) $\int_0^1 (1 + t^2 + t^4)\, dt$

c) $\int_0^1 \sqrt{1 + 4t^2}\, dt$ d) $\int_0^1 \sqrt{5}(1 + t^2 + t^4)\, dt$

Section 16.3

_____ **1.** For $f(x, y) = x^2 - y^2$ and C, the parabola $y = x^2$ from $(0, 0)$ to $(2, 4)$,
$$\int_C \nabla f \cdot d\mathbf{r} =$$

    a) 0   b) 2   c) -2 d) -12

_____ **2.** Which curve is simple but not closed?

    a)   b)

    c)   d)

_____ **3.** True or False:
$\mathbf{F}(x, y) = (x^2 + 2y^2)\mathbf{i} + (y^2 + 2x^2)\mathbf{j}$ is conservative.

_____ **4.** True or False:
Let $\mathbf{F}(x, y) = 2x \cos y\mathbf{i} - x^2 \sin y\mathbf{j}$ and C be the cubic $y = x^3$ curve
for $0 \le x \le 2$. Then $\int_C \mathbf{F} \cdot d\mathbf{r}$ is independent of path.

_____ **5.** Let C be the line from $(1, 0)$ to $(0, 1)$ and $\mathbf{F}(x, y) = y\mathbf{i} + (x + y)\mathbf{j}$. Then
$$\int_C \mathbf{F} \cdot d\mathbf{r} =$$

    a) $\frac{1}{2}$ b) 1   c) 2   d) 4

Section 16.4

_____ 1. True or False:

Suppose C in the figure at the right is piecewise smooth and **F** has continuous partials. By Green's Theorem,

$$\oint_C \mathbf{F} \cdot d\mathbf{r} = \iint_D f(x, y) \, dA.$$

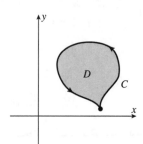

_____ 2. If D is a region enclosed by a closed curve C, such as that pictured in question 1, and $\mathbf{F}(x, y) = P(x, y)\mathbf{i} + Q(x, y)\mathbf{j}$ is conservative, then

$$\int_C \mathbf{F} \cdot d\mathbf{r} =$$

a) 1

b) 0

c) $\iint_D \left(\dfrac{\partial Q}{\partial y} - \dfrac{\partial P}{\partial x} \right) dA$

d) It cannot be determined from the given information.

_____ 3. Let C be the curve that bounds the rectangle at the right. For $\mathbf{F}(x, y) = (x^3 + y)\mathbf{i} + (2xy)\mathbf{j}$, write a double integral for $\oint_C \mathbf{F} \cdot d\mathbf{r}$.

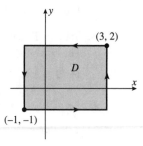

a) $\iint_D (2y - 1) \, dA$

b) $\iint_D (2xy - x^2 y) \, dA$

c) $\iint_D (2x - 3x^2) \, dA$

d) $\iint_D (x^3 + y - 2xy) \, dA$

_____ 4. The area of a region D enclosed by a curve C, such as that pictured in question 1, may be evaluated using the line integral:

a) $\dfrac{1}{2} \oint_C x \, dy$

b) $\dfrac{1}{2} \oint_C y \, dx$

c) $\dfrac{1}{2} \oint_C x \, dy + y \, dx$

d) $\dfrac{1}{2} \oint_C x \, dy - y \, dx$

Section 16.5

____ **1.** Find the curl of $\mathbf{F}(x, y, z) = (x^2 + y^2)\mathbf{i} + (xz)\mathbf{j} + (yz)\mathbf{k}$.

     a) $-z\mathbf{i} + 2y\mathbf{j} + x\mathbf{k}$ b) $(z - x)\mathbf{i} + (z - 2y)\mathbf{k}$

     c) $(x - z)\mathbf{i} + 2x\mathbf{j} + 2y\mathbf{k}$ d) $y\mathbf{k}$

____ **2.** True or False:
    If $\mathbf{F}$ is conservative, then curl $\mathbf{F} = \mathbf{0}$.

____ **3.** Find div $\mathbf{F}$ for $\mathbf{F}(x, y, z) = (x^2 + y^2)\mathbf{i} + (xz)\mathbf{j} + (yz)\mathbf{k}$.

     a) $2z - x - y$ b) $2x + 3y + 2z$

     c) $2x + y$ d) $y + x$

____ **4.** True or False:
    div curl $\mathbf{F} = 0$.

____ **5.** One of the vector forms of Green's Theorem says that $\oint_C \mathbf{F} \cdot d\mathbf{r} =$

     a) $\iint\limits_D \text{curl } \mathbf{F} \, dA$ b) $\iint\limits_D \text{div } \mathbf{F} \, dA$

     c) $\iint\limits_D (\text{curl } \mathbf{F}) \cdot \mathbf{k} \, dA$ d) $\iint\limits_D (\text{div } \mathbf{F}) \cdot \mathbf{k} \, dA$

____ **6.** One of the vector forms of Green's Theorem says that if $\mathbf{n}$ is the outer normal vector to C, then $\int_C \mathbf{F} \cdot \mathbf{n} \, ds =$

     a) $\iint\limits_D \text{curl } \mathbf{F} \, dA$ b) $\iint\limits_D \text{div } \mathbf{F} \, dA$

     c) $\iint\limits_D (\text{curl } \mathbf{F}) \cdot \mathbf{k} \, dA$ d) $\iint\limits_D (\text{div } \mathbf{F}) \cdot \mathbf{k} \, dA$

Section 16.6

1. A parametrization of the cylinder in the figure is:

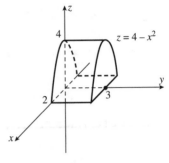

a) $x = u, y = v, z = v - u^2, -2 \le u \le 2, 0 \le v \le 3$
b) $x = u, y = 3, z = 4 - v^2, -2 \le u \le 2, 0 \le v \le 3$
c) $x = u, y = v, z = 4 - u^2, -2 \le u \le 2, 0 \le v \le 3$
d) $x = u, y = v, z = u - v^2, -2 \le u \le 2, 0 \le v \le 2$

2. Find the normal vector to the tangent plane at the point corresponding to $(u, v) = (1, 2)$ on the surface given by $x = 2u, y = u^2 + v^2, z = 3v$.

a) $6\mathbf{i} + 6\mathbf{j} - 8\mathbf{k}$
b) $-6\mathbf{i} - 6\mathbf{j} + 8\mathbf{k}$
c) $6\mathbf{i} - 6\mathbf{j} + 8\mathbf{k}$
d) $-6\mathbf{i} + 6\mathbf{j} - 8\mathbf{k}$

3. The equation of the tangent plane corresponding to $(1, 2)$ to the surface S parameterized by $\mathbf{r}(u, v) = (u + v)\mathbf{i} + (u^2 - v^2)\mathbf{j} + (u^2 + v^2)\mathbf{k}$ is:

a) $16(x - 3) - 2(y + 3) - 6(z - 5) = 0$
b) $8(x - 3) - 2(y + 3) - 0(z - 5) = 0$
c) $16(x + 3) - 6(y - 3) - 2(z - 5) = 0$
d) $8(x + 3) - 6(y + 3) - 2(z - 5) = 0$

4. A surface S is parameterized by $x = \sin u, y = \sin v, z = \cos v$ for $(u, v) \in D$, $0 \le u \le \frac{\pi}{2}, 0 \le v \le \frac{\pi}{2}$. A double integral for the surface area of S is:

a) $\iint_D \sqrt{(\cos u \cos v)^2 + (\sin u \sin v)^2} \, dA$
b) $\iint_D \sqrt{(\cos u \sin v)^2 + (\cos v \sin u)^2} \, dA$

c) $\iint_D \cos u \, dA$
d) $\iint_D \sqrt{3} \, dA$

5. Write an iterated integral for the area of that part of the plane $z = x$ that is above the region D at the right.

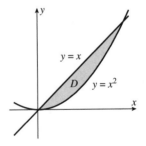

a) $\int_0^1 \int_x^{x^2} \sqrt{x^2 + 1} \, dy \, dx$

b) $\int_0^1 \int_{x^2}^x \sqrt{x^2 + 1} \, dy \, dx$

c) $\int_0^1 \int_{x^2}^x \sqrt{2} \, dy \, dx$

d) $\int_0^1 \int_y^{\sqrt{y}} \sqrt{y + 1} \, dx \, dy$

_____ 1. Find an iterated integral for the surface integral $\iint_S f(x, y, z)\, dS$ where
$f(x, y, z) = x + y$ and the surface S is given by
$\mathbf{r}(u, v) = (u + v)\mathbf{i} + (u - v)\mathbf{j} + (uv)\mathbf{k}, 0 \le u \le 1, 0 \le v \le 1$.

a) $\displaystyle\int_0^1\int_0^1 2u\, \sqrt{2u^2 + 2v^2}\, du\, dv$

b) $\displaystyle\int_0^1\int_0^1 \sqrt{2u^2 + 2v^2}\, du\, dv$

c) $\displaystyle\int_0^1\int_0^1 2u\, \sqrt{2u^2 + 2v^2 + 4}\, du\, dv$

d) $\displaystyle\int_0^1\int_0^1 \sqrt{2u^2 + 2v^2 + 4}\, du\, dv$

_____ 2. Find an iterated integral for $\iint_S \mathbf{F} \cdot d\mathbf{S}$ where S is the surface
$\mathbf{r}(u, v) = u^2\mathbf{i} + v^2\mathbf{j} + uv\mathbf{k}, 0 \le u \le 1, 0 \le v \le 1$, and $\mathbf{F}(x, y, z) = x\mathbf{i} + y\mathbf{j} + 2z\mathbf{k}$.

a) $\displaystyle\int_0^1\int_0^1 2uv\, du\, dv$

b) $\displaystyle\int_0^1\int_0^1 4u^2v^2\, du\, dv$

c) $\displaystyle\int_0^1\int_0^1 (2u^2 + 2v^2)\, du\, dv$

d) $\displaystyle\int_0^1\int_0^1 (2u^2 + 2v^2 + 4u^2v^2)\, du\, dv$

_____ 3. Let $\mathbf{F}(x, y, z) = (2x + 3y + z)\mathbf{i} + (z - 2y)\mathbf{j} + (x + 3y^2 + xz + yz)\mathbf{k}$ and the
surface S be that portion of the graph of $g(x, y) = xy + y$ above the
triangular region $D: 0 \le x \le 2, 0 \le y \le x$. Find $\iint_S \mathbf{F} \cdot d\mathbf{S}$.

a) 2 b) $2\sqrt{2}$ c) $\sqrt{2}$ d) $\dfrac{\sqrt{2}}{2}$

_____ 1. Let $\mathbf{F}(x, y, z) = P\mathbf{i} + Q\mathbf{j} + R\mathbf{k}$ have continuous first partial derivatives and S be a smooth, simple, connected, and positively oriented surface bounded by a simple closed curve C.

By Stokes' Theorem, $\int_C \mathbf{F} \cdot d\mathbf{r} =$

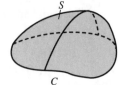

a) $\iint_S \text{curl } \mathbf{F} \cdot d\mathbf{r}$

b) $\iint_S \text{curl } \mathbf{F} \cdot d\mathbf{S}$

c) $\iint_S \text{div } \mathbf{F} \cdot d\mathbf{r}$

d) $\iint_S \text{div } \mathbf{F} \cdot d\mathbf{S}$

_____ 2. Let $\mathbf{F}(x, y, z) = z\mathbf{i} - x\mathbf{j} + y\mathbf{k}$, let C be the circle $x^2 + y^2 = 36$, $z = 0$, and let S be the surface given by $z = 36 - x^2 - y^2$ and $z \geq 0$. Use Stokes' Theorem to convert $\int_C \mathbf{F} \cdot d\mathbf{r}$ to a surface integral and then rewrite the surface integral as a double integral.

a) $\int_{-6}^{6} \int_{-\sqrt{36-u^2}}^{\sqrt{36-u^2}} (2u + 2v - 1) \, dv \, du$

b) $\int_{-6}^{6} \int_{-\sqrt{36-u^2}}^{\sqrt{36-u^2}} (2u - 2v - 1) \, dv \, du$

c) $\int_{-6}^{6} \int_{-\sqrt{36-u^2}}^{\sqrt{36-u^2}} (2u + 2v + 1) \, dv \, du$

d) $\int_{-6}^{6} \int_{-\sqrt{36-u^2}}^{\sqrt{36-u^2}} (2u - 2v + 1) \, dv \, du$

_____ 3. Let $\mathbf{F}(x, y, z) = z\mathbf{i} + x\mathbf{j} - y\mathbf{k}$ and the curve C be the boundary of that portion of the surface S above the rectangular region D: $0 \leq x \leq 1$, $0 \leq y \leq 2$. Use Stokes' Theorem to convert $\iint_S \text{curl } \mathbf{F} \cdot d\mathbf{S}$ to a line integral and then rewrite the line integral as a double integral.

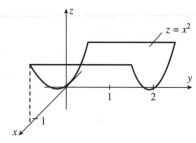

a) $\int_0^2 \int_0^1 (2u + 1) \, dv \, du$

b) $\int_0^2 \int_0^1 (1 - 2u) \, dv \, du$

c) $\int_0^1 \int_0^2 (2u + 1) \, dv \, du$

d) $\int_0^2 \int_0^1 (1 - 2u) \, dv \, du$

Section 16.9

_____ 1. Let $\mathbf{F}(x, y, z) = P\mathbf{i} + Q\mathbf{j} + R\mathbf{k}$ have continuous first partial derivatives and E be a simple solid with boundary surface S that has a positive orientation. By the Divergence Theorem, $\iint_S \mathbf{F} \cdot d\mathbf{S} =$

a) $\iiint_E \text{div } \mathbf{F} \cdot d\mathbf{r}$

b) $\iiint_E \text{div } \mathbf{F} \cdot \mathbf{k} \, ds$

c) $\iiint_E \text{div } \mathbf{F} \cdot dV$

d) $\iint_S \text{div } \mathbf{F} \cdot d\mathbf{S}$

_____ 2. Let S be the surface of the closed cylindrical "half can" and $\mathbf{F}(x, y, z) = 2xy^2\mathbf{i} + xz^2\mathbf{j} + yz\mathbf{k}$. Write a triple integral for $\iint_S \mathbf{F} \cdot d\mathbf{S}$.

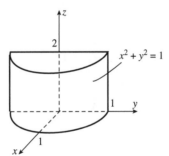

a) $\int_0^2 \int_0^1 \int_{-\sqrt{1-x^2}}^{\sqrt{1-x^2}} (4xy + 2xz + y) \, dy \, dx \, dz$

b) $\int_0^2 \int_0^1 \int_{-\sqrt{1-x^2}}^{\sqrt{1-x^2}} (2y^2 - z) \, dy \, dx \, dz$

c) $\int_0^2 \int_{-1}^1 \int_0^{\sqrt{1-y^2}} (4xy + 2xz + y) \, dx \, dy \, dz$

d) $\int_0^2 \int_{-1}^1 \int_0^{\sqrt{1-y^2}} (2y^2 + y) \, dx \, dy \, dz$

Section 17.1

____ **1.** True or False:

$(x + y)y'' + (x - 10)y' + xy + 10 = 0$ is homogeneous.

____ **2.** True or False:

If y_1 and y_2 are linearly independent solutions to $P(x)y'' + Q(x)y' + R(x)y = 0$, then all solutions have the form $y = c_1y_1 + c_2y_2$.

____ **3.** The general solution to $y'' - 6y' + 8y = 0$ is:

a) $y = e^x(c_1 \cos 4x + c_2 \sin 4x)$
b) $y = e^{4x}(c_1 \cos x + c_2 \sin x)$
c) $y = c_1e^{2x} + c_2e^{4x}$
d) $y = c_1e^{2x} + c_2xe^{4x}$

____ **4.** The solution to $y'' - 10y' + 25y = 0, y(0) = 5, y(1) = 0$ is:

a) $y = 5e^{5x} - xe^{5x}$
b) $y = 5e^{5x} - 5xe^{5x}$
c) $y = e^{5x} - 5xe^{5x}$
d) $y = 5e^{5x} + xe^{5x}$

____ **5.** True or False:

An initial-value problem specifies $y(x_0) = y_0$ and $y(x_1) = y_1$ in the statement of the problem.

____ **6.** Solve $y'' - 2y' + 2 = 0, y(0) = 2, y'(0) = 1$.

a) $y = e^{-x}(\cos x - 2 \sin x)$
b) $y = e^{-x}(2 \cos x - \sin x)$
c) $y = e^x(\cos x - 2 \sin x)$
d) $y = e^x(2 \cos x - \sin x)$

_____ **1.** If $y_c(x)$ is the general solution to $ay'' + by' + cy = 0$ and $y_p(x)$ is a particular solution to $ay'' + by' + cy = G(x)$, then the general solution to the nonhomogeneous equation is:

a) $y(x) = y_c(x) + y_p(x)$
b) $y(x) = y_c(x) - y_p(x)$
c) $y(x) = y_p(x) - y_c(x)$
d) $y(x) = y_c(x)y_p(x)$

_____ **2.** By the method of undetermined coefficients, a particular solution to $y'' - 5y' + 6y = e^{4x}$ will have the form:

a) $y_p(x) = Ae^x$
b) $y_p(x) = A_1e^x + A_2e^{6x}$
c) $y_p(x) = A_1e^{2x} + A_2e^{3x}$
d) $y_p(x) = Ae^{4x}$

_____ **3.** By the method of variation of parameters, a particular solution to $y'' - 2y' + y = 3e^{4x}$ is $u_1y_1 + u_2y_2$, where $y_1 = e^x$, $y_2 = xe^x$, and:

a) $u_1 = -xe^{3x} + e^{3x}$, $u_2 = -e^{3x}$
b) $u_1 = -xe^{3x}$, $u_2 = e^{3x}$
c) $u_1 = -xe^{3x}$, $u_2 = xe^{3x} - e^{3x}$
d) $u_1 = -xe^{3x} + \frac{1}{3}e^{3x}$, $u_2 = e^{3x}$

_____ **4.** Use the method of undetermined coefficients to solve $y'' - 6y' + 5y = -6e^{2x}$.

a) $y = c_1e^x + c_2e^{5x} - 4e^{2x}$
b) $y = c_1e^x + c_2e^{5x} + 2e^{2x}$
c) $y = c_1e^{-x} + c_2e^{-5x} - 4e^{2x}$
d) $y = c_1e^{-x} + c_2e^{-5x} + 2e^{2x}$

_____ **5.** Use the method of variation of parameters to solve $y'' - 2y' = 4e^{2x}$.

a) $y = c_1 + (c_2 + 2x + 1)e^{2x}$
b) $y = c_1 + (c_2 + 2x - 1)e^{2x}$
c) $y = c_1 + c_2e^{2x} + (2x - 1)$
d) $y = c_1 + c_2e^{2x} + (2x + 1)$

Section 17.3

_____ **1.** A series circuit has a resistor with $R = 15 \ \Omega$, an inductor with $L = 2$, a capacitor with $C = 0.005$ F and a 6-V battery. An equation for the charge Q at time t is:

a) $2Q'' + 200Q' + 15Q = 6$

b) $2Q'' + 6Q' + 15Q = 200$

c) $2Q'' + 15Q' + 200Q = 6$

d) $2Q'' + 15Q' + 6Q = 200$

_____ **2.** A spring with a 10-kg mass is immersed in a fluid with damping constant 60. To maintain the spring stretched 3 m beyond its equilibrium requires a force of 150 N. The general solution to the equation modeling the motion is:

a) $x = c_1 e^t + c_2 e^{5t}$

b) $x = c_1 e^{-t} + c_2 e^{-5t}$

c) $x = e^{-t}(c_1 \cos 5t + c_2 \sin 5t)$

d) $x = e^{-5t}(c_1 \cos t + c_2 \sin t)$

Section 17.4

_____ 1. Find the recursive relation among the coefficients for the series solution

$$y = \sum_{n=0}^{\infty} c_n x^n \text{ to } y'' = 2y.$$

a) $c_n = \dfrac{2c_{n-2}}{n(n-1)}$

b) $c_n = \dfrac{2^n c_{n-2}}{n(n-1)}$

c) $c_n = \dfrac{2c_{n-1}}{n(n-1)}$

d) $c_n = \dfrac{2^n c_{n-1}}{n(n-1)}$

_____ 2. The series solution to $y' = xy$ is:

a) $y = \sum_{n=0}^{\infty} \dfrac{c_0}{2^n n!} x^{2n}$

b) $y = \sum_{n=0}^{\infty} \dfrac{c_0}{n!} x^{2n}$

c) $y = \sum_{n=0}^{\infty} \dfrac{2^n c_0}{n!} x^{2n}$

d) $y = \sum_{n=0}^{\infty} \dfrac{n! c_0}{2^n} x^{2n}$

Answers

■ CHAPTER 10

Section 10.1
1. c
2. d
3. a

Section 10.2
1. d
2. c
3. b
4. b
5. a
6. b

Section 10.3
1. d
2. c
3. b
4. b
5. d
6. b

Section 10.4
1. a
2. d
3. d

Section 10.5
1. c
2. b
3. a

Section 10.6
1. c
2. b
3. d
4. c

■ CHAPTER 11

Section 11.1
1. b
2. Always
3. Sometimes
4. Always
5. False
6. a
7. c
8. d

Section 11.2
1. d
2. True
3. False
4. True
5. b
6. False
7. c
8. True

Section 11.3
1. c
2. False
3. False
4. False
5. True
6. b

Section 11.4
1. Never
2. True
3. False
4. True
5. b

Section 11.5
1. True
2. True
3. b
4. False

Section 11.6
1. False
2. True
3. True
4. a
5. b
6. True

Section 11.7
1. True
2. False
3. True
4. True
5. True
6. True

Section 11.8
1. Sometimes
2. True
3. c
4. b
5. c

Section 11.9
1. c
2. c
3. c
4. d
5. a

Section 11.10
1. c
2. d
3. c
4. a
5. b
6. True
7. c
8. b
9. b
10. a
11. d
12. b

Section 11.11
1. d
2. b
3. b

■ CHAPTER 12

Section 12.1
1. a
2. c
3. d
4. d
5. d

Section 12.2
1. b
2. c
3. d

4. False
5. c
6. d

Section 12.3
1. b
2. a
3. d
4. b
5. d
6. b

Section 12.4
1. d
2. False
3. b
4. True
5. d
6. c

Section 12.5
1. d
2. True
3. a
4. d
5. c
6. a

Section 12.6
1. b
2. d
3. c
4. b
5. b
6. True

Answers

■ CHAPTER 13

Section 13.1
1. a
2. a
3. b
4. b
5. False

Section 13.2
1. a
2. False
3. c
4. c

Section 13.3
1. a
2. d
3. d
4. True
5. a
6. True
7. a

Section 13.4
1. b
2. c
3. c
4. a

■ CHAPTER 14

Section 14.1
1. c
2. b
3. b
4. b
5. c

Section 14.2
1. True
2. a
3. d
4. d
5. False

Section 14.3
1. b
2. c
3. b
4. a
5. Sometimes
6. d

Section 14.4
1. False
2. a
3. c
4. c

Section 14.5
1. d
2. a
3. a
4. c

Section 14.6
1. c
2. True
3. d
4. b
5. c
6. d

Section 14.7
1. a
2. b
3. b
4. a
5. False

Section 14.8
1. a
2. a

■ CHAPTER 15

Section 15.1
1. c
2. d
3. b
4. Sometimes
5. False

Section 15.2
1. b
2. False
3. b

Section 15.3
1. d
2. False
3. b
4. b

Section 15.4
1. b
2. d

Section 15.5
1. a
2. c
3. False
4. False

Section 15.6
1. c

Section 15.7
1. False
2. a
3. a
4. b

Section 15.8
1. a
2. c
3. d
4. a

Section 15.9
1. a
2. d
3. d
4. b

Section 15.10
1. d
2. b

Answers

■ CHAPTER 16

Section 16.1
1. d
2. c
3. b
4. True

Section 16.2
1. b
2. True
3. a
4. False
5. a

Section 16.3
1. d
2. b
3. False
4. True
5. a

Section 16.4
1. False
2. b
3. a
4. d

Section 16.5
1. b
2. True
3. c
4. True
5. c
6. b

Section 16.6
1. c
2. c
3. a
4. c
5. c

Section 16.7
1. c
2. b
3. a

Section 16.8
1. b
2. a
3. c

Section 16.9
1. c
2. d

■ CHAPTER 17

Section 17.1
1. False
2. True
3. c
4. b
5. False
6. d

Section 17.2
1. a
2. d
3. d
4. b
5. b

Section 17.3
1. c
2. b

Section 17.4
1. a
2. a